AF802894

ADVANCED RESEARCH METHODS IN FOOD PROCESSING TECHNOLOGIES

Innovations in Agricultural and Biological Engineering

ADVANCED RESEARCH METHODS IN FOOD PROCESSING TECHNOLOGIES

Edited by

Junaid Ahmad Malik, PhD

Megh R. Goyal, PhD, PE

Preeti Birwal, PhD

Ritesh B. Watharkar, PhD

AAP | APPLE ACADEMIC PRESS

First edition published 2024

Apple Academic Press Inc.
1265 Goldenrod Circle, NE,
Palm Bay, FL 32905 USA

760 Laurentian Drive, Unit 19,
Burlington, ON L7N 0A4, CANADA

CRC Press
2385 NW Executive Center Drive,
Suite 320, Boca Raton FL 33431

4 Park Square, Milton Park,
Abingdon, Oxon, OX14 4RN UK

© 2024 by Apple Academic Press, Inc.

Apple Academic Press exclusively co-publishes with CRC Press, an imprint of Taylor & Francis Group, LLC

Reasonable efforts have been made to publish reliable data and information, but the authors, editors, and publisher cannot assume responsibility for the validity of all materials or the consequences of their use. The authors, editors, and publishers have attempted to trace the copyright holders of all material reproduced in this publication and apologize to copyright holders if permission to publish in this form has not been obtained. If any copyright material has not been acknowledged, please write and let us know so we may rectify in any future reprint.

Except as permitted under U.S. Copyright Law, no part of this book may be reprinted, reproduced, transmitted, or utilized in any form by any electronic, mechanical, or other means, now known or hereafter invented, including photocopying, microfilming, and recording, or in any information storage or retrieval system, without written permission from the publishers.

For permission to photocopy or use material electronically from this work, access www.copyright.com or contact the Copyright Clearance Center, Inc. (CCC), 222 Rosewood Drive, Danvers, MA 01923, 978-750-8400. For works that are not available on CCC please contact mpkbookspermissions@tandf.co.uk

Trademark notice: Product or corporate names may be trademarks or registered trademarks and are used only for identification and explanation without intent to infringe.

Library and Archives Canada Cataloguing in Publication

Title: Advanced research methods in food processing technologies / edited by Junaid Ahmad Malik, PhD, Megh R. Goyal, PhD, PE, Preeti Birwal, PhD, Ritesh B. Watharkar, PhD.
Names: Malik, Junaid Ahmad, 1987- editor. | Goyal, Megh R., editor. | Birwal, Preeti, editor. | Watharkar, Ritesh B., editor.
Series: Innovations in agricultural and biological engineering.
Description: First edition. | Series statement: Innovations in agricultural and biological engineering | Includes bibliographical references and index.
Identifiers: Canadiana (print) 20230475213 | Canadiana (ebook) 20230475248 | ISBN 9781774913482 (hardcover) | ISBN 9781774913499 (softcover) | ISBN 9781003369394 (ebook)
Subjects: LCSH: Food industry and trade—Technological innovations.
Classification: LCC TP370 .A38 2024 | DDC 664—dc23

Library of Congress Cataloging-in-Publication Data

Names: Malik, Junaid Ahmad, 1987- editor. | Goyal, Megh R., editor. | Birwal, Preeti, editor. | Watharkar, Ritesh B., editor.

Title: Advanced research methods in food processing technologies / edited by Junaid Ahmad Malik, Megh R. Goyal, Preeti Birwal, Ritesh B. Watharkar.

Description: Palm Bay, FL : Apple Academic Press, 2024. | Series: Innovations in agricultural and biological engineering | Includes bibliographical references and index.

Identifiers: LCCN 2023027572 (print) | LCCN 2023027573 (ebook) | ISBN 9781774913482 (hardback) | ISBN 9781774913499 (paperback) | ISBN 9781003369394 (ebook)

Subjects: LCSH: Food industry and trade--Technological innovations.

Classification: LCC TP371 .A357 2024 (print) | LCC TP371 (ebook) | DDC 664--dc23/eng/20230703

LC record available at https://lccn.loc.gov/2023027572

LC ebook record available at https://lccn.loc.gov/2023027573

ISBN: 978-1-77491-348-2 (hbk)
ISBN: 978-1-77491-349-9 (pbk)
ISBN: 978-1-00336-939-4 (ebk)

ABOUT THE BOOK SERIES: INNOVATIONS IN AGRICULTURAL AND BIOLOGICAL ENGINEERING

Under this book series, Apple Academic Press Inc. is publishing book volumes over a span of 8–10 years in the specialty areas defined by the American Society of Agricultural and Biological Engineers (<asabe.org>). Apple Academic Press Inc. aims to be a principal source of books in agricultural and biological engineering. We welcome book proposals from readers in areas of their expertise.

The mission of this series is to provide knowledge and techniques for agricultural and biological engineers (ABEs). The book series offers high-quality reference and academic content on agricultural and biological engineering (ABE) that is accessible to academicians, researchers, scientists, university faculty and university-level students, and professionals around the world.

Agricultural and biological engineers ensure that the world has the necessities of life, including safe and plentiful food, clean air and water, renewable fuel and energy, safe working conditions, and a healthy environment by employing knowledge and expertise of the sciences, both pure and applied, and engineering principles. Biological engineering applies engineering practices to problems and opportunities presented by living things and the natural environment in agriculture.

ABE embraces a variety of the following specialty areas (<asabe.org>): aquaculture engineering, biological engineering, energy, farm machinery and power engineering, food, and process engineering, forest engineering, information, and electrical technologies, soil, and water conservation engineering, natural resources engineering, nursery, and greenhouse engineering, safety, and health, and structures and environment.

For this book series, we welcome chapters on the following specialty areas (but not limited to):

- Academia to industry to end-user loop in agricultural engineering;
- Agricultural mechanization;
- Aquaculture engineering;
- Biological engineering in agriculture;

- Biotechnology applications in agricultural engineering;
- Energy source engineering;
- Farm to fork technologies in agriculture;
- Food and bioprocess engineering;
- Forest engineering;
- GPS and remote sensing potential in agricultural engineering;
- Hill land agriculture;
- Human factors in engineering;
- Impact of global warming and climatic change on agriculture economy;
- Information and electrical technologies;
- Irrigation and drainage engineering;
- Nanotechnology applications in agricultural engineering;
- Natural resources engineering;
- Nursery and greenhouse engineering;
- Potential of phytochemicals from agricultural and wild plants for human health;
- Power systems and machinery design;
- Robot engineering and drones in agriculture;
- Rural electrification;
- Sanitary engineering;
- Simulation and computer modeling;
- Smart engineering applications in agriculture;
- Soil and water engineering;
- Micro-irrigation engineering;
- Structures and environment engineering;
- Waste management and recycling;
- Any other focus areas.

For more information on this series, readers may contact:

Megh R. Goyal, PhD, PE
Book Series Senior Editor-in-Chief: Innovations in Agricultural and Biological Engineering
E-mail: goyalmegh@gmail.com

OTHER BOOKS ON AGRICULTURAL AND BIOLOGICAL ENGINEERING BY APPLE ACADEMIC PRESS, INC.

Management of Drip/Trickle or Micro Irrigation
Megh R. Goyal, PhD, PE, Senior Editor-in-Chief
Evapotranspiration: Principles and Applications for Water Management
Megh R. Goyal, PhD, PE and Eric W. Harmsen, PhD Editors

Book Series: Research Advances in Sustainable Micro Irrigation
Senior Editor-in-Chief: Megh R. Goyal, PhD, PE

- **Volume 1:** Sustainable Micro Irrigation: Principles and Practices
- **Volume 2:** Sustainable Practices in Surface and Subsurface Micro Irrigation
- **Volume 3:** Sustainable Micro Irrigation Management for Trees and Vines
- **Volume 4:** Management, Performance, and Applications of Micro Irrigation Systems
- **Volume 5:** Applications of Furrow and Micro Irrigation in Arid and Semi-Arid Regions
- **Volume 6:** Best Management Practices for Drip Irrigated Crops
- **Volume 7:** Closed Circuit Micro Irrigation Design: Theory and Applications
- **Volume 8:** Wastewater Management for Irrigation: Principles and Practices
- **Volume 9:** Water and Fertigation Management in Micro Irrigation
- **Volume 10:** Innovation in Micro Irrigation Technology

Book Series: Innovations and Challenges in Micro Irrigation
Senior Editor-in-Chief: Megh R. Goyal, PhD, PE

- Engineering Interventions in Sustainable Trickle Irrigation: Water Requirements, Uniformity, Fertigation, and Crop Performance
- Management Strategies for Water Use Efficiency and Micro Irrigated Crops: Principles, Practices, and Performance

- Micro-Irrigation Engineering for Horticultural Crops: Policy Options, Scheduling, and Design
- Micro-Irrigation Management: Technological Advances and Their Applications
- Micro-Irrigation Scheduling and Practices
- Performance Evaluation of Micro-Irrigation Management: Principles and Practices
- Potential of Solar Energy and Emerging Technologies in Sustainable Micro-Irrigation
- Principles and Management of Clogging in Micro-Irrigation
- Sustainable Micro-Irrigation Design Systems for Agricultural Crops: Methods and Practices

Book Series: Innovations in Agricultural and Biological Engineering
Senior Editor-in-Chief: Megh R. Goyal, PhD, PE

- Advanced Research Methods in Food Processing Technologies
- Advances in Food Process Engineering: Novel Processing, Preservation and Decontamination of Foods
- Advances in Green and Sustainable Nanomaterials: Applications in Energy, Biomedicine, Agriculture, and Environmental Science
- Advances in Sustainable Food Packaging Technology
- Analytical Methods for Milk and Milk Products, 2-volume set:
    - Volume 1: Sampling Methods, Chemical and Compositional Analysis
    - Volume 2: Physicochemical Analysis of Concentrated, Coagulated and Fermented Products
    - Volume 3: Microbiological Analysis, Isolation, and Characterization
- Biological and Chemical Hazards in Food and Food Products: Prevention, Practices, and Management
- Bioremediation and Phytoremediation Technologies in Sustainable Soil Management, 4-volume set:
    - Volume 1: Fundamental Aspects and Contaminated Sites
    - Volume 2: Microbial Approaches and Recent Trends
    - Volume 3: Inventive Techniques, Research Methods, and Case Studies
    - Volume 4: Degradation of Pesticides and Polychlorinated Biphenyls

- Dairy Engineering: Advanced Technologies and Their Applications
- Developing Technologies in Food Science: Status, Applications, and Challenges
- Emerging Technologies in Agricultural Engineering
- Engineering Interventions in Agricultural Processing
- Engineering Interventions in Foods and Plants
- Engineering Practices for Agricultural Production and Water Conservation: An Interdisciplinary Approach
- Engineering Practices for Management of Soil Salinity: Agricultural, Physiological, and Adaptive Approaches
- Engineering Practices for Milk Products: Dairyceuticals, Novel Technologies, and Quality
- Enzyme Inactivation in Food Processing: Technologies, Materials, and Applications
- Field Practices for Wastewater Use in Agriculture: Future Trends and Use of Biological Systems
- Flood Assessment: Modeling and Parameterization
- Food Engineering: Emerging Issues, Modeling, and Applications
- Food Process Engineering: Emerging Trends in Research and Their Applications
- Food Processing and Preservation Technology: Advances, Methods, and Applications
- Food Technology: Applied Research and Production Techniques
- Functional Dairy Ingredients and Nutraceuticals: Physicochemical, Technological, and Therapeutic Aspects
- Handbook of Research on Food Processing and Preservation Technologies, 5-volume set:
    - Volume 1: Nonthermal and Innovative Food Processing Methods
    - Volume 2: Nonthermal Food Preservation and Novel Processing Strategies
    - Volume 3: Computer-Aided Food Processing and Quality Evaluation Techniques
    - Volume 4: Design and Development of Specific Foods, Packaging Systems, and Food Safety
    - Volume 5: Emerging Techniques for Food Processing, Quality, and Safety Assurance
- Modeling Methods and Practices in Soil and Water Engineering
- Nanotechnology and Nanomaterial Applications in Food, Health, and Biomedical Sciences

- Nanotechnology Applications in Agricultural and Bioprocess Engineering: Farm to Table
- Nanotechnology Applications in Dairy Science: Packaging, Processing, and Preservation
- Nanotechnology Horizons in Food Process Engineering, 3-volume set:
    - Volume 1: Food Preservation, Food Packaging and Sustainable Agriculture
    - Volume 2: Scope, Biomaterials, and Human Health
    - Volume 3: Trends, Nanomaterials, and Food Delivery
- Nonthermal Light-Based Technologies in Food Processing
- Nonthermal Plasma Technology for Food and Food Products
- Novel and Alternative Methods in Food Processing: Biotechnological, Physicochemical, and Mathematical Approaches
- Novel Dairy Processing Technologies: Techniques, Management, and Energy Conservation
- Novel Processing Methods for Plant-Based Health Foods: Extraction, Encapsulation and Health Benefits of Bioactive Compounds
- Novel Strategies to Improve Shelf-Life and Quality of Foods: Quality, Safety, and Health Aspects
- Phytochemicals and Medicinal Plants in Food Design: Strategies and Technologies for Improved Healthcare
- Processing of Fruits and Vegetables: From Farm to Fork
- Processing Technologies for Milk and Milk Products: Methods, Applications, and Energy Usage
- Quality Control in Fruit and Vegetable Processing: Methods and Strategies
- Scientific and Technical Terms in Bioengineering and Biological Engineering
- Soil and Water Engineering: Principles and Applications of Modeling
- Soil Salinity Management in Agriculture: Technological Advances and Applications
- State-of-the-Art Technologies in Food Science: Human Health, Emerging Issues and Specialty Topics
- Sustainable and Functional Foods from Plants
- Sustainable Biological Systems for Agriculture: Emerging Issues in Nanotechnology, Biofertilizers, Wastewater, and Farm Machines
- Sustainable Green Nanomaterials: Synthesis, Characterization, and Engineering Applications

- Sustainable Nanomaterials for Biomedical Engineering: Impacts, Challenges, and Future Prospects
- Sustainable Nanomaterials for Biosystems Engineering: Trends in Renewable Energy, Environment, and Agriculture
- Technological Interventions in Dairy Science: Innovative Approaches in Processing, Preservation, and Analysis of Milk Products
- Technological Interventions in Management of Irrigated Agriculture
- Technological Interventions in the Processing of Fruits and Vegetables
- Technological Processes for Marine Foods, From Water to Fork: Bioactive Compounds, Industrial Applications, and Genomics
- The Chemistry of Milk and Milk Products: Physicochemical Properties, Therapeutic Characteristics, and Processing Methods

ABOUT SENIOR-EDITOR-IN-CHIEF

Megh R. Goyal, PhD, PE

Retired Professor in Agricultural and Biomedical Engineering, University of Puerto Rico, Mayaguez Campus; Senior Acquisitions Editor, Biomedical Engineering and Agricultural Science, Apple Academic Press, Inc.

Megh R. Goyal, PhD, PE, is currently a Retired Professor of Agricultural and Biomedical Engineering from the General Engineering Department at the College of Engineering at the University of Puerto Rico–Mayaguez Campus (UPRM); and Senior Acquisitions Editor and Senior Technical Editor-in-Chief for Agricultural and Biomedical Engineering for Apple Academic Press Inc. During his professional career, he has worked as a Soil Conservation Inspector; Research Assistant at Haryana Agricultural University and Ohio State University; Research Agricultural Engineer/Professor at the Department of Agricultural Engineering of UPRM; and Professor of Agricultural and Biomedical Engineering in the General Engineering Department of UPRM. He spent a one-year sabbatical leave in 2002–2003 at the Biomedical Engineering Department of Florida International University, Miami, USA.

Dr. Goyal was the first agricultural engineer to receive the professional license in agricultural engineering from the College of Engineers and Surveyors of Puerto Rico. In 2005, he was proclaimed the "Father of Irrigation Engineering in Puerto Rico for the Twentieth Century" by the American Society of Agricultural and Biological Engineers, Puerto Rico Section, for his pioneering work on micro irrigation, evapotranspiration, agroclimatology, and soil and water engineering.

During his professional career of 52 years, he has received many awards, including Scientist of the Year, Membership Grand Prize for the American Society of Agricultural Engineers Campaign, Felix Castro Rodriguez Academic Excellence Award, Man of Drip Irrigation by the Mayor of Municipalities of Mayaguez/Caguas/Ponce and Senate/Secretary of Agriculture of ELA, Puerto Rico, and many others. He has been recognized as one of the experts "who rendered meritorious service for the development of

[the] irrigation sector in India" by the Water Technology Centre of Tamil Nadu Agricultural University in Coimbatore, India, and ASABE (American Society of Agricultural and Biological Engineers), which was bestowed on him the 2018 Netafim Microirrigation Award.

Dr. Goyal has authored more than 200 journal articles and edited more than 100 books. AAP has published many of his books, including *Management of Drip/Trickle or Micro Irrigation; Evapotranspiration: Principles and Applications for Water Management;* ten-volume set on *Research Advances in Sustainable Micro Irrigation.* He has also authored the textbooks *Elements of Agroclimatology* (Spanish) by UNISARC, Colombia, and two *Bibliographies on Drip Irrigation.* Dr. Goyal has also developed several book series with AAP, including Innovations in Agricultural & Biological Engineering (with over 60 titles in the series to date), Innovations and Challenges in Micro Irrigation; and Innovations in Plant Science for Better Health: From Soil to Fork.

Dr. Goyal received his BSc degree in Engineering from Punjab Agricultural University, Ludhiana, India, and his MSc and PhD degrees from the Ohio State University, Columbus, Ohio, USA. He also earned a Master of Divinity degree from the Puerto Rico Evangelical Seminary, Hato Rey, Puerto Rico, USA.

Readers may contact him at goyalmegh@gmail.com.

ABOUT THE EDITORS

Junaid Ahmad Malik, PhD
Lecturer, Department of Zoology, Government Degree College, Bijbehara, Kashmir, Jammu and Kashmir, India

Junaid Ahmad Malik, PhD, is a Lecturer with the Department of Zoology at Government Degree College, Bijbehara, Kashmir (J&K), India, and is actively involved with teaching and research activities. He has more than eight years of research experience. His areas of interest are ecology, soil macrofauna, wildlife biology, conservation biology, etc. Dr. Malik has published over 20 research articles and technical papers in international peer-reviewed journals and has authored and edited books, book chapters, and more than 10 popular editorial articles. He also serves as an editor and reviewer of several journals. He has participated in several state, national, and international conferences, seminars, workshops, and symposia and has more than 20 conference papers to his credit. He is a life member of the Society for Bioinformatics and Biological Sciences.

Dr. Malik received a BSc (2008) in Science from the University of Kashmir, Srinagar, J & K; MSc (2010) in Zoology from Barkatullah University, Bhopal, Madhya Pradesh, India; and PhD (2015) in Zoology from the same university. He completed his BEd program in 2017 at the University of Kashmir, Srinagar, J&K.

Preeti Birwal, PhD
Scientist (Processing and Food Engineering), Department of Processing and Food Engineering, College of Agricultural Engineering and Technology, Punjab Agricultural University, Ludhiana, Punjab, India

Preeti Birwal, PhD, is working as a Scientist (processing and food engineering) in the Department of Processing and Food Engineering at the College of Agricultural Engineering and Technology at Punjab Agricultural University, Ludhiana,

Punjab, India. She is currently working in the area of nonthermal food preservation, fermented beverages, food packaging, and technology of millet-based beer. She has served at Jain (Deemed to be University), Bangalore, as a member of the board of examiners and placements. She has participated at several national and international conferences and seminars and has delivered lectures as a resource person on doubling farmers' income through dairy technology training sponsored by the Directorate of Extension, Ministry of Agriculture and Farmers Welfare, Government of India. Dr. Birwal has published research papers, an edited book, book chapters, popular articles, conference papers, abstracts, and editorial opinions. She is advising several MTech scholars in food technology and has successfully guided five postgraduate students for their dissertation work. She also serves as an external examiner for various Indian state agricultural universities and as an editor and reviewer of several journals. Dr. Birwal has been named outstanding reviewer of the month by the online journal *Current Research in Nutrition and Food Science.* She has successfully completed AUTOCAD 2D & 3D certification. She is a life member of IDEA. She graduated with a degree in Dairy Technology from ICAR–National Dairy Research Institute, Karnal, India, holds a master's degree in Food Process Engineering and Management from NIFTEM, Haryana and PhD (Dairy Engineering) from ICAR–NDRI, Bangalore, India. She has received several fellowships from MHRD, Nestle India, GATE, and UGC-RGNF.

Ritesh B. Watharkar, PhD
Assistant Professor,
College of Agriculture and Food Business Management,
MIT-ADT University, Rajbaugh,
Maharashtra–India

Ritesh B. Watharkar, PhD, is currently working as an Assistant Professor in the College of Agriculture and Food Business Management at MIT-ADT University, Maharashtra, India.

He received his BTech degree in Food Technology from Dr. Balasaheb Sawant Konkan Agriculture University, Dapoli, Maharashtra, India; his MTech in Food Processing and Engineering from the Karunya University, India; his PhD in Food Engineering and Technology from Tezpur University (A Central University), Assam, India.

He was formerly a Senior Research Fellow under a funded project by the Ministry of Food Processing Industries, New Delhi (2013–2015), and

Assistant Professor at K.K. Wagh Food Tech College (2011–2013); at Shramshakti College of Food Technology, affiliated by Mahatma Phule Agricultural University (2019–2020).

During his professional career, he received awards such as Senior Research Fellow under the UNESCO/China Great Wall Fellowship Program and worked for one year in the School of Biosystem Engineering and Food Science at Zhejiang University, Hangzhou, China. He was also awarded for best popular article from *Krishidut Magazine,* Maharashtra, India. Currently, he is a life member of the Association of Food Scientists and Technologists, India. He has published more than 10 journal articles and attended more than 10 national and international conferences. He has also published more than 10 popular articles in the *Marathi Literacy Magazine*.

CONTENTS

CONTRIBUTORS

Mohammed Shafiq Alam
Principal Scientist, Department of Processing and Food Engineering, College of Agricultural Engineering and Technology, Punjab Agricultural University (PAU), Ludhiana, Punjab, India

Farhana Mehraj Allai
Assistant Professor, Department of Food Technology, Islamic University of Science and Technology, Awantipora, Kashmir, India; and PhD Research Scholar, Department of Postharvest Engineering and Technology, Aligarh Muslim University, Aligarh, Uttar Pradesh, India

Cintia Da Silva Araújo
Post-Doctoral Fellow, Department of Food Engineering, Federal University of Espírito Santo, Alegre, Brazil

Syed Amir Ashraf
Assistant Professor, Department of Clinical Nutrition, University of Hael, KSA, Hael, Saudi Arabia

Z. R. Azaz Ahmad Azad
Associate Professor, Department of Postharvest Engineering and Technology, Aligarh Muslim University, Aligarh, Uttar Pradesh, India

Shashikant V. Bhandari
Head and Professor, Department of Pharmaceutical Chemistry, AISSMS College of Pharmacy, Pune, Maharashtra, India

Harinderjeet Kaur Bhullar
PhD Research Scholar, Department of Food Science and Technology, Punjab Agricultural University, Ludhiana, Punjab, India

Preeti Birwal
Scientist, Department of Processing and Food Engineering, College of Agricultural Engineering and Technology, Punjab Agricultural University, Ludhiana, Punjab, India

B. B. Borse
Professor AcSIR and Senior Principal Scientist, Department of Spice and Flavor Science (SFS), CFTRI, Mysuru, Karnataka, India

K. L. Claudia
PhD Research Scholar, Department of Food Process Engineering, Agricultural Engineering College and Research Institute, Tamil Nadu Agricultural University (TNAU), Coimbatore, Tamil Nadu, India

Jefferson Luiz Gomes Corrêa
Professor, Department of Food Science, Federal University of Lavras, Lavras, Brazil

Lavanya Devaraj
PhD Research Scholar, Department of Food Engineering and Center for Excellence in Grain Science, National Institute of Food Technology and Entrepreneurship Management (NIFTEM), Thanjavur (formerly Indian Institute of Food Processing Technology), Ministry of Food Processing Industries (MoFPI), Government of India, Tamil Nadu, India

M. Dharini
PhD Research Scholar, Center of Excellence in Non-Thermal Processing, National Institute of Food Technology and Entrepreneurship Management (NIFTEM), Thanjavur (formerly Indian Institute of Food Processing Technology), Ministry of Food Processing Industries (MoFPI), Government of India, Pudukkottai Road, Thanjavur, Tamil Nadu, India

Soumya Doodi
Assistant Professor, Department of Food Technology, Faculty of Life and Allied Health Sciences (FLAHS), Ramaiah University of Applied Sciences (RUAS), Bangalore, Karnataka, India

Pranjal Dutta
MTech Student, Department of Agricultural Engineering, Assam University, Silchar, Cachar District, Assam, India

Gitanjali Gautam
PhD Research Scholar, Department of Food Engineering and Technology, School of Engineering, Tezpur University, Tezpur, Napaam, Sonitpur, Assam, India

Mainao Alina Gayary
Assistant Professor, Department of Food Engineering and Technology, Central Institute of Technology, Kokrajhar, Assam, India

Nileema S. Gore
Assistant Professor, MGMs Institute of Biosciences and Technology, Mahatma Gandhi Mission University, Aurangabad, Maharashtra, India

N. A. Nanje Gowda
Assistant Professor and Head, Department of Food Technology, Faculty of Life and Allied Health Sciences (FLAHS), Ramaiah University of Applied Sciences (RUAS), Bangalore, Karnataka, India

Megh R. Goyal
Senior Editor-in-Chief (Agriculture and Biomedical Engineering) for AAP, Retired Professor in Agricultural and Biomedical Engineering, University of Puerto Rico-Mayaguez, Mayaguez, Puerto Rico, USA

Chennappa Gurikar
Assistant Professor, Department of Food Technology, Faculty of Life and Allied Health Sciences (FLAHS), Ramaiah University of Applied Sciences (RUAS), Bangalore, Karnataka, India

V. Harsha
PhD Research Scholar, Department of Agriculture and Food Engineering and Technology, Indian Institute of Technology (IIT), Kharagpur, West Bengal, India

Irineu Petri Júnior
Professor, Department of Engineering, Federal University of Lavras, Lavras, Brazil

Baldev Singh Kalsi
PhD Research Scholar, Department of Processing and Food Engineering, College of Agricultural Engineering and Technology, Punjab Agricultural University (PAU), Ludhiana, Punjab, India

Varsha Kanojia
Scientist, Department of Processing and Food Engineering, College of Agricultural Engineering and Technology (COAET), Punjab Agricultural University (PAU), Ludhiana, Punjab, India

Amarjeet Kaur
Senior Milling Technologist, Department of Food Science and Technology, Punjab Agricultural University (PAU), Ludhiana, Punjab, India

Gurveer Kaur
Scientist, Department of Processing and Food Engineering, College of Agricultural Engineering and Technology, Punjab Agricultural University (PAU), Ludhiana, Punjab, India

Roshna Khawas
MTech Student, Department of Agricultural Engineering, Assam University, Silchar, Cachar District, Assam, India

Pranali V. Kuthe
PhD Research Scholar, Pharmaceutical Chemistry, AISSMS College of Pharmacy, Pune, Maharashtra, India

Leandro Levate Macedo
Post-Doctoral Fellow, Department of Food Science, Federal University of Lavras, Lavras, Brazil

Charu Lata Mahanta
Professor, Department of Food Engineering and Technology, School of Engineering, Tezpur University, Napaam, Sonitpur, Assam, India

R. Mahendran
Associate Professor and Director, Center of Excellence in Non-Thermal Processing, National Institute of Food Technology and Entrepreneurship Management (NIFTEM), Thanjavur (formerly Indian Institute of Food Processing Technology), Ministry of Food Processing Industries (MoFPI), Government of India, Tamil Nadu, India

Nikhil Kumar Mahnot
Assistant Professor, Department of Bioengineering and Technology, Gauhati University Institute of Science and Technology (GUIST), Gauhati, Assam, India

Junaid Ahmad Malik
Lecturer, Department of Zoology, Government Degree College, Bijbehara, Kashmir, Jammu and Kashmir, India

Nikitha Modupalli
PhD Research Scholar, Department of Food Engineering and Center for Excellence in Grain Science, National Institute of Food Technology and Entrepreneurship Management (NIFTEM), Thanjavur (formerly Indian Institute of Food Processing Technology), Ministry of Food Processing Industries (MoFPI), Government of India, Tamil Nadu, India

Amit Kumar Mukherjee
Assistant Professor, Department of Food Technology, Haldia Institute of Technology, West Bengal, India

Mohan Naik
Assistant Professor, Department of Food Engineering, National Institute of Food Technology and Entrepreneurship Management (NIFTEM), Thanjavur (formerly Indian Institute of Food Processing Technology), Ministry of Food Processing Industries (MoFPI), Government of India, Tamil Nadu, India

Sadaf Nazir
Assistant Professor, Department of Food Technology, Institute of Engineering and Technology, Bundelkhand University, Jhansi, Uttar Pradesh, India

Anand Kumar Pandey
Assistant Professor, Department of Biotechnology Engineering, Institute of Engineering and Technology, Bundelkhand University, Jhansi, Uttar Pradesh, India

Pragya Pandey
Assistant Professor, Department of Food Technology and Nutrition, School of Agriculture, Lovely Professional University, Jalandhar, Punjab, India

Kaninika Paul
Assistant Professor, Department of Food Technology and Nutrition, School of Agriculture, Lovely Professional University, Jalandhar, Punjab, India

Yashaswini Japannanayakanakote Puttalingappa
Assistant Professor, Department of Food Engineering, National Institute of Food Technology and Entrepreneurship Management (NIFTEM), Thanjavur (formerly Indian Institute of Food Processing Technology), Ministry of Food Processing Industries (MoFPI), Government of India, Tamil Nadu, India

Ratish K. Ramanan
MTech Student, National Institute of Food Technology and Entrepreneurship Management (NIFTEM), Thanjavur (formerly Indian Institute of Food Processing Technology), Ministry of Food Processing Industries (MoFPI), Government of India, Tamil Nadu, India

Soumya Rathore
Assistant Professor, P.G. Department of Food Technology, RBS College, Agra, Uttar Pradesh, India

Utsav Rathore
Research Scholar, Department of Food Technology and Nutrition, School of Agriculture, Lovely Professional University, Jalandhar, Punjab, India

E. J. Rifna
MTech Student, National Institute of Food Technology and Entrepreneurship Management (NIFTEM), Thanjavur (formerly Indian Institute of Food Processing Technology), Ministry of Food Processing Industries (MoFPI), Government of India, Tamil Nadu, India

M. S. Roopesh
Assistant Professor, Department of Agricultural, Food and Nutritional Science, University of Alberta, Edmonton, Canada

Bhagwan K. Sakhale
Associate Professor, Department of Chemical Technology, Dr. Babasaheb Ambedkar Marathwada University, Aurangabad, Maharashtra, India

Aniket P. Sarkate
Assistant Professor, Department of Chemical Technology, Dr. Babasaheb Ambedkar Marathwada University, Aurangabad, Maharashtra, India

Sandhya Singh
Senior Scientist, Department of Processing and Food Engineering, College of Agricultural Engineering and Technology, Punjab Agricultural University (PAU), Ludhiana, Punjab, India

Prem Prakash Srivastav
Associate Professor, Agricultural and Food Engineering Department, Indian Institute of Technology (IIT), Kharagpur, West Bengal, India

Alphy Nisa Steaphen
PhD Research Scholar, Department of Microbiology, Sree Sankara College, Kalady, Kerala, India

Anjali Sudhakar
PhD Research Scholar, Agricultural Processing and Structures, ICAR–Central Institute of Agricultural Engineering, Bhopal, Madhya Pradesh, India

C. K. Sunil
Assistant Professor, Department of Food Engineering and Center for Excellence in Grain Science, National Institute of Food Technology and Entrepreneurship Management (NIFTEM), Thanjavur (formerly Indian Institute of Food Processing Technology), Ministry of Food Processing Industries (MoFPI), Government of India, Tamil Nadu, India

T. Supriya
MSc Student, JSS College of Arts, Commerce and Science College, Mysuru, Karnataka, India

Barinderjeet Singh Toor
PhD Research Scholar, Department of Food Science and Technology, Punjab Agricultural University, Ludhiana, Punjab, India

Cinu Varghese
PhD Research Scholar, Center for Rural Development and Innovative Sustainable Technology, Indian Institute of Technology (IIT), Kharagpur, West Bengal, India

N. Venkatachalapathy
Professor and Head, Department of Food Engineering and Center for Excellence in Grain Science (NIFTEM), Thanjavur (formerly Indian Institute of Food Processing Technology), Ministry of Food Processing Industries (MoFPI), Government of India, Tamil Nadu, India

Wallaf Costa Vimercati
PhD Research Fellow, Department of Food Science, Federal University of Lavras, Lavras, Brazil,

Ritesh B. Watharkar
Assistant Professor, College of Management, Food and Agriculture Management Department, MIT Art, Design and Technology University, Pune, Maharashtra, India

Insha Zahoor
Post-Doctoral Fellow, Department of Neurology, Henry Ford Hospital, Detroit, Michigan, USA

ABBREVIATIONS

2-DCB	2-dodecylcyclobutanone
2-TCB	2-tetradecylcyclobutanone
3DFP	3D food printing
AA	ascorbic acid
AAP	Apple Academic Press Inc.
ABEs	agricultural and biological engineer
AC	alternating current
AC	antioxidant capacity
ACP	atmospheric cold plasma
Ag	silver
Al_2O_3	aluminum oxide
AM	additive manufacturing
ANF	anti-nutritional factor
APPJ	atmospheric pressure plasma jet
Ar	argon
ASABE	American Society of Agricultural and Biological Engineers
BLG	β-lactoglobulin
BO	biaxial orientation
BOD	biochemical oxygen demand
BSA	bovine serum albumin
BT	barium titanate
CAD	computer-aided design
CD	cyclodextrin
CF	cutting force
CFU	colony forming unit
CIP	cleaning in place
ClO_2	chlorine dioxide
CMC	carboxymethyl cellulose
COD	chemical oxygen demand
CP	cold plasma
Cr	chromium
CW	cooperative extractor
DBD	dielectric barrier discharge

DC	direct current
DNA	deoxyribonucleic acid
DPPH	2,2-diphenyl-1-picrylhydrazyl
DSC	differential scanning calorimeter
e.m.f.	electromotive force
ELISA	enzyme-linked immunosorbent assay
EMAT	electromagnetic acoustic transducers
EMC	ethyl methylcellulose
EPA	Environmental Protection Agency
ET	ethanol treatment
EVOH	ethylene vinyl alcohol
FAO	Food and Agriculture Organization
FAV	fruits and vegetables
FBD	fluidized bed drying
FCC	Federal Communications Commission
FDA	Food and Drug Administration
FDM	fused deposition modeling
Fe	iron
GA	glycyrrhizic acid
GA	gum Arabic
GABA	gamma-aminobutyric acid
GCO	green coffee oil
GDA	gliding arc discharge
GHz	gigahertz
GIT	gastrointestinal tract
GMF	genetically modified food
GPP	grapefruit-peel-phenolic
GRAS	generally recognized as safe
H_2O_2	hydrogen peroxide
HHP	high hydrostatic pressure
HIPE	high internal phase emulsion
HPLC	high-performance liquid chromatography
HPMC	hydroxy propyl methyl cellulose
HPP	high-pressure processing
HPU	high-power ultrasound
HTST	high-temperature, short-time
HVACP	high voltage atmospheric cold plasma
Hz	hertz
IF	immersion freezing

IgE	immunoglobulins E
IgG	immunoglobulins
IMD	intermittent microwave drying
ISM	industrial, scientific, and medical applications
J	joule
KCl	potassium chloride
kGy	kilo gray
kHz	kilohertz
kV	kilo volt
L and W Yield Meter	locker and wild yield meter
LDPE	low density polyethylene
LOD	limit of detection
LOX	lipoxygenase
MAE	microwave assisted extraction
MAP	modified atmosphere packaging
MD	microwave discharge
MDA	malondialdehyde
Mg	magnesium
MHz	megahertz
MIRINZ	Meat Industry Research Institute of New Zealand
MIT	Massachusetts Institute of Technology
MP	myofibrillar protein
MPa	megapascal
MS	manosonication
MS	mass spectrometry
MTS	manothermosonication
MW	microwave
MWAD	microwave-assisted air drying
MWE	microwave-assisted extraction
MWSD	microwave-enhanced spouted bed drying
N_2	nitrogen molecule
NBS	natural bioactive substances
NDT	non-destructive testing
Ni	nickel
NLCs	nanostructured lipid carriers
nm	nanometer
NMR	nuclear magnetic resonance
NOS	nitrogen species
NPs	nanoparticles

NT	nanotechnology
NTP	non-thermal plasma
O_2	oxygen molecule
OD	osmotic dehydration
OTR	oxygen transmission rate
PA	polyamide
PAW	plasma activated water
PCL	polycaprolactone
PCR	polymerase chain reaction
PEF	pulsed electric field
PEG	polyethylene glycol
PEO	polyethylene oxide
PET	polyethylene terephthalate
PJ	plasma jet
PL	pulsed-light
PLA	polylactic acid
PME	pectin methylesterase
PN	lead meta niobate
PO/POD	peroxidase
PP	polypropylene
ppb	parts per billion
PPO	polyphenol oxidase
PTFE	polyterafluroethylene
PVA	polyvinyl alcohol
PVP	polyvinylpyrrolidone
PZT	lead zirconate titanate
RF	radiofrequency
RFD	radiofrequency discharge
RFH	radiofrequency heating
ROS	reactive oxygen species
RSM	response surface methodology
RT-PCR	reverse transcription-polymerase chain reaction
SD	Sprague Dawley
SEM	scanning electron microscope
SERS	surface-enhanced Raman spectroscopy
SFE	supercritical fluid extraction
SHASAM	selective hot air sintering and melting
SiO_2	silicon dioxide
SKIT	soybean Kunitz-type tryps in inhibitor

SLNs	solid lipid nanoparticles
SLS	selective laser sintering
SMD	surface micro-discharge
SPP	sweet potato puree
SSP	salt soluble protein
TA	titratable acidity
TBA	thiobarbituric acid
TEAC	Trolox equivalent antioxidant capacity
TP	thermal plasma
TPC	total phenolic contents
TS	thermosonication
TSS	total soluble solids
UAE	ultrasound-assisted extraction
UHT	ultra-high temperature
UMAE	ultrasonic microwave-assisted extraction
US	ultrasonic
USAC	ultrasound-assisted compaction
USC-CO_2	ultrasound-assisted supercritical carbon dioxide
USDA	United States Department of Agriculture
USDA-FSIS	United States Department of Agriculture–Food Safety and Inspection Service
USFE	ultrasound-assisted supercritical fluid extraction
UV	ultraviolet
V	voltage
VI	vacuum impregnation
W	watt
WHC	water holding capacities
WHO	World Health Organization
ZnO	zinc oxide

PREFACE

Food processing and preservation have been part of human lives for over a millennia. Preservation is essential for better storage of food and to meet the hunger demand when growing and harvesting conditions are not suitable. There are many traditional technologies present for food preservation. Though extension of shelf-life can be achieved through conventional methods there are many negative impacts on foods' nutritional and sensorial properties. Researchers and industrialists have investigated new methods for food preservation and processing. Advanced technologies like cold plasma (CP), ultrasonic treatment, microwave (MW), heating, electrospinning, encapsulation, and thermosonication (TS) are there, and have proven to be better processing technologies. Technologies are reported to have extensive use in food processing. They are useful for processing fruits, vegetables, eggs, meat, nutraceuticals, milk, and milk products.

This book mainly aims to provide knowledge about the present status of the studies and research cases of advanced food processing and preservation technologies. The book will motivate students, academicians, industrialists, and research scholars to opt for advanced technologies and come out with more innovative products and formulations for the world. The book will encourage engineers to develop even better food processing and preservation equipment.

Each chapter covers the principles, history, design, development, and application of several advanced food processing technologies.

The book comprises of three parts. Part I: Cold Plasma and Ultrasound Processing of Foods, includes introduction to food processing technologies, applications, and potential of cold plasma technology in the food industry, impact of cold plasma on physical surface modifications of foods and food products, ultrasound processing of foods and food products: an overview, potential of ultrasound technology in nutraceuticals and pharmaceuticals, principles of ultrasonication processing in the food industry, potential of thermosonication in the food industry, role of ultrasound-assisted mechanical shear force in preservation of foods; Part II: Microwave Processing of Foods describe applications of microwave technology in food processing, potential of microwave heating in food processing, intermittent microwave drying (IMD) of strawberry fruits; Part III: Research Advances in Processing of

Foods discusses high pressure processing of foods and food products: an overview, applications of electrospinning technology in foods: an overview, encapsulation techniques and its applications in the food industry, impact of freezing and thawing processes on textural properties of food products, 3D printing of foods: concepts, applications, and prospects, potential of Enzyme-Linked Immunosorbent Assay (ELISA) in food authentication, applications of nanotechnology (NT) in food processing.

We introduce this book volume under the book series *Innovations in Agricultural and Biological Engineering*. This book volume is an excellent source of information and a significant source of reference material for researchers, students, scientists, traders, growers, industries, and processors for food preservation and processing.

This book has been shaped excellently by the outstanding efforts of contributing authors and editors. Their names have been indicated in each chapter and mentioned in the contributor's list. We are indeed grateful to the authors for their contribution, knowledge in the field, commitment to work on time, and enthusiasm. We believe this book to be supportive and helpful to the food preservation and processing community.

We give genuine thanks to the editorial office of Apple Academic Press, Inc. for their enormous simplification and propositions right through this assignment, alongside the AAP production staff for typesetting the whole manuscript and for bringing out this quality production.

We hereby appeal to our readers for their productive suggestions that may form the basis for improving future editions.

We take this opportunity to wholeheartedly thank our families for their motivation, moral support, and blessings in counteracting every obstacle coming our way. Also, we express our gratitude to our respective institutes for their support during the composing of this publication. Last but not least, we wish to thank our spouses for their understanding and patience throughout this project.

—***Editors***

PART I

COLD PLASMA AND ULTRASOUND PROCESSING OF FOODS

CHAPTER 1

INTRODUCTION TO FOOD PROCESSING TECHNOLOGIES

SOUMYA DOODI, N. A. NANJE GOWDA, and
CHENNAPPA GURIKAR

ABSTRACT

Novel nonthermal food processing technologies (such as ohmic heating, ozonation, infrared heating, microwave (MW) heating, and pulsed electric fields (PEFs)) have already been commercially adopted in the food industry to deal with issues of rising competition, globalization, and increasingly dynamic and diverse consumer demands. Cold plasma (CP), plasma-activated water (PAW), surface Raman spectroscopy, radiofrequency (RF) heating, and chlorine dioxide (ClO_2) are still in the research stage and will make their way into commercial applications in the future. Food production and the supply system are under tremendous pressure due to the gap between academic research and industrial applications of food processing technologies. A close collaboration between university researchers and decision-makers in food industries is the need of the hour for new technological advancements and the creation of better food processing techniques for the future.

1.1 INTRODUCTION

Heating is one of the oldest and most traditional food processing and preservation techniques. Heating effectively preserves food by eliminating enzymes and pathogenic and spoilage bacteria. The "State of Food Security and

Advanced Research Methods in Food Processing Technologies. Junaid Ahmad Malik, Megh R. Goyal, Preeti Birwal, & Ritesh B. Watharkar (Eds.)
© 2024 Apple Academic Press, Inc. Co-published with CRC Press (Taylor & Francis)

Nutrition" suggests that the global population is predicted to cross 10 billion in 2050, thus requiring a 50% increase in global food production to meet the nutritional requirements [55]. The vision of the entire world community is to provide safe, nutritious, and high-quality food to the emerging population, which has always been a pressing concern for food supply system. Despite significant amounts of food waste being generated during processing and storage, it is predicted that more than 820 million people would be deficit of food [27].

The concept of food preservation has evolved throughout to obtain quality and safe food products with extended shelf-life [14]. Consumers demand high-quality and safe food with increased nutritional levels of vitamins, minerals, and antioxidants. Thermal sterilization/pasteurization techniques ensure effective elimination of microbes, but it results in significant loss of thermo-labile components and has a negative impact on taste, physicochemical, and nutritional qualities [47]. Current barriers, such as high investment prices, poor control of factors connected with process operation, and lack of regulatory permission have slowed the use of these technologies on a large scale [15].

The extent to which nonthermal food processing a preservation method (i.e., ohmic heating, pulsed electric field (PEF), pulsed-light (PL), supercritical carbon dioxide, high hydrostatic pressure (HHP), ultrasound (US) and irradiation) are being adapted is uncertain. As a result, the experts have surveyed to determine the extent of its use in the United States (US), as well as the major reasons for food manufacturers to develop certain specific food processing technologies. By retaining unstable bioactive components and modifying enzyme activity, nonthermal food processing and preservation techniques can fulfill microbiological food safety criteria, while also increasing the product's physical, sensory, and nutritional properties [2].

This chapter elucidates the scientific information on the use of novel food processing and quality analysis techniques (such as: cold plasma (CP), plasma activated water (PAW), surface enhanced Raman spectroscopy (SERS), radiofrequency (RF) heating, and chlorine dioxide (ClO_2) technology) in the food industry.

1.2 NOVEL TECHNOLOGIES

Food scientists have started to investigate on various novel processing and microbial inactivation techniques alternate to thermal food processing that would produce a safer product with superior nutritional content and sensory qualities. The most researched processing and quality analysis techniques currently have been presented in this section.

1.2.1 COLD PLASMA (CP) TECHNOLOGY

This is an effective decontamination and shelf-life extension capability; and is an alternative tool for conventional food processing methods. With its unique, nonthermal, low-cost, adaptable, and eco-friendly approach, it can be used in the food industry for enzyme inactivation [38], toxin elimination, food packaging decontamination, and wastewater treatment, etc. [50]. In the food processing industry, CP has proven to be effective against food-borne pathogenic bacteria, such as: *E. coli, S. aureus, S. typhimurium,* and *L. monocytogenes* [72].

1.2.1.1 MECHANISM OF CP

Plasma can be produced by ionizing gases (such as: helium, nitrogen, and argon), visible light, UV radiation, Electric or electromagnetic fields, electrical, thermal, X-ray, and radioactive (gamma radiation), electromagnetic waves, which are widely used to generate CP (Figure 1.1) [41]. The versatility of CP generating sources allows for unique designs that are compatible with modern food industries. "Dielectric barrier discharge (DBD)" and "jet plasma" are the most extensively used types of plasma generating techniques in the food industry [16, 20]. The principles of various CP producing devices are briefly summarized in Table 1.1.

The plasma jet device consists of two concentric metal electrodes: the inner electrode is usually connected to a high RF power source, which causes the working gas to ionize and exit the nozzle in a jet fashion. Two metal electrodes are used in DBD devices, one of which is covered by a dielectric barrier. The dielectric barrier acts as a stabilizing agent, preventing arc transitions and facilitating the development of a large number of microdischarges for uniform treatment. When the packaged food is placed between two electrodes, the head space gas is ionized, which creates reactive species. This type of processing method assists in scaling up of the technique into a continuous pilot scale system with increased antibacterial efficiency, and reduction in the cross contamination [16, 42].

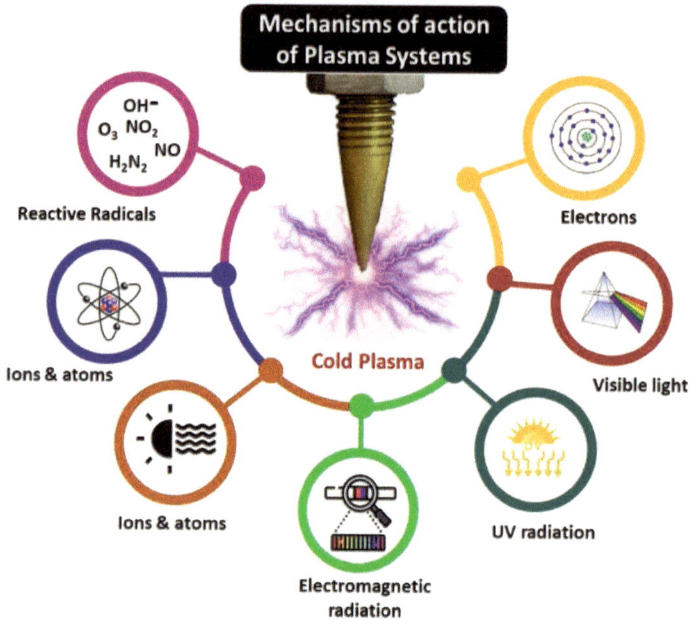

FIGURE 1.1 Mechanism of cold plasma system.

TABLE 1.1 Plasma Generation Principles of Various CP Devices

Cold Plasma Producing Devices	Working Mechanism
Corona discharge	Plasma is formed when an electric field exists at the tip of a pointed electrode, and it is a self-contained discharge.
Dielectric barrier discharge (DBD)	Plasma is formed when two electrodes (Cu, Alu, Brass, Steel) are surrounded by a layer of dielectric barrier material (quart, PP, ceramic, or glass).
	A voltage supply (AC or DC current) can be used to work at high pressures (104–106 Pa).
Microwave discharge (MD)	Plasma is generated by a magnetron in an open, closed, or resonant configuration contained in a cooling system. It makes use of microwaves (2.45 GHz (gigahertz)) at 50–1,000 W power range.
Plasma jet (PJ)	Plasma is generated by the energized electrode and flows into a different surface. DBD jets and similar jets, single electrode jets, and dielectric electrode jets are the four most prevalent forms of plasma jets.
Radiofrequency discharge (RFD)	Plasma is formed at the tip of an electrode needle and is extended to the outside ring electrodes, which are coupled to RF voltage electrodes (13.56 MHz).

1.2.1.2 EFFICACY

CP inactivates bacterial spores, fungi, and biofilms by attacking these with radical species (OH^- and NO^-), which cause the volatile molecules (such as: CO_2 and H_2O) to be adsorbed onto the surface. The cell membrane is subjected to high intensity electric fields to create an electrostatic tension within the cell, causing cell death by creating wounds on the surface [11].

Inert gasses (such as N_2, He, and Ar) have been frequently used in CP for several reasons, including high thermal conductivity, a broad UV emission spectrum, and atmospheric pressure discharge voltage. However, air seems to be the better option because other options are not cost-effective [37]. The rate at which active species are driven towards the target location is influenced by gas flow rates. In addition to velocity, the gas flow rate influences the retention time, discharge operation, and mass transfer procedure, thus certain short-lived active plasma species may not reach the sample effectively [39]. The flow rate is directly proportional to the impact and potential of reactive species to react; and any increase in the flow rate will also increase the collision and reaction time [66].

The short-lived reactive plasma species are transported into the cell through external cell membrane and causes its damage (Figure 1.2); however, the penetration depth is restricted since each element in liquid matter interacts with applied plasma. As bacteria and components are modified to demonstrate efficient antibacterial capacity while preserving other dietary ingredients, optimization is critical [58].

1.2.1.3 FUTURE TRENDS AND SCOPE OF CP

The knowledge of reactive chemical plasma species, inactivation procedures, and other unfavorable issues influence the future of CP in the food industry. UV-induced and chemical effects on radicals and reactive plasma species are important components of its antibacterial action; however, a better understanding of its mechanism of action is required before it can be optimized for specific food processing applications.

CP is a chemical-free processing technique; making it a good alternative for more organic food processing research. The impact of CP processing on taste, flavor, aroma, and other sensory aspects needs in-depth investigation for better understanding. It must also be assessed if any chemical residues are left over, which in turn can cause toxicological effects. The evidence on insignificant effect of CP on various food product categories will enable regulatory authorities to authorize and raise public awareness for its approval for broader applications.

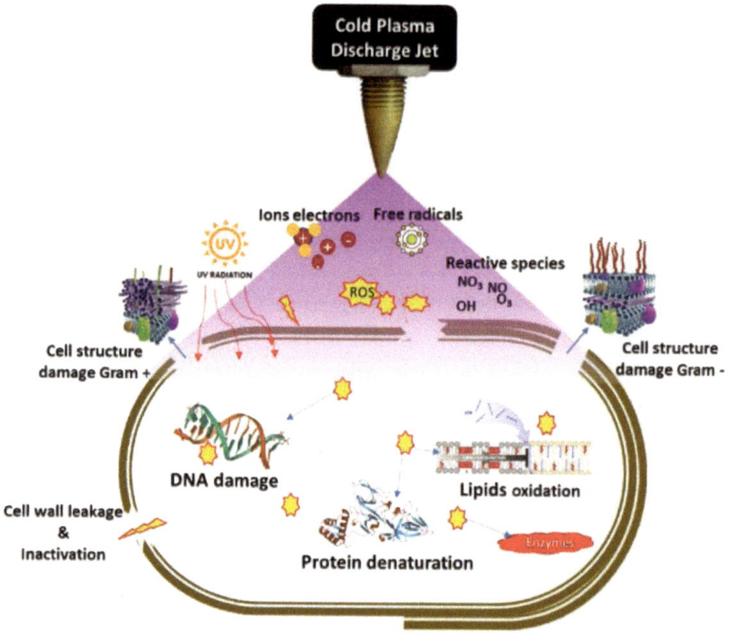

FIGURE 1.2 Bacterial cell damage (internal and external) caused by cold plasma.

1.2.2 PLASMA-ACTIVATED WATER (PAW)

Plasma-activated water (PAW) is an advancement of CP treatment technique with the capability to overcome the limitations of CP equipment and acting directly on food surface for its decontamination [19, 63]. It is a cutting-edge eco-friendly technology that has gained a lot of interest recently due to its ability to decontaminate food from both biological and chemical origins when compared to other common chemical sanitizers [62]. The uses of PAW technique are being studied extensively in seed germination, to enhance plant growth, to retard various pests and diseases, and for several aspects of agriculture [19].

1.2.2.1 MECHANISM

Plasma is capable of generating both electric and magnetic fields, ultra-violet (UV) rays, electrons, charged particles, reactive oxygen, and nitrogen species (NOS) [5, 51]. PAW is produced primarily through atmospheric cold plasma (ACP) discharge in any of three forms, such as: direct discharge within liquids, indirect discharge in the gas phase over the liquid surface and multiphase discharges (such as discharges in bubbles within contacting liquid sprays) [9, 22]. The most common plasma sources to generate PAW are plasma jet, gliding arc discharge (GDA), surface micro-discharge (SMD), and DBD [3, 6, 22].

Several plasma–water reaction systems (such as GDA (Figure 1.3), pulsed corona discharge, DBD and plasma steamer discharge) are frequently used to generate PAW to facilitate these plasma–water interactions [21, 43]. APPJ (Atmospheric pressure plasma jet) uses a discharge plasma that extends beyond the surrounding atmosphere of the plasma generation area and it assists in strong interaction with the water molecules. APPJ and DBD are most commonly used, because of their ability to produce a diverse range of reactive species in PAW [7, 21, 30].

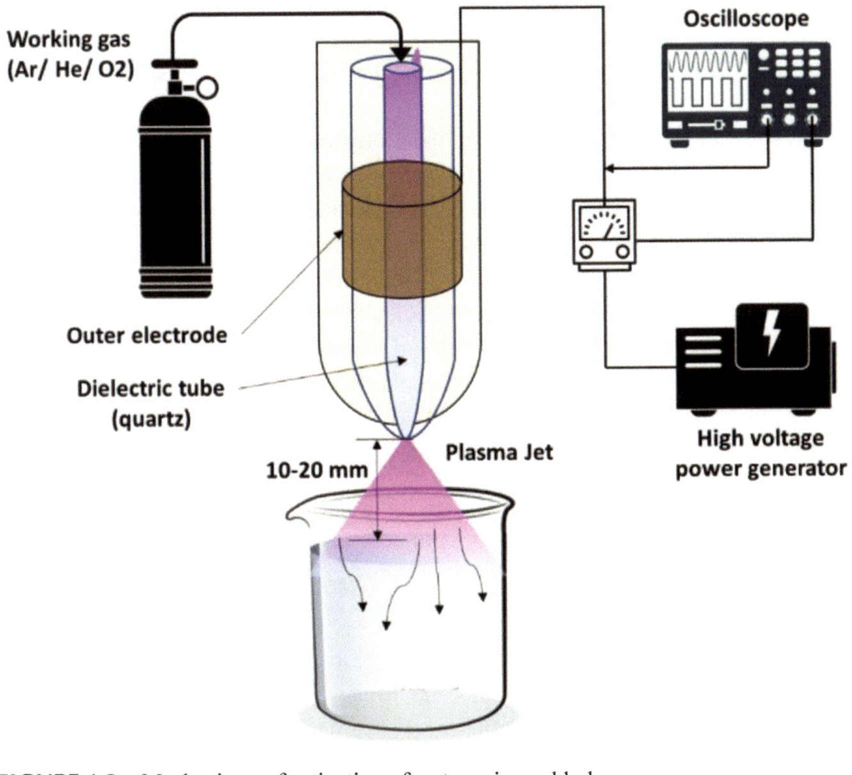

FIGURE 1.3 Mechanisms of activation of water using cold plasma.

The plasma generator has two major components, such as: a copper spring and a quartz cylindrical tube. The quartz tube inner wall is connected to the high-voltage electrode (copper spring). The external wall of the quartz tube is enmeshed inside a copper mesh, which functions as a ground electrode. The discharge is achieved approximately about 110 mm in length, at atmospheric air pressure by maintaining a flow rate of 2.5 L/min. The reactive CP species produced in the gas phase are then transported from the gas phase to the liquid phase with an enhanced rate of mass transfer using an aerator. Small diameter bubbles (1 mm) are blown into the reaction cylinder, which were expected to operate as micro-reactors, permitting the transport of active species in the air into liquid [67, 70].

PAW is utilized in place of chemical substances (like chlorine) in a variety of culinary and industrial processes. PAW has the power to inactivate harmful bacteria and prevent cross contamination for one period of time, after which the chemical that affects bacteria inactivation will dissipate [48].

1.2.2.2 EFFICACY

The characteristics of the working, gas composition, humidity, and flow rate have a significant impact on PAW efficiency [69]. The distance between the plasma discharge tip and the water surface is another parameter to be considered while generating PAW above the water surface. Extensive studies are in progress to understand the influence of probe distances (from 1 to 4 mm) on the generation of active species specifically N_2 and OH near the water surface, and it has been concluded that increasing the gap between probe tip and water will increase the electron density thus decreasing OH^- concentration [28]. In contrast, PAW in combination with low temperature can enhance the inactivation efficiency, while the high temperatures can exacerbate the extent of microbial destruction [5, 32, 69].

1.2.2.3 PAW: FUTURE TRENDS AND SCOPE FOR DEVELOPMENT

Despite the fact that various PAW generation devices have been developed over the years, the high cost and the low yield are major constraints for application of PAW to enhance the food safety and quality. The production of PAW in large volumes at cost effective way is a key challenge, which needs to address in future research to enhance the PAW application. In addition, transition from "He" to a more simpler and cost-effective plasma gas (air) is also critical, along with a better understanding of PAW stability over time. The current PAW production devices do not meet food processing requirements, and the manufacturing cost of PAW is higher (one euro per 30 L of PAW) than the chemical disinfectants, which is also a limiting factor for applications of PAW in the food industries.

1.2.3 SURFACE ENHANCED RAMAN SPECTROSCOPY (SERS)

Surface-Enhanced Raman Spectroscopy (SERS: Figure 1.4) can analyze trace elements (such as pesticide residues, toxicants, abandoned pharmaceuticals), and food additives in complex food matrices. SERS has evolved as a better and efficient technique for quick and accurate detection of multiple analytes in food products compared to other detection techniques, because it offers high sensitivity and quick "fingerprinting identification" ability [33, 60]. Noble metal nanostructures (such as: silver or gold or copper) use an electromagnetic and chemical enhancement techniques to enhance or amplify the Raman signals of molecules in these metal nanostructures [33, 34, 65].

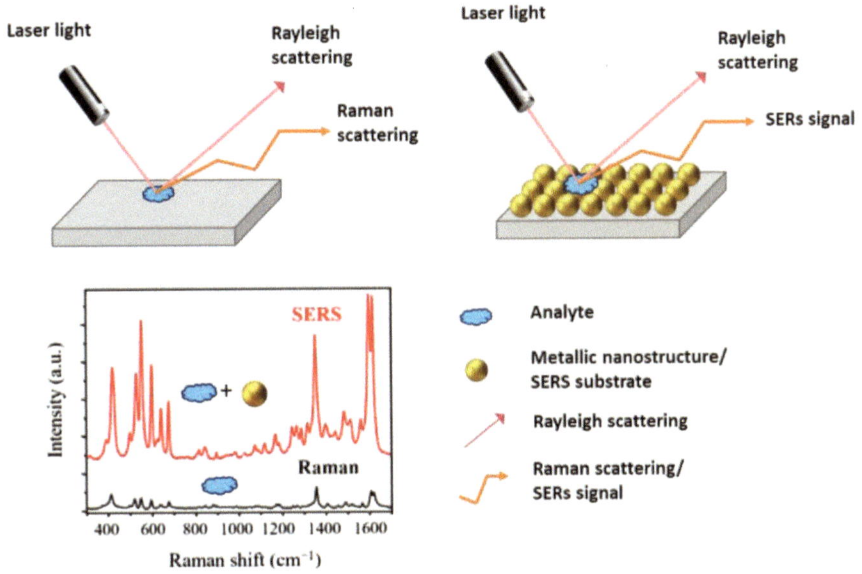

FIGURE 1.4 Mechanism of surface enhanced Raman spectroscopy (SERS).

1.2.3.1 MECHANISM OF SERS

Raman spectroscopy works on the basis of a spectrum scattering to analyze the food samples based on vibrations of molecules, rotation of molecules, or other low-frequency modes. When a single frequency laser beam passes through a molecule, it creates polarization, which can lead to the formation of higher energy intermediate "virtual state" that rapidly emits photons as scattered radiation (Figure 1.4). In general, both the scattered and incident photons will have similar frequency, which is termed as Rayleigh scattering; however, Raman scattering accounts only small proportion of scattered photons, which occurs only when one out of every 100 scattered photons have a different energy than the incident photons [22].

SERS employs a unique surface enhancement technique to improve the analysis of Raman scattering signals [22, 68]. The features of SERS substrates (such as food additives, food toxicants, etc.) are examined, and it focuses on the chemical properties of the food [61]. To achieve the preeminent SERS effect, it is critical to select an appropriate wavelength for the incident laser. For example, 633 nm is closer to crystal violet's maximum absorption wavelength (max. of 590 nm that had a stronger enhancement effect than 780 nm [31]. Similarly, 633 nm laser would perform superior than the 780 nm laser in SERS analysis of malachite green, offering an ultra-high sensitivity to detect chemicals as low as 1 ppb or below [22, 71]. The total SERS effect is a result of multiple enhancement processes, with the relative degree of enhancement induced by each resonance that depends on the changes in experimental parameters [22].

1.2.3.2 EFFICACY

SERS methodologies are of two types depending on their effectiveness in enhancing SERS spectra: (i) The most commonly used technique is to enhance the effectiveness, shape, size, composition, and SERS-active nanostructures self-assembly, such as, metal nanoparticles (NPs); and (ii) The other technique is to develop an effective target separation or concentration processes for removing target molecules from complex matrixes more quickly and efficiently [10, 24].

For the sensitive detection of two food chemical hazards, a SERS microsystem has been developed. To enrich argon NPs and analytic compounds, this system employs a porous microfluidic matrix composed of packed silica microspheres. In addition, two optical fibers are inserted and aligned into the microfluidic system to increase automation. The limits of detection (LOD) for melamine and a fungicide (called thiram) were 63 ppb and 50 ppb, respectively [32, 64].

1.2.3.3 SERS: FUTURE TRENDS AND SCOPE FOR DEVELOPMENT

SERS has shown a huge potential for rapid analysis and detection of trace compounds; however, the technology is still in infancy stage to be employed for commercial applications [22]. Significant research advances have been achieved to enhance the sensitivity, selectivity, and repeatability in a variety of food matrices. Faster and better resolution chemical pictures are possible with strong Raman imaging microscopes, which can provide critical information in *in-situ* study; and faster and higher resolution chemical images are conceivable. Still, several challenges need to be addressed before it can be used for realistic food analysis.

1.2.4 PASTEURIZATION VIA RADIOFREQUENCY (RF) HEATING

Radiofrequency heating (RFH) is an effective pasteurization technique for maintaining food safety and prolong the shelf-life, since it provides rapid and volumetric heating. The faster heating rate of RFH allows the product quality to be maintained while reducing the heat treatment time [25]. RFH generates heat within the food, which is then distributed from a heated medium to a food sample for more uniform heating, whereas the heat is transferred from a heated medium to a food sample in the traditional method, resulting in non-uniform heat transfer [26].

1.2.4.1 MECHANISM OF RADIOFREQUENCY HEATING (RFH)

RFH is a non-ionizing radiation heating method because a very low energy (<10 eV) is generally produced to ionize the biological atoms. Despite the fact that RF waves have a frequency range of 1–300 MHz, the US Federal Communications Commission (FCC) permits three frequencies of 13.56, 27.12, and 40.68 MHz for industrial, scientific, and medical applications (ISM), respectively [40, 44].

Electromagnetic energy is absorbed by water, salts, polar, and charged molecules in the food and is converted to heat. The product is fed during RFH inside the RF chamber, and an alternating electric field is applied between two electrodes (Figure 1.5). An alternating electric field heats the food due to the friction among spinning polar dielectric molecules and moving ions [35]. The heat transfer by conduction and convection depends on the food matrix and might prolong the heating. As a result, the physio-chemical changes can also occur significantly in thermally processed foods, resulting in changes in sensory, textural properties and loss in the nutritive value [54].

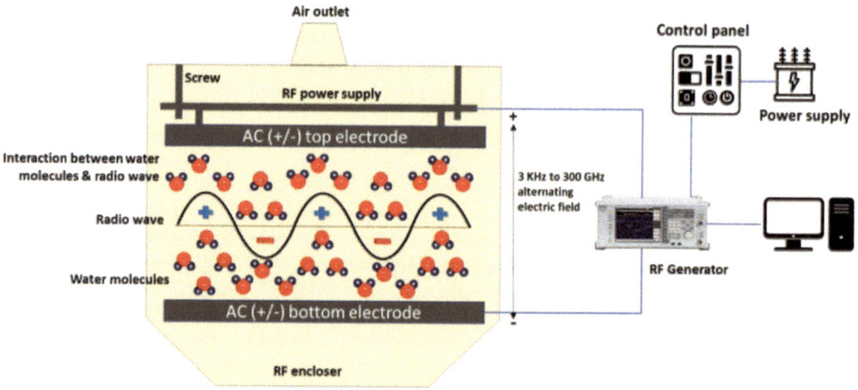

FIGURE 1.5 Mechanism of radiofrequency heating of foods.

1.2.4.2 EFFICACY

RFH has the potential to replace conventional heating for semi-solid and solid foods having low to moderate thermal conductivities [23]. With limited applications, RFH has been used in the food industries, where two RF applications have already been commercialized: biscuit post-baking and meat thawing/tempering, while RF drying and disinfestation are partially marketed processes. The combination of hot air circulation and RF treatment has been found to be effective for drying and is frequently used to reduce operational costs. Furthermore, RFH is a proven alternative method to fumigation, chemical treatment to inactivate pests in food products. The food industry has recently commercialized RF postharvest disinfestation of agricultural samples [26].

Furthermore, heating of low moisture meals is slow and ineffective due to their poor thermal diffusivities. RFH can overcome the limits of conventional heating by providing volumetric heat with deep penetration into the food [26].

1.2.4.3 RADIOFREQUENCY HEATING (RFH): FUTURE TRENDS AND SCOPE FOR DEVELOPMENT

RFH provides several advantages, such as the capability to instantly penetrate up to 20 cm or more into food, more consistent and efficient heating, and less unwanted changes (such as: decreased food quality or sensory changes). RF pasteurization has yet to be commercialized due to the considerable limitations of RFH (i.e., non-uniformity in temperature distribution). Computer simulations, on the other hand, can provide an understanding of heating mechanisms by solving the governing equations and demonstrating the 3D distribution of the selected parameters, such as: temperature and electric field distribution with the least amount of time, labor, and experimental work.

Microwaves (MWs) may be phased out in favor of more efficient RF cooking and heating equipment. Current MW equipment includes a single-button option that controls MW cycling over time for specified foods. Consider an RF oven with single-button control that allows to change the frequency, duration, and cycle to increase the quality and health benefits of a certain food. Because of the research findings with RFH in the past, it is apparent that RFH is a highly appealing processing strategy for producing safe and high-quality food items.

1.2.5 PACKAGING SOLUTIONS WITH CHLORINE DIOXIDE (CLO_2)

Chlorine dioxide (ClO_2) is a chemical compound formed by reactions of chlorine and oxygen. Despite the fact that it has the word "chlorine" in its name, it is not due to the presence of chlorine (bleach), as most people think [4]. ClO_2 has been used as a powerful disinfectant for packaging materials, metal cans, and sterilizing medical instruments across the world. It finds its application in food processing for decontamination and shelf-life enhancement of perishable products by inactivating bacteria, yeasts, molds, and some viruses [4, 29].

1.2.5.1 MECHANISM

ClO_2 has a higher penetrating capacity than the aqueous solutions, and ClO_2 gas is frequently used for food decontamination. ClO_2 gas diffuses to oxidize proteins on the cell surface resulting in failure in cellular metabolism [45, 56]. When hydrochloric acid (strong acid) is combined with sodium chlorite or sodium chlorate, a salt-based compound known as ClO_2 is formed in a chemical reactor, which is often employed as a surface disinfectant. The gas can be used either indirectly (by absorbing it in water before applying it to the disinfected product) or directly (by soaking it in water before applying it to the disinfected product) through direct or indirect contact with a food without any negative consequences. ClO_2 gas can be sprayed as gas on the surface that can to sanitize the food equipment, or even disseminate into the air to minimize odors or to disinfect the spoilage-causing organisms in the ambient [4, 29].

1.2.5.2 EFFICACY

The duration of ClO_2 exposure with food or surface, the concentration of targeted pathogen, and type of food being treated are important factors that affect the performance of ClO_2 treatment. Soft rot in fruits can be reduced by 80–90% by exposing fruits to small concentration of ClO_2 gas (2 mg/kg of fruit) for two hours [36]. Furthermore, ClO_2 treatment does not produce any chemical residue on the surface of treated food [4, 59]. The use of ClO_2 gas in processed products (such as: sliced, diced, and boiled products) has been approved by the US Food and Drug Administration (FDA); while the US Environmental Protection Agency (EPA) has approved the use of ClO_2 gas only for raw potatoes, tomatoes, and melons.

1.3 SUMMARY

The food sector needs novel technologies to handle the difficulties of rising competition, globalization, and increasingly dynamic and diverse consumer demands. Novel nonthermal food processing and preservation technologies are superior for processing and preserving food quality than thermally processed food products. This chapter also documents the knowledge of these novel techniques, their working mechanisms, efficacy, limitations, and their future scope. More detailed research is needed to understand the potential capability of novel techniques discussed in this chapter. Newer concepts (such as cold plasma, PAW, surface Raman spectroscopy, RFH, and ClO_2) are still in the research stage and will make their way into commercial applications in the near future.

KEYWORDS

- **chlorine dioxide packaging**
- **cold plasma technology**
- **food processing**
- **novel technologies**
- **plasma activated water**
- **radiofrequency pasteurization**
- **surface-enhanced Raman spectroscopy**

REFERENCES

1. Altemimi, A., Aziz, S. N., Al-Hilphy, A. R., Lakhssassi, N., Watson, D. G., & Ibrahim, S. A., (2019). Critical review of radiofrequency (RF) heating applications in food processing. *Food Quality and Safety, 3*(2), 81–91.
2. Amaral, G. V., Silva, E. K., Cavalcanti, R. N., Cappato, L. P., Guimaraes, J. T., Alvarenga, V. O., Esmerino, E. A., et al., (2017). Dairy processing using supercritical carbon dioxide technology, theoretical fundamentals, quality, and safety aspects. *Trends in Food Science and Technology, 64*(1), 94–101.
3. Andrasch, M. J., Stachowiak, O., Schluter, U., Schnabel, & Ehlbeck, J., (2017). Scale-up to pilot plant dimensions of plasma processed water generation for fresh-cut lettuce treatment. *Food Packaging and Shelf Life, 14*(1), 40–45.
4. Arnold, N., Yang, L., & Boyer, R. R., (2019). *How is Chlorine Dioxide Gas Used to Process Foods*. Virginia Cooperative Extension, Virginia State University. https://vtechworks.lib.vt.edu/bitstream/handle/10919/95919/FST-325.pdf?sequence=1&isAllowed=y (accessed on 09 February 2023).
5. Bai, Y., Muhammad, A. I., Hu, Y., Koseki, S., Liao, X., Chen, S., & Ding, T., (2020). Inactivation kinetics of *Bacillus cereus* spores by plasma activated water (PAW). *Food Research International, 13*(1), 109041.
6. Barrionuevo, J. M., Berrone, P., & Ricart, J. E., (2012). Smart cities, sustainable progress. *IESE Insight, 14*(14), 50–57.
7. Bourke, P., Ziuzina, D., Han, L., Cullen, P., & Gilmore, B., (2017). Microbiological interactions with cold plasma. *Journal of Applied Microbiology, 123*(2), 308–324.

8. Cappato, L. P., Ferreira, M. V. S., Guimarães, J. T., Portela, J. B., Costa, A. L. R., Freitas, M. Q., Cunha, R. L., et al., (2017). Ohmic heating in dairy processing: Relevant aspects for safety and quality. *Trends in Food Science and Technology, 62*(1), 104–112.
9. Ceriani, E., Marotta, E., Schiorlin, M., Ren, X., Ceretta, C., Gobbo, R., & Paradisi, C., (2018). A versatile prototype plasma reactor for water treatment supporting different discharge regimes. *Journal of Physics D: Applied Physics, 51*(27), 274001.
10. Chen, J., Park, B., Huang, Y. W., Zhao, Y., & Kwon, Y., (2018). Label-free SERS detection of *Salmonella typhimurium* on DNA aptamer modified AgNR substrates. *J. Food Meas. Charact., 11*(1), 1773–1779.
11. Coutinho, N. M., Silveira, M. R., Rocha, R. S., Moraes, J., Ferreira, M. V. S., Pimentel, T. C., & Cruz, A. G., (2018). Cold plasma processing of milk and dairy products. *Trends in Food Science and Technology, 74*(1), 56–68.
12. Dimitrakellis, P., Giannoglou, M., Xanthou, Z. M., Gogolides, E., Taoukis, P., & Katsaros, G., (2021). Application of plasma-activated water as an antimicrobial washing agent of fresh leafy produce. *Plasma Processes and Polymers, 18*(12), 2100030.
13. Ellis, D. A., Martin, J. W., De Silva, A. O., Mabury, S. A., Hurley, M. D., Sulbaek, A. M. P., & Wallington, T. J., (2004). Degradation of fluorotelomer alcohols: A likely atmospheric source of perfluorinated carboxylic acids. *Environmental Science & Technology, 38*(12), 3316–3321.
14. Ellis, D. I., Muhamadali, H., Haughey, S. A., Elliott, C. T., & Goodacre, R., (2015). Point-and-shoot: Rapid quantitative detection methods for on-site food fraud analysis–moving out of the laboratory and into the food supply chain. *Analytical Methods, 7*(22), 9401–9414.
15. Fellows, P. J., (2017). *Food Processing Technologies: Principles and Practices* (p. 223). Cambridge, M A: Wood head Publishing, Elsevier.
16. Fox, J., & Gutiérrez, J. M., (2018). *Snake Venom Metalloproteinases*. MDPI.
17. Gao, G., Ren, P., Cao, X., Yan, B., Liao, X., Sun, Z., & Wang, Y., (2016). Comparing quality changes of cupped strawberry treated by high hydrostatic pressure and thermal processing during storage. *Food and Bioproducts Processing, 100*, 221–229.
18. Gómez-López, V. M., Rajkovic, A., Ragaert, P., Smigic, N., & Devlieghere, F., (2009). Chlorine dioxide for minimally processed produce preservation: A review. *Trends in Food Science & Technology, 20*(1), 17–26.
19. Guo, D., Liu, H., Zhou, L., Xie, J., & He, C., (2021). Plasma-activated water production and its application in agriculture. *Journal of the Science of Food and Agriculture, 101*(12), 891899.
20. Han, Y., Cheng, J. H., & Sun, D. W., (2019). Activities and conformation changes of food enzymes induced by cold plasma: A review. *Critical Reviews in Food Science and Nutrition, 59*(5), 794–811.
21. Herianto, S., Hou, C. Y., Lin, C. M., & Chen, H. L., (2021). Nonthermal plasma-activated water: A comprehensive review of this new tool for enhanced food safety and quality. *Comprehensive Reviews in Food Science and Food Safety, 20*(1), 583–626.
22. Huang, Y., Wang, X., Lai, K., Fan, Y., & Rasco, B. A., (2020). Trace analysis of organic compounds in foods with surface-enhanced Raman spectroscopy: Methodology, progress, and challenges. *Comprehensive Reviews in Food Science and Food Safety, 19*(2), 622–642.
23. Jeong, S. G., & Kang, D. H., (2014). Influence of moisture content on inactivation of *Escherichia coli* O157: H7 and *Salmonella enterica* serovar typhimurium in powdered

red and black pepper spices by radio-frequency heating. *International Journal of Food Microbiology, 176*, 15–22.
24. Jiang, Y., Sun, D. W., Pu, H., & Wei, Q., (2018). Surface enhanced Raman spectroscopy (SERS): A novel reliable technique for rapid detection of common harmful chemical residues. *Trends in Food Science & Technology, 75*, 10–22.
25. Jiao, Y., Tang, J., & Wang, S., (2014). A new strategy to improve heating uniformity of low moisture foods in radio frequency treatment for pathogen control. *J. Food Eng., 141*, 128–138.
26. Jiao, Y., Tang, J., Wang, Y., & Koral, T. L., (2018). Radio-frequency applications for food processing and safety. *Annu. Rev. Food Sci. Technol., 9*, 105–127.
27. Katsigiannis, A. S., Bayliss, D. L., & Walsh, J. L., (2021). Cold plasma decontamination of stainless-steel food processing surfaces was assessed using an industrial disinfection protocol. *Food Control, 121*, 107543.
28. Kinandana, A. W., Sumariyah, S., & Nur, M. (2018). Analysis of pH on plasma activated medium in aqueous solution in cold atmospheric pressure plasma jets. *International Journal of Innovative Research in Advanced Engineering, 5*, 22–25.
29. Kingsley, D. H., Pérez-Pérez, R. E., Niemira, B. A., & Fan, X., (2018). Evaluation of gaseous chlorine dioxide for the inactivation of Tulane virus on blueberries. *International Journal of Food Microbiology, 273*, 28–32.
30. Kong, M., Ganguly, B., & Hicks, R., (2012). Plasma jets and plasma bullets. *Plasma Sources Science and Technology, 21*(3), 030201.
31. Li, Y., Li, F., Tang, J., Zhang, R., Wang, Y., Koral, T., & Jiao, Y., (2018). Radio frequency tempering uniformity investigation of frozen beef with various shapes and sizes. *Innov. Food Sci. Emerg. Technol., 48*, 42–55.
32. Liao, W., & Lu, X., (2016). Determination of chemical hazards in foods using surface-enhanced Raman spectroscopy coupled with advanced separation techniques. *Trends in Food Science & Technology, 54*, 103–113.
33. Lin, Z., & He, L., (2019). Recent advance in SERS techniques for food safety and quality analysis: A brief review. *Current Opinion in Food Science, 28*, 82–87.
34. Liu, Y., Zhou, H., & Hu, Z., (2017). Label and label-free based surface-enhanced Raman scattering for pathogen bacteria detection: A review. *Biosensors and Bioelectronics, 94*, 131–140.
35. Llave, Y., Terada, Y., Fukuoka, M., & Sakai, N., (2014). Dielectric properties of frozen tuna and analysis of defrosting using a radio-frequency system at low frequencies. *Journal of Food Engineering, 139*, 1–9.
36. Mahovic, M. J., Joel, D. T., & Jerry, A. B., (2007). Applications of chlorine dioxide gas for control of bacterial soft rot in tomatoes. *Plant Disease, 91*(10), 1316–1320.
37. Misra, N. N., & Jo, C., (2017). Applications of cold plasma technology for microbiological safety in the meat industry. *Trends in Food Science & Technology, 64*, 74–86.
38. Misra, N. N., Pankaj, S. K., Segat, A., & Ishikawa, K., (2016). Cold plasma interactions with enzymes in foods and model systems. *Trends Food Sci. Technol., 55*, 39–47.
39. Nishime, T. M. C., Borges, A. C., Koga-Ito, C. Y., Machida, M., Hein, L. R. O., & Kostov, K. G., (2017). Nonthermal atmospheric pressure plasma jet applied to inactivation of different microorganisms. *Surface and Coating Technology, 312*, 19–24.
40. Ozturk, S., Kong, F., & Singh, R. K., (2020). Evaluation of enterococcus faecium NRRL B-2354 as a potential surrogate of *Salmonella* in packaged paprika, white pepper and cumin powder during radio frequency heating. *Food Control, 108*, 106833.

41. Pankaj, S. K., & Kenner, K. M., (2017). Cold plasma: Background, applications and current trends. *Current Opinion in Food Science, 16,* 49–52.
42. Pankaj, S. K., & Keener, K. M., (2017). Cold plasma applications in food packaging. In: *Reference Module in Food Science* (pp. 46–64). Elsevier: Amsterdam, The Netherlands.
43. Perinban, S., Orsat, V., & Raghavan, V., (2017). Nonthermal plasma liquid interactions in food processing: A review. *Comprehensive Reviews in Food Science and Food Safety, 18*(6), 1985–2008.
44. Piyasena, P., Dussault, C., Koutchma, T., Ramaswamy, H. S., & Awuah, G. B., (2003). Radio frequency heating of foods: Principles, applications and related properties—A review. *Critical Reviews in Food Science and Nutrition, 43*(6), 587–606.
45. Ran, Y., Qingmin, C., & Maorun, F., (2019). Chlorine dioxide generation method and its action mechanism for removing harmful substances and maintaining quality attributes of agricultural products. *Food and Bioprocess Technology, 12*(7), 1110–1122.
46. Ronit, M., Anika, S., & Anubhav, P. S., (2018). Recent developments in cold plasma decontamination technology in the food industry, *Trends in Food Science & Technology,* 1–37.
47. Roobab, U., Aadil, R. M., Madni, G. M., & Bekhit, A. E., (2018). The impact of nonthermal technologies on the microbiological quality of juices: A review. *Compr. Rev. Food Sci. F., 17,* 437–457.
48. Royintarat, T., Boonyawan, D., Seesuriyachan, P., & Wattanutchariya, W., (2018). A comparison of plasma activated water techniques for bacteria inactivation. *Journal of Biomaterials and Tissue Engineering, 8*(6), 887–891.
49. Sarangapani, C., Misra, N. N., Milosavljevic, V., Bourke, P., O'Regan, F., & Cullen, P. J., (2016). Pesticide degradation in water using atmospheric air cold plasma. *J. Water Process Eng., 9,* 225–232.
50. Schnabel, U., Sydow, D., Schluter, O., Andrasch, M., & Ehlbeck, J., (2015). Decontamination of fresh-cut iceberg lettuce and fresh mung bean sprouts by nonthermal atmospheric pressure plasma processed water (PPW). *Modern Agricultural Science and Technology, 1*(1), 23–39.
51. Scholtz, V., Pazlarova, J., Souskova, H., Khun, J., & Julak, J., (2015). Nonthermal plasma-A tool for decontamination and disinfection. *Biotechnology Advances, 33*(62), 1108–1119.
52. Segat, A., Misra, N. N., Cullen, P. J., & Innocente, N., (2016). Effect of atmospheric pressure cold plasma (ACP) on activity and structure of alkaline phosphatase. *Food and Bioproducts Processing, 98,* 181–188.
53. Sharma, B., Frontiera, R. R., Henry, A. I., Ringe, E., & Van, D. R. P., (2012). SERS: Materials, applications, and the future. *Materials Today, 15,* 16–25.
54. Siefarth, C., & Buettner, A., (2014). The aroma of goat milk: Seasonal effects and changes through heat treatment. *Journal of Agricultural and Food Chemistry, 62*(49), 11805–11817.
55. Siefarth, C., Tran, T. B., Mittermaier, P., Pfeiffer, T., & Buettner, A., (2014). Effect of radio frequency heating on yoghurt, II: Microstructure and texture. *Foods (Basel, Switzerland), 3,* 369–393.
56. Sun, X., Baldwin, E., & Bai, J., (2019). Applications of gaseous chlorine dioxide on postharvest handling and storage of fruits and vegetables: A review. *Food Control, 95,* 18–26.

57. Surowsky, B., Bußler, S., & Schlüter, O. K., (2016). Cold plasma interactions with food constituents in liquid and solid food matrices; Chapter 7. In: Misra, N. N., Schlüter, O. K., & Cullen, P. J., (eds.), *Cold Plasma in Food and Agriculture: Fundamentals and Application* (pp. 179–204). London: Elsevier.
58. Tchoursinov, Y., Kovaliova, O., Kalyna, V., Mykolenko, S., & Khomuk, N. I., (2020). *Phyto-Feed Additives Production: Technological Aspects and Biological Value* (p. 11). Scientific Bulletin, University of Agronomic Sciences and Veterinary Medicine of Bucharest. http://elartu.tntu.edu.ua/bitstream/lib/33064/1/Biotechnologies_VolXXIV-2%282020%29_220pag%20BT1%20%281%29%20%282%29.pdf (accessed on 09 February 2023).
59. Trinetta, V., Vaidya, N., Linton, R., & Morgan, M., (2011). Evaluation of chlorine dioxide gas residues on selected food produce. *Journal of Food Science, 76*(1), 11–15.
60. Wang, C., Madiyar, F., Yu, C., & Li, J., (2017). Detection of extremely low concentration waterborne pathogen using a multiplexing self-referencing SERS microfluidic biosensor. *Journal of Biological Engineering, 11*, 9.
61. Wang, P., Wu, L., Lu, Z., Li, Q., Yin, W., Ding, F., & Han, H., (2017). Gecko-inspired nanotentacle surface-enhanced Raman spectroscopy substrate for sampling and reliable detection of pesticide residues in fruits and vegetables. *Analytical Chemistry, 89*, 2424–2431.
62. Wu, V. C. H., & Rioux, A., (2010). A simple instrument-free gaseous chlorine dioxide method for microbial decontamination of potatoes during storage. *Food Microbiology, 27*, 179–184.
63. Yang, B., Zhong, C. S., Wang, W. Z., Liao, Z. J., Chen, R., & Wang, Y., (2019). Advances in the application of plasma-activated water in agriculture. *Agri Eng., 9*, 101–107.
64. Yazdi, S. H., & White, I. M., (2012). Optofluidic surface enhanced Raman spectroscopy microsystem for sensitive and repeatable on-site detection of chemical contaminants. *Analytical Chemistry, 84*(18), 7992–7998.
65. Zhang, X., Dai, Z., & Zhang, X., (2016). Recent progress in the fabrication of SERS substrates based on the arrays of polystyrene nanospheres. *Science China Physics, Mechanics and Astronomy, 59*, 126801.
66. Zhang, H., Ma, D., Qiu, R., Tang, Y., & Du, C., (2017). Nonthermal plasma technology for organically contaminated soil remediation: A review. *Chemical Engineering Journal, 313*, 157–170.
67. Zhang, T., Zhou, R., Wang, P., Mai-Prochnow, A., McConchie, R., Li, W., & Cullen, P. J., (2021). Degradation of cefixime antibiotic in water by atmospheric plasma bubbles: Performance, degradation pathways and toxicity evaluation. *Chemical Engineering Journal, 421*, 127730.
68. Zhao, H., Hasi, W., Li, N., Sha, X., Lin, S., & Han, S., (2019). In situ analysis of pesticide residues on the surface of agricultural products via surface-enhanced Raman spectroscopy using a flexible Au-Ag–PDMS substrate. *New Journal of Chemistry, 43*, 13075–13082.
69. Zhao, Y. M., Patange, A., Sun, D. W., & Tiwari, B., (2020). Plasma-activated water: Physicochemical properties, microbial inactivation mechanisms, factors influencing antimicrobial effectiveness, and applications in the food industry. *Comprehensive Reviews in Food Science and Food Safety, 19*(6), 3951–3979.

70. Zhou, R., Zhou, R., Wang, P., Luan, B., Zhang, X., Fang, Z., & Bazaka, K., (2019). Microplasma bubbles: Reactive vehicles for biofilm dispersal. *ACS Applied Materials & Interfaces, 11*(23), 20660–20669.
71. Zhu, J., Liu, M., Li, J., Li, X., & Zhao, J., (2018). Multi-branched gold nanostars with fractal structure for SERS detection of the pesticide thiram. *Spectrochimica Acta Part A: Molecular and Biomolecular Spectroscopy, 189,* 586–593.
72. Ziuzina, D., Patil, S., Cullen, P. J., Keener, K. M., & Bourke, P., (2014). Atmospheric cold plasma inactivation of *Escherichia coli, Salmonella entericaserovar, Typhimurium* and *Listeria monocytogenes* inoculated on fresh produce. *Food Microbiol., 42*, 109–116.

CHAPTER 2

APPLICATIONS OF COLD PLASMA TECHNOLOGY IN THE FOOD INDUSTRY

KANINIKA PAUL, PRAGYA PANDEY, UTSAV RATHORE, and N. A. NANJE GOWDA

ABSTRACT

Cold plasma (CP) technology consists of completely or moderately ionized gas containing neutral particles, free radicals, ions as well as electrons. On the basis of temperature, this plasma can be characterized as hot or thermal plasma (TP) and cold or non-thermal plasma (NTP). In food industries, CP has potential for its applications in microbial inhibition, enzyme inactivation, removal of pesticides, etc. In CP, the reactive species formed cause the effect on microbial cell that leads to the final inactivation of the microbial load. These reactive species cause damage to the cellular composition, damage to proteins by oxidizing them and also by damaging the DNA (deoxyribonucleic acid) of the cells thus leading to denaturation of microbial cell. Owing to these effects of the CP, this novel technology can be efficiently applied in treating polluted/contaminated water, sterilization of foods and packaging materials, and edible oil hydrogenation, reduction in the severity of food allergens, elimination of anti-nutritive factors, modifying seed sprouting process and controlling of food waste in the industries.

2.1 INTRODUCTION

Food safety and quality with respect to preservation of food is a major area of concern in today's world. The popular/commonly used technologies for food

Advanced Research Methods in Food Processing Technologies. Junaid Ahmad Malik, Megh R. Goyal, Preeti Birwal, & Ritesh B. Watharkar (Eds.)
© 2024 Apple Academic Press, Inc. Co-published with CRC Press (Taylor & Francis)

preservation namely pasteurization, high heat sterilization, dehydration, and vaporization, can assure the biological safety or constancy of their foods, but can abolish some of the food elements, specifically the temperature subtle polyphenols and vitamins, which are interrelated with the excellence of the food. In addition to the accumulative desire for fine quality foods, the non-thermal technologies are associated with insignificant or zero changes on food flavors and crucial nutrients [45]. As a result, in recent years' non-thermal technologies in food processing have been investigated comprehensively. Non-thermal technologies have the potential capability to moderately, or entirely, switch the customary deep-rooted preservation practices [31].

Non-thermal technologies can make sure the sensate eminence and dietary standards of food in shorter processing time and at low temperature environment and still be able to cast-off enrich food safety and prolong the shelf life of food product [45]. The sterilization equipment associated with non-thermal technologies primarily state to the succeeding two traits [33]: (i) altering the plasmalemma arrangement to eradicate the controlling mechanism of the microbe; and (ii) extinguishing the chromosomal material to source metabolic ailments in the microorganism. This technology is also useful for extraction, drying, and additional mass transfer practices. Furthermore, the desire for more green technologies has expanded international attentiveness to lessen water and energy utilization in the course of food processing.

Similarly, novel non-thermal methodologies with respect to preservation approaches were studied and reinvigorated to deplete environmental effect, lessen food-processing expenditures, and build-up significance into the products sustaining the physicochemical properties, prolonging usable life, and guaranteeing the food preservation of the matrices [30]. At present, the existence of pathogenic and microorganisms' deterioration has turn out to be a task to attain a nontoxic and quality matrix. Non-thermal technologies of current industrial applications have unveiling efficacy in preventing deterioration and contagious microorganisms' growth in fish and meat products [5].

Innovative foodstuff processing technologies (such as: supercritical fluid, ohmic heating, high-pressure processing (HPP), ultrasound, intense pulse lights ozone, irradiation, pulsed electric field (PEF) extraction, non-thermal plasma (NTP)), or grouping of these technologies enact negligible effect on the food quality and also on Mother Nature consequently, which meet the benchmarks for non-polluting technology. Owing to this, the NTP technologies are also known as green technologies [5].

This chapter focuses on aspects related to the novel non-thermal technology of cold plasma (CP) and its applications in food industries.

2.2 WHAT IS COLD PLASMA (CP)?

Amid the non-thermal technologies, one of the developing non-thermal technologies in the field of food processing and preservation is CP. The term, "plasma" has been coined by Langmuir in 1925. The three conventional forms of matter are gases, solid, and liquid. Apart from solid, liquid, and gas, the fourth stage of the matter is plasma [21]. The escalation in the kinetic energy of molecules in the solid-state leads to the nature change from solid to liquid and likewise from liquid to gaseous form of matter. Further escalation in the kinetic energy of the molecules in gaseous form will lead to the conversion of neutral atoms or molecules into electrically charged atoms and molecules of the gas molecules thus, generating the fourth stage of matter that is plasma. In other words, plasma can be defined as a gas that is to completely or moderately ionized and constitute collaboration of numerous of sub-atomic, molecular bodies as well as quanta of electromagnetic radiation [26].

Plasma can be categorized as hot or thermal plasma (TP), or cold or NTP centered on the condition of thermodynamic stability of the ions and electrons. The number of atom collisions per unit time increases when the pressure is increased [40]. This causes kinetic energy to be distributed between electrons and heavier particles, resulting in "thermal plasma." The temperature of the electrons, gas, and ions in TP is nearly similar in this state [43]. CP, in contrast to TP, has a far greater electron temperature than the universal gas temperature and the establishing ions, as of this reason it is also known as disequilibrium plasma. Electron temperature, ion temperature, and gas temperature are all measured in the CP state. They're called "non-thermal/non-equilibrium plasma" or "cold plasma" since the gas temperature is low (about room temperature) [27]. CP, in simple terms, is an ionized gas produced under atmospheric or low-pressure circumstances.

The capacity of CP to produce a unique blend of organically active components namely reactive oxygen species (ROS) and reactive nitrogen species (NOS) at ambient temperature exists as key property that allows it to be used safely on biological materials, including food. CP technology has been proposed in several recent research as a unique method for maintaining food safety in a variety of food goods [16].

The sterilizing property of plasma was patented in 1968 [20]. The eradication of 106 dormant bacteria in the innermost surface of vials by argon plasma by pulsated radiofrequency (RF) field atmospheric plasma took less than a second [20]. Furthermore, Ashman & Menashi (1972), Boucher (1980) and Bithell (1982) filed many patents claiming that electric charge release within precise gases can be a key factor in thorough sterilization [1–3].

2.3 MECHANISM OF COLD PLASMA (CP) GENERATION

Plasma biological reactions can be classified into two types depending on reactions: (a) standardized gas-phase reactions; and (b) varied reactions in which plasma comes into contact with a solid or liquid intermediary [43]. These varied reactions are further divided into three subcategories [27]: (i) material is removed from the surface in the primary sub-classification, which is referred to as ablation or etching; (ii) the stuff is attached to the compact surface in the form of a slender film detected in the course of chemical action of plasma, by a method referred to as plasma enriched chemical vapor installation; and (iii) the plasma exposure results in physical and chemical modification of the substrate surface only without addition or removal of any material.

Electrical discharge methods (Figure 2.1) can be used to create a plasma state, namely: (i) dielectric barrier discharge (DBD); (ii) glow discharge; (iii) corona discharge; (iv) microwave discharge (MD); (v) high frequency discharge; and (vi) pulse discharge. The type of discharge is governed by the rhythm of the power source (alternating current (AC) and direct current (DC)), gas pressure (atmospheric pressure, low-pressure) and the geometry of the electrodes, etc. [6].

2.3.1 CORONA DISCHARGE

Corona discharge is associated with the electrodes having highly curved and pointed tips (sharp, pin-point tips) or small diameter wires. A vigorous electric field is generated at bend surfaces of the electrodes on application of high voltage to the device [6]. As a result of the imposed high voltage at air pressure and ambient temperature, the electric field surrounding the electrode becomes uneven. As a consequence of the focused electric field, bland gas present in the area of the bend electrode surface is ionized, resulting in the generation of corona discharge plasma from the dielectric disintegration [6].

2.3.2 GLOW DISCHARGE

Glow discharge is a long-lasting discharge created on exposure of low pressure (1–1,000 Pa) gas to direct voltage ranging from 100 V to several kV. In a glow discharge, the electrons possess higher energy than that of neutral gases and ions [23]. As a result, glow discharge plasma is classified as non-thermal/non-equilibrium plasma. Glow discharge, in comparison to other types of discharge, is spatially highly uniform and capable of producing plasma at low temperatures and in enormous volumes [23].

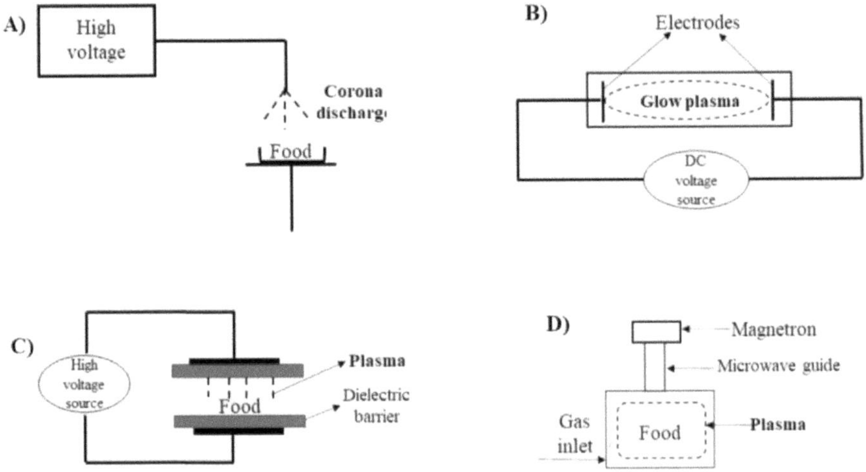

FIGURE 2.1 Schematic representation of: (A) corona discharge plasma; (B) glow plasma; (C) dielectric barrier discharge plasma; and (D) microwave discharge plasma.

2.3.3 DIELECTRIC BARRIER DISCHARGE (DBD)

A dielectric barrier discharge (DBD) equipment consists of solid insulating material (glass, plastic, silicon, ceramic) that is used to cover two electrodes present in the device. The insulator covering the electrodes has the effect of suppressing an arc discharge, which would otherwise result in a hot plasma. As DBD occurs in a short amount of time, the temperature of the neutral particles and ions does not rise. DBD is produced by alternating high-voltage pulses with frequencies > 10 Hz [39]. An afterglow discharge is a sort of DBD in which high-heat plasma created on the inside of a tube composed of ceramic or glass is forced outside of the tube as it cools. An atmospheric pressure plasma torch (plasma jet) is being developed that uses AC high-voltage discharge (of several kHz) along with gases such as N_2, He, O_2 and air for the generation of CP [39].

2.3.4 PULSE DISCHARGE

In case of pulse discharge, a sequence of short, pulsed discharges along with a sharp increase in voltage is used to generate the CP. Unlike electrons, heavy ions are not accelerated under these conditions. Even at air pressure, a stable discharge can be achieved using this pulse discharge approach. The transition into an arc discharge can be avoided by using a voltage with a very short duration. High-voltage pulse power sources with short pulses and high repetition have recently been created [34].

2.3.5 HIGH-FREQUENCY DISCHARGE

A radiofrequency (RF) power supply at around 13 MHz is commonly used to create a high-frequency discharge. The shape of the electrodes or antenna classifies this discharge mechanism. RF power is applied between facing electrodes in a capacitively linked plasma [18]. The electrodes are then surrounded by plasma. The RF power is supplied to a coiled antenna in the inductively coupled plasma. Inside the coil, high frequency plasma is created. A large volume of dense plasma can be created from a high frequency discharge at low air pressure and utilizing minimal heating. Disinfection of a material can be carried out employing this method without having any adverse effect on the material. In addition, the nature of plasma generation can be modified according to the size of the object to be treated [18].

2.3.6 MICROWAVE DISCHARGE (MD)

Electromagnetic radiation with frequency ranging from 300 MHz–300 GHz is generally used to create a MD. Electrodes aren't required for this form of discharge, and plasma can be created with antennas or a waveguide, however shielding is required. This approach can produce large-diameter plasma discharge at low gas pressure [35].

2.4 EFFECTS OF COLD PLASMA (CP) ON FOODS AND FOOD PRODUCTS

2.4.1 EFFECT OF PLASMA ON MICROBIAL CELLS

Effects of CP on food matrices are presented in Table 2.1. The cell interactions and plasma ions are mostly responsible for the outcome of plasma treatment on microbial cells [4, 41]. The continuous oxidative impacts on the exterior surface of microbial cells have been linked to the sensitive species in plasma. Plasma's outcome is highly dependent on the presence of water; peak outcome in a moist organism was comparable to nethermost in a dry organism [19]. The evidence that the plasma reactive species mutilate the genetic material in the chromosomes [12] is driving the use of plasma in inactivation of biological cells. Plasma causes the formation of malondialdehyde (MDA) in microbial cells, which leads to DNA adduct formation and cell mutilation [12].

When reactive species combine with water, hydroxyl (OH^-) ions form, and they are extremely reactive and damaging to cells. These OH^- radicals are created in the humidifying layer encircling the DNA molecule and are answerable for 90% of DNA deterioration [10]. These hydroxyl radicals then react with the neighboring organic components causing sequence oxidation and consequently damaging the DNA molecules, cellular membranes, and auxiliary cell components. The lipid bi-layer of a microbial cell is more sensitive to atomic oxygen, which being highly reactive in nature can degrade lipids, proteins, and DNA in cells [10]. The mutilation of the two-fold bonds in the lipid bi-layer source hampered molecular transport into and outside of the cell. The attack of ROS on the bacterial cell surface disrupts the membrane lipids as well [17]. At the cellular level, Figure 2.2 depicts the impact of CP on food matrices.

2.4.2 EFFECTS OF PLASMA ON ENDOGENOUS ENZYMES

Plasma not only acts in isolation to microbes, but also in complex biological combinations such as enzymes [9]. Enzymatic browning occurred during post-harvest treatment and storage, is the most common cause of fruit and vegetable spoilage. This enzymatic browning is caused by endogenous enzymes, particularly polyphenol oxidase and peroxidase (PO), which oxidize phenols in the presence of H_2O_2, resulting in off flavors [38]. Through

TABLE 2.1 Effects of Cold Plasma on Food Matrices

Target Compounds in Food	Source and Gas Used in Plasma Generation	Effects
Alkaline phosphatase (milk)	DBD; atmospheric air	Inhibition of enzyme activity.
Bacteria inoculated stainless steel surface	DBD; air	Reduction in microbial load.
Banana starch	Corona electrical discharge; air	Increase in crystallinity, gelatinization temperature and surface roughness.
Beef Jerky	Flexible thin-layer plasma system; ambient air.	Change in peroxide content and color parameters.
Blueberries	Atmospheric pressure plasma jet.	Antimicrobial efficacy.
Cherry tomato	Alternating corona discharge plasma jet; air at atmospheric pressure.	Increase in shelf life.
Chicken breast	DBD; atmospheric cold plasma.	Increase in rehydration ratio and shelf life (at 4°C).
Corn	DBD; air	Antimicrobial efficacy.
Fresh cut pear	In-package atmospheric cold plasma.	Reduction in respiration rate, peroxidase, and pectin methyl esterase activities.
Lipoxygenase (extra virgin olive oil)	DBD; argon.	Inhibition of enzyme activity.
MAP packaged ready-to-eat ham	DBD; air.	Antimicrobial efficacy.
Peanut and dry defatted peanut flour	DBD; air.	Allergen degradation in terms of reducing binding capacity of allergen.
Peroxidase in sliced tomatoes	DBD; helium, air.	Reduction in enzyme activity.
Pumpkin seed coat	Cold plasma jet; helium or argon.	Modification of seed coats surface properties along with enhancement in water uptake, seed germination and seed growth.
Rapeseed seed	Corona discharge plasma jet; air.	Exposure of plasma for >3 min had adverse effects on sensory characteristics.

Source: Adapted from Ref. [41].

oxidation events mediated by atomic and oxygen free radicals, enzymes are rendered inactive. After making use of CP, Dobrynin et al. scrutinized a lessening in trypsin's enzymatic activity [9]. The authors claimed that breakage of peptide bonds in plasma could change the 3D configuration of proteins in trypsin enzymes [9].

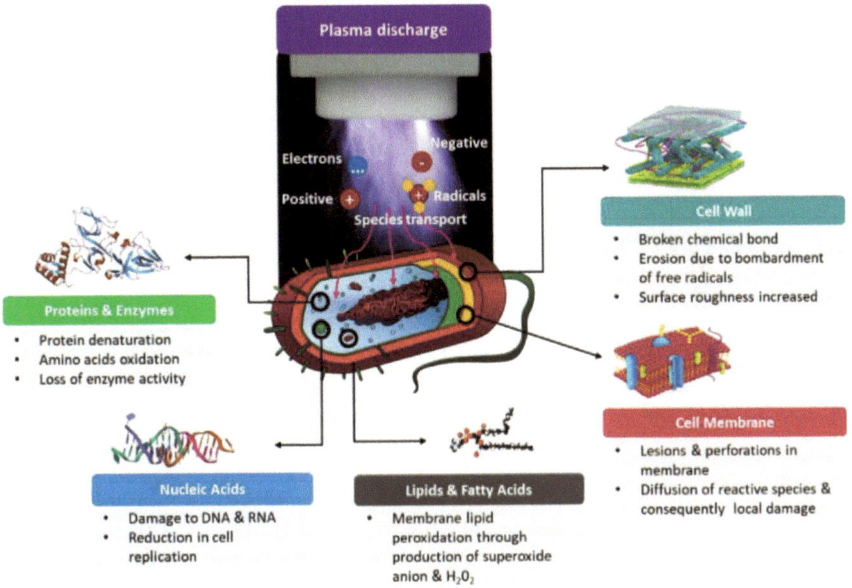

FIGURE 2.2 Effects of cold plasma on cellular level.

2.4.3 *EFFECTS ON STARCH GRANULES*

Altered starches are structured to increase their usefulness as food condiments in a variety of meal preparations by adjusting their intended qualities. The usage of changed starches was found to improve the ductility and mechanical properties of starch films. The natural polymers available for the manufacture of changed starches are sensitive to chemicals like strong acid for surface etching [28]. As a result, mild chemicals are used to change smooth hydrophobic surfaces into rough hydrophilic surfaces. CP technology is a substitute for parched etching for surface change along with removal of biopolymers, according to Pashkuleva et al. [28], who noted that oxygen-containing compounds such as hydroxyl, carbonyl, and carboxyl increase surface hydrophilicity, which can change the characteristics of starch granules [28].

2.4.4 EFFECTS ON SEED GERMINATION

Several writers have found that treating seeds with plasma technology can result in early sprouting. This can be accomplished by penetrating the seed coat with plasma particles and reliably controlling the cells within the seed [29]. The increase in sprouting rate is attributed to an increase in oxygen and moisture transmission through the seed coat, which alters the sprouting rate. It's possible that the interaction of cells with plasma resulted in DNA mutilation, cell wall rupture, growth factor augmentation, and protein structure changes that result in enzymatic activity. All of these factors would have a cumulative effect on seed sprouting rates [7].

2.4.5 EFFECTS ON FOOD PACKAGING MATERIAL

CP is used to sanitize food packaging material on the surface, where the likelihood of a shadow effect is minimal because the plasma wanders all over the surface [8]. In the case of packing materials, plasma dispensation is well known for causing surface variations or modifications. It enhances surface action drives like cleaning, coating, printing, and adhesive bonding [26]. Low-temperature gas plasma decontamination provides for quick and safe decontamination of packaging materials while preserving the material qualities [44]. CP can also be used to purify heat-susceptible packaging materials such as polycarbonate and polythene.

2.5 POTENTIAL OF COLD PLASMA (CP) IN THE FOOD INDUSTRY

2.5.1 APPLICATIONS OF COLD PLASMA (CP) FOR FOOD SAFETY

Due to bacterial toxins, foodborne viruses, pesticide residues, and mycotoxins, pathogenic microorganisms constitute one of the biggest life-threatening food safety challenges [42]. Furthermore, the ability of bacteria to form biofilms, internalize infected cells within a host tissue and produce exceptionally resistant spores frequently obscures or contradicts food-cleaning techniques [32]. Owing to the non-thermal nature and recognized capability to improve bacteriological safety and preserve quality physiognomies of an extensive variety of foods in a short processing time, CP technology has been confirmed as a possible substitute for conservative approaches [32].

2.5.2 INACTIVATION OF FOODBORNE PATHOGENIC MICROORGANISMS

For *Salmonella* sp. inhibition, CP was created by combining a RF argon plasma jet with microwave (MW)-produced air plasma [14. The formation of reactive nitrogen and oxygen species in CP effectively inactivates bacteria. Reactive NOS is able to build up on microbial surfaces and effortlessly infiltrate across cell membranes resulting in a reduction in intracellular pH. This intracellular pH affects several factors such as reaction rates, enzyme activity, protein stability and nucleic acid structure, to name a few [14].

2.5.3 COLD PLASMA (CP) CONTROL OF BACTERIAL BIOFILMS

Bacterial biofilms are complex surface formed by adhered consortium of bacteria bound to the surface of self-produced polymer matrices comprising of polysaccharides, peptides, and extracellular DNA [11, 36]. In terms of biofilm susceptibility to CP processes, Jahid et al. [15] confirmed that in 15 seconds of plasma treatment bring down planktonic populations of *Aeromonas hydrophila* by more than 5 log. On the other hand, 5 min treatment resulted in significant reductions in bacterial biofilm populations on lettuce leaves [46].

2.5.4 APPLICATION IN FRUITS AND VEGETABLES (FAV)

Softening is a common textural variation associated with plant cell walls that results in a loss in fruit and vegetable quality [25]. Delaying or preventing fruit softening is an efficient way to extend the shelf life of the fruit. CP has been effective in preserving the fruit's shine. CP treatments can prevent fruit softening in a variety of methods, including preserving fruit firmness, suppressing enzyme activity, inactivating fungal infections, and lowering respiration rates [37]. Thus, among all postharvest talents, CP is a viable approach that has a lot of promise in terms of maintaining fruit quality and preventing softening [37].

2.5.5 APPLICATION IN MEAT INDUSTRY

Chicken breast meat is particularly perishable in fresh poultry products. Several authors have reported on the effectiveness of in-package CP treatment in cleaning chicken breast surfaces and extending shelf life [13]. In-package CP is a reliable solution for outmoded chemical sterilization of chicken surfaces since it is an energy-efficient and chemical-free sanitization method. The oxidation of lipids caused by ROS has been a limiting factor in the use of CP to meat commodities, prompting extensive research employing low oxygen, enhanced atmospheres [24]. The effect of an in-package high voltage CP therapy combined with MAP gas application has been investigated to see if this combined method may be used in meat product safety situations [22].

2.6 SUMMARY

The CP technology has been considered a viable alternative to both thermal and non-thermal technologies. Despite having plasma generation equipment, more improvement in plasma generation systems is required in terms of improved design, dependable action, and economic feasibility. There is also a need for research into prototype development, plasma treatment of liquefied meals, and the use of CP in conjunction with other technologies. CP combined with MAP technology could offer a new path to technological advancement.

KEYWORDS

- **cold plasma**
- **enzyme inactivation**
- **food preservation**
- **fruits and vegetables**
- **meat products**
- **microbial inactivation**
- **non-thermal technology**
- **packaging material**

REFERENCES

1. Ashman, L. E., & Menashi, W. P., (1972). US Patent, Patent No. 3,701,628.
2. Bithell, R. M., (1982). US Patent, Patent No. 4,321,232.

3. Boucher, R. M., (1980). US Patent, Patent No. 4, 207,286.
4. Bourke, P., Ziuzina, D., Han, L., Cullen, P. J., & Gilmore, B. F., (2017). Microbiological interactions with cold plasma. *Journal of Applied Microbiology, 123*(2), 308–324.
5. Chakka, A. K., Sriraksha, M. S., & Ravishankar, C. N., (2021). Sustainability of emerging green non-thermal technologies in the food industry with food safety perspective: A review. *LWT, 151*, E-article 112140.
6. Chang, J. S., Lawless, P. A., & Yamamoto, T., (1991). Corona discharge processes. *IEEE Transactions on Plasma Science, 19*(6), 1152–1166.
7. Dey, A., Rasane, P., Choudhury, A., Singh, J., Maisnam, D., & Rasane, P., (2016). Cold plasma processing: A review. *Journal of Chemical and Pharmaceutical Sciences, 9*, 2980–2984.
8. Dhanapal, A., Rajamani, L., & Banu, M., (2012). Edible films from polysaccharides. *Food Science and Quality Management, 3*, 9–17.
9. Dobrynin, D., Fridman, G., Friedman, G., & Fridman, A., (2009). Physical and biological mechanisms of direct plasma interaction with living tissue. *New Journal of Physics, 11*(11), E-article 115020.
10. Gavahian, M., & Cullen, P. J., (2020). Cold plasma as an emerging technique for mycotoxin-free food: Efficacy, mechanisms, and trends. *Food Reviews International, 36*(2), 193–214.
11. Giaouris, E., Heir, E., Hébraud, M., Chorianopoulos, N., Langsrud, S., Møretrø, T., Habimana, O., et al., (2013). Attachment and biofilm formation by foodborne bacteria in meat processing environments: Causes, implications, role of bacterial interactions and control by alternative novel methods. *Meat Science, 97*, 298–309.
12. Han, L., Patil, S., Boehm, D., Milosavljević, V., Cullen, P. J., & Bourke, P., (2016). Mechanisms of inactivation by high-voltage atmospheric cold plasma differ for *Escherichia coli* and *Staphylococcus aureus*. *Applied and Environmental Microbiology, 82*(2), 450–458.
13. Han, L., Ziuzina, D., Heslin, C., Boehm, D., Patange, A., Sango, D. M., Valdramidis, V. P., et al., (2016). Controlling microbial safety challenges of meat using high voltage atmospheric cold plasma. *Frontiers in Microbiology, 7*, 977–983.
14. Hertwig, C., Reineke, K., Ehlbeck, J., Knorr, D., & Schlüter, O., (2015). Decontamination of whole black pepper using different cold atmospheric pressure plasma applications. *Food Control, 55*, 221–229.
15. Jahid, I. K., Han, N., & Ha, S. D., (2014). Inactivation kinetics of cold oxygen plasma depend on incubation conditions of *Aeromonas hydrophila* biofilm on lettuce. *Food Research International, 55*, 181–189.
16. Jiafeng, J., Xin, H., Ling, L. I., Jiangang, L., Hanliang, S., Qilai, X., & Yuanhua, D., (2014). Effect of cold plasma treatment on seed germination and growth of wheat. *Plasma Science and Technology, 16*(1), 54–58.
17. Liao, X., Muhammad, A. I., Chen, S., Hu, Y., Ye, X., Liu, D., & Ding, T., (2019). Bacterial spore inactivation induced by cold plasma. *Critical Reviews in Food Science and Nutrition, 59*(16), 2562–2572.
18. Lister, G. G., Li, Y. M., & Godyak, V. A., (1996). Electrical conductivity in high-frequency plasmas. *Journal of Applied Physics, 79*(12), 8993–8997.
19. Lu, H., Patil, S., Keener, K. M., Cullen, P. J., & Bourke, P., (2014). Bacterial inactivation by high-voltage atmospheric cold plasma: Influence of process parameters and effects on cell leakage and DNA. *Journal of Applied Microbiology, 116*(4), 784–794.

20. Menashi, W. P., (1968). *Treatment of Surfaces*. US Patent, Patent No. 3,383,163.
21. Misra, N. N., & Jo, C., (2017). Applications of cold plasma technology for microbiological safety in the meat industry. *Trends in Food Science & Technology, 64*, 74–86.
22. Misra, N. N., Yepez, X., Xu, L., & Keener, K., (2019). In-package cold plasma technologies. *Journal of Food Engineering, 244*, 21–31.
23. Moon, S. Y., Choe, W., & Kang, B. K., (2004). A uniform glow discharge plasma source at atmospheric pressure. *Applied Physics Letters, 84*(2), 188–190.
24. Moutiq, R., Misra, N. N., Mendonca, A., & Keener, K., (2020). In-package decontamination of chicken breast using cold plasma technology: Microbial, quality and storage studies. *Meat Science, 159*, E-article 107942.
25. Pan, Y. W., Cheng, J. H., & Sun, D. W., (2021). Inhibition of fruit softening by cold plasma treatments: Affecting factors and applications. *Critical Reviews in Food Science and Nutrition, 61*(12), 1935–1946.
26. Pankaj, S. K., Bueno-Ferrer, C., Misra, N. N., Milosavljević, V., O'donnell, C. P., Bourke, P., & Cullen, P. J., (2014). Applications of cold plasma technology in food packaging. *Trends in Food Science & Technology, 35*(1), 5–17.
27. Pankaj, S. K., Wan, Z., & Keener, K. M., (2018). Effects of cold plasma on food quality: A review. *Foods, 7*(1), 4–10.
28. Pashkuleva, I., Lopez-Perez, P. M., & Reis, R. L., (2008). Surface modification for natural-based biomedical polymers. In: *Natural-based Polymers for Biomedical Applications* (pp. 163–192). Cambridge, Woodhead Publishing Limited.
29. Percival, M., (1996). *Antioxidants: Clinical Nutrition Insights*. Advanced Nutrition Publications, Inc. NUT031-1/96.
30. Picart-Palmade, L., Cunault, C., Chevalier-Lucia, D., Belleville, M. P., & Marchesseau, S., (2019). Potentialities and limits of some non-thermal technologies to improve sustainability of food processing. *Frontiers in Nutrition, 5*, 1–18.
31. Putnik, P., Kresoja, Ž., Bosiljkov, T., Jambrak, A. R., Barba, F. J., Lorenzo, J. M., & Kovačević, D. B., (2019). Comparing the effects of thermal and non-thermal technologies on pomegranate juice quality: A review. *Food Chemistry, 279*, 150–161.
32. Raybaudi-Massilia, R. M., Mosqueda-Melgar, J., Soliva-Fortuny, R., & Martin-Belloso, O., (2009). Control of pathogenic and spoilage microorganisms in fresh-cut fruits and fruit juices by traditional and alternative natural antimicrobials. *Comprehensive Reviews in Food Science and Food Safety, 8*, 157–180.
33. Rosario, D. K., Rodrigues, B. L., Bernardes, P. C., & Conte-Junior, C. A., (2021). Principles and applications of non-thermal technologies and alternative chemical compounds in meat and fish. *Critical Reviews in Food Science and Nutrition, 61*(7), 1163–1183.
34. Sakudo, A., Misawa, T., Shimizu, N., & Imanishi, Y., (2014). N_2 gas plasma inactivates influenza virus mediated by oxidative stress. *Frontiers in Bioscience (Elite edition), 6*, 69–79.
35. Sakudo, A., Misawa, T., & Yagyu, Y., (2020). Equipment design for cold plasma disinfection of food products; Chapter 10. In: Bermudez-Aguirre, D., (ed.), *Advances in Cold Plasma Applications for Food Safety and Preservation* (pp. 289–307). San Diego–USA: Academic Press.
36. Sharma, G., Rao, S., Bansal, A., Dang, S., Gupta, S., & Gabrani, R., (2014). Pseudomonas aeruginosa biofilm: Potential therapeutic targets. *Biologicals, 42*, 1–7.
37. Sharma, R. R., Reddy, S. V. R., & Sethi, S., (2018). Cold plasma technology for surface disinfection of fruits and vegetables; Chapter 10. In: Siddiqui, M. W., (ed.), *Postharvest Disinfection of Fruits and Vegetables* (pp. 197–209). San Diego – USA: Academic Press.

38. Surowsky, B., Fischer, A., Schlueter, O., & Knorr, D., (2013). Cold plasma effects on enzyme activity in a model food system. *Innovative Food Science & Emerging Technologies, 19*, 146–152.
39. Teschke, M., Kedzierski, J., Finantu-Dinu, E. G., Korzec, D., & Engemann, J., (2005). High-speed photographs of a dielectric barrier atmospheric pressure plasma jet. *IEEE Transactions on Plasma Science, 33*(2), 310–311.
40. Thirumdas, R., Sarangapani, C., & Annapure, U. S., (2015). Cold plasma: A novel non-thermal technology for food processing. *Food Biophysics, 10*(1), 1–11.
41. Uçar, Y., Ceylan, Z., Durmuş, M., Tomar, O., & Teçinkaya, T., (2021). Application of cold plasma technology in the food industry and its combination with other emerging technologies. *Trends In Food Science & Technology, 114*, 355–371.
42. Van, B. S., Habib, I., Jacxsens, L., De Vocht, M., Baert, L., Van De, P. E., Rajkovic, A., et al., (2013). Food safety issues in fresh produce: Bacterial pathogens, viruses and pesticide residues indicated as major concerns by stakeholders in the fresh produce chain. *Food Control, 32*, 190–197.
43. Varilla, C., Marcone, M., & Annor, G. A., (2020). Potential of cold plasma technology in ensuring the safety of foods and agricultural produce: A review. *Foods, 9*(10), 1435.
44. Vesel, A., & Mozetic, M., (2012). Surface modification and ageing of PMMA polymer by oxygen plasma treatment. *Vacuum, 86*(6), 634–637.
45. Zhang, Z. H., Wang, L. H., Zeng, X. A., Han, Z., & Brennan, C. S., (2019). Non-thermal technologies and its current and future application in the food industry: A review. *International Journal of Food Science & Technology, 54*(1), 1–13.
46. Ziuzina, D., Han, L., Cullen, P. J., & Bourke, P., (2015). Cold plasma inactivation of internalized bacteria and biofilms for *Salmonella enterica Serovar typhimurium, Listeria monocytogenes* and *Escherichia coli. International Journal of Food Microbiology, 210*, 53–61.

CHAPTER 3

APPLICATIONS OF COLD PLASMA TECHNOLOGY IN FOOD PROCESSING

VARSHA KANOJIA

ABSTRACT

Cold plasma (CP) technology has been expanded to the food industry as a potent tool for non-thermal processing in the recent decade, with a variety of uses. This chapter provides an overview of recent research on CP applications in the food industries. CP has also been used in fields such as hydrogenation of edible oils, food allergy mitigation, inactivation of anti-nutritional factors (ANFs), seed germination performance tailoring, and effluent control.

3.1 INTRODUCTION

Cold plasma (CP) technology is different from the conventional processing technologies for being non-thermal and non-destructive method, yet cost-intensive, adaptable, and eco-friendly. As per the recent trends in the food applications, the usage of CP has been suggested for microbial decontamination of food, owing to its ample amount of reactive oxygen species (ROS) in quasi-neutral plasma gas, toxin degradation [39], inactivation of enzymes [44], removal of toxins, processing of packaging materials by improving the barrier properties and increasing the antimicrobial properties [47], and wastewater treatment [14].

Regarding the areas of food decontamination, CP method has proven to be efficient against many pathogenic microorganisms, i.e., *Salmonella Typhimurium*, *E. coli* [39], *Listeria monocytogenes*, *Staphylococcus aureus*,

Advanced Research Methods in Food Processing Technologies. Junaid Ahmad Malik, Megh R. Goyal, Preeti Birwal, & Ritesh B. Watharkar (Eds.)
© 2024 Apple Academic Press, Inc. Co-published with CRC Press (Taylor & Francis)

etc. There are still unexploited areas for its application; therefore, this paves a trail for researchers to point out the research options and also figure out the future arenas. Being a non-thermal, non-destructive technology, the main focus has been on the positive impact of this technique in the food processing sector; however, not enough focus has been given on its impact on food attributes and overall quality factors.

This chapter amalgamates current trends related to potential of CP technology in the food processing industry, its effect on the overall attributes of the food products and its future prospects in food industry.

3.2 COLD PLASMA (CP) TECHNOLOGY

3.2.1 FUNDAMENTALS OF COLD PLASMA (CP) TECHNOLOGY

Aside from the three existing forms of matter (solid, liquid, and gas), plasma is a fourth state of matter (Figure 3.1). In 1928, Irving Langmuir introduced the term "Plasma," which is partially or completely ionized state of gas. Plasma indicates a neutral gas made up of free electrons, molecules atoms and ions. It is obtained by an electrical discharge [15], in which a partially or completely ionized gas is formed. The compounds and molecules in plasma state are designated as light species (photons and electrons) or "heavy" species (remaining constituents) [38, 39]. The process parameters and gas in use will determine the nature of the plasma [46].

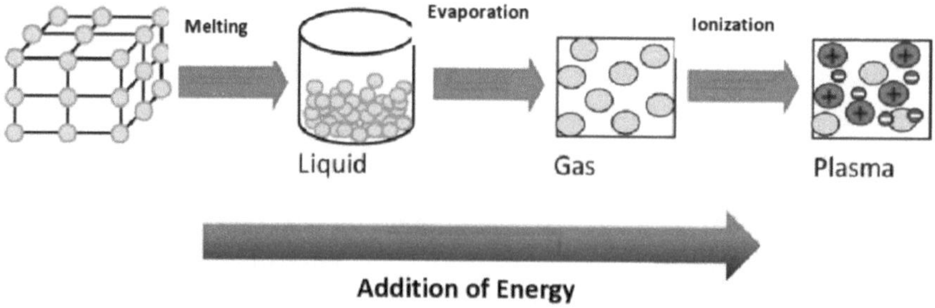

FIGURE 3.1 The four states of matter.

Source: Adapted from Ref. [71]

The change in phase of matter from solid into liquid and then into gaseous state occurs when the energy input is increased. Ionization of gas molecules starts beyond a certain level in gaseous state, which yields the plasma state

[10]. Based on the thermal equilibrium, plasma may be found either in non-equilibrium, low temperature, glow discharge type of plasma or equilibrium TP. In equilibrium TP, all the components lie in a thermodynamic equilibrium, i.e., the temperatures of all components are same in limited areas of plasma. Low temperature plasma is categorized into TP (quasi-equilibrium plasma), and non-thermal plasma (NTP) (non-equilibrium plasma). The NTP that has heavier species and electrons in non-equilibrium thermal condition is known as CP. On the basis of pressure, plasma is further classified as low-pressure, atmospheric pressure and high-pressure plasma. The atmospheric pressure plasma is generated at the normal atmospheric pressure; therefore, it eradicates the necessity of cost-intensive reaction chambers to maintain the pressure.

3.2.2 GENERATION AND SOURCES OF PLASMA SYSTEM

For the generation of CP, there are some common techniques, such as: plasma jets, DBDs, corona discharges, and microwave discharge (MD) at atmospheric pressure.

3.2.2.1 DIELECTRIC BARRIER DISCHARGE (DBD) PLASMA GENERATION

When two metal electrodes are spaced apart using a dielectric substance (such as plastic, quartz, or ceramic) at a discharge gap ranging from 100 mm to several cm, DBD plasma is created (Figure 3.2). The dielectric substance prevents sparks from forming owing to charge mobility. Simple operation, use of various gasses, reduced gas flow rate, versatility due to varied electrode geometries, and uniform discharge igniting across several meters are all advantages of DBD. The method, however, necessitates high ignition voltages of 10 kV.

Typical applications include ozone generation, UV generation in exciter lamps and CO_2 lasers. In 1857, Siemens carried out experiments on DBD, which has been explored to have number of applications, such as bacterial inactivation, living tissue sterilization, surface treatment, angiogenesis, and exciter formation. DBD contains two metal electrodes, which are flat and enclosed by a dielectric material (Figure 3.2). There is a gas moving around these two electrodes that are ionized to form the plasma. One of these electrodes is a high voltage electrode and another one is an earthed electrode. Here, the generation of "Plasma" requires very high voltages. The configuration of the electrodes has many distinctions, although they follow the same concept.

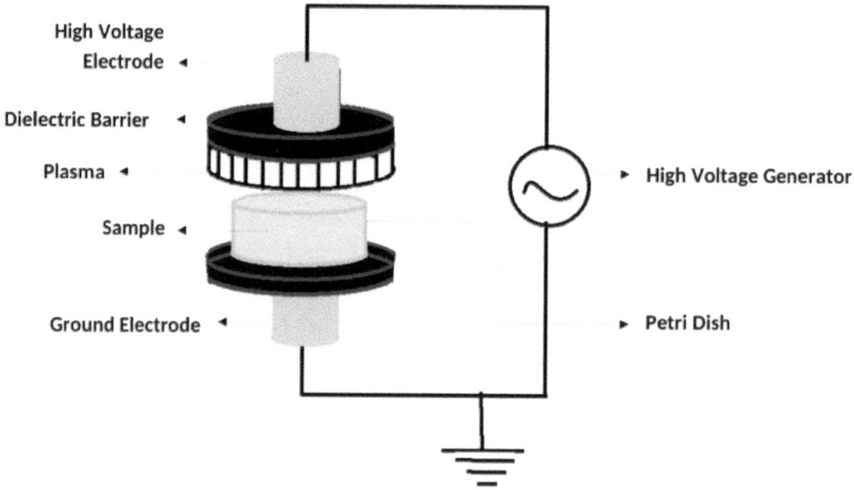

FIGURE 3.2 Generation of dielectric barrier discharge (DBD) plasma.
Source: Adapted from Ref. [72]

The shape of electrodes may be cylindrical, instead of being flat; and instead of covering both electrodes with a dielectric material, only one of these electrodes may cover the same. Recently, the floating electrode DBD plasma was developed [18] that is identical to the earlier DBD, made of two electrodes, i.e., first one is insulated high voltage electrode and another one is active electrode. In floating electrode DBD, the second electrode is not connected with the earth and stands at as a floating potential. This second active electrode is a human or animal skin or organ. The active electrode should be near to another electrode (i.e.,< 3 mm distance) to produce the discharge. This electrode had been used for sterilization of living tissues and to deactivate the bacterial cells [18, 26].

3.2.2.2 PLASMA JET

3.2.2.2.1 Radiofrequency (RF) Plasma Jets

Radiofrequency (RF) plasma jet is a type of atmospheric pressure plasma jet (APPJ), which is used for bacterial sterilization [53], and it made of two coaxial electrodes (Figure 3.3). There is a gas (i.e., the mixtures of oxygen, helium, and other gases) between the two electrodes, which moves at very high rate. The discharge is created by applying RF power of 50–100 W to the central electrode at 13.56 MHz; however, the outer electrode is earthed.

The reactive substance that exits through nozzle at high velocity, lands over the area to be treated. APPJ is also applied for inactivation of many micro-organisms [1, 17, 24].

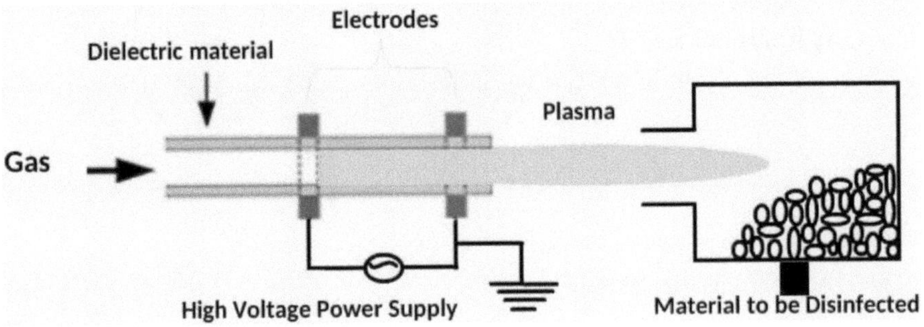

FIGURE 3.3 Atmospheric pressure plasma jet (APPJ).

Source: Adapted from Ref. [73]

In 2002, a very small atmospheric plasma jet (named as plasma needle) was developed [56]. Later in 2004, an advanced type of plasma jet was produced [27]. In the previous model, the needle was hemmed in a box because of which the samples were needed to be put inside the box. However, in latest model, the plasma needle comprises of metal strand having diameter of 0.3 mm with sharp tip inside a Perspex tube. Helium is the most frequently used gas, since it has high thermal conductivity. The mixture of gas and air produces a micro-discharge at the tip of needle. The plasma glow generated is of 2 mm diameter.

3.2.2.3 PULSED DIRECT CURRENT (DC)-DRIVEN PLASMA JETS

Manufacturing of a small jet, known as plasma pencil (Figure 3.4), consisting of dielectric cylindrical tube having diameter of 2.5 cm was achieved. In this tube, two disk electrodes of equal diameter are installed. Both electrodes are placed at the distance of about 0.3 cm to 10 cm from each other [31].

The electrodes consist of a thin copper ring connected to the disk. Sub-microsecond high voltage pulses are applied between two electrodes to create the plasma; and a gas is passed through the holes of electrodes. High voltage pulse generator is used to supply the power to the electrodes. This plasma pencil has been found effective in the treatment of Leukemia cells, *P. gingivalis,* and *E. coli* [4, 31]. Also, a plasma jet by applying the principle of DBD was developed [16, 64, 69].

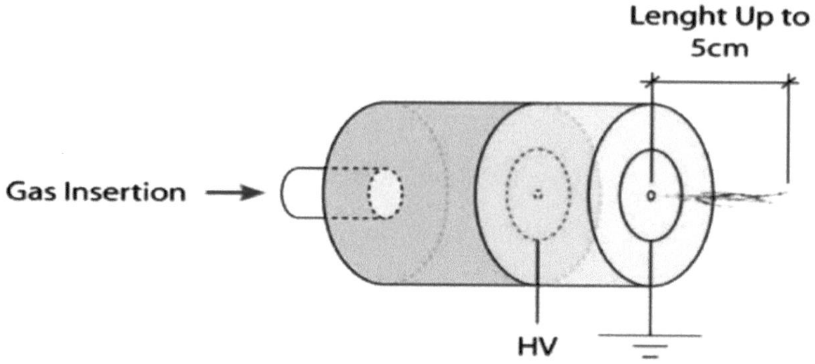

FIGURE 3.4 The plasma pencil [31].
Source: Adapted from Ref. [31]

3.2.2.4 CORONA DISCHARGE

When a high voltage is utilized, a corona discharge plasma results from the high electric field that surrounds an electrically conductive space region (Figure 3.5). The high electric field around the needle point or wire causes electrical breakdown and ionization of the gas surrounding the point, resulting in plasma, which shoots from the point or wire like a fountain spray.

The amount, temperature, and density of free electrons in a system determine the plasma type. The discharge is so constrained that the fabric's residence time in the plasma is too short for commercial use. Furthermore, the applied power is severely limited by the wire's cross-sectional capacity and its ability to distribute the heat created during treatment.

3.3 CURRENT APPLICATIONS OF COLD PLASMA (CP) METHODS IN PROCESSING OF FOODS AND FOOD PRODUCTS

Originally, CP was developed for revolutionizing the printing and adhesion properties of polymers, modification of textiles [14] and also for sterilization and disinfection of equipment surfaces [60]; however, recently, CP has gained lot of endorsement in the food industries.

3.3.1 MICROBIAL DECONTAMINATION OF FOOD PRODUCTS

The antimicrobial property of CP technology has been documented by several research studies. It has been proven to be most beneficial non-thermal preservative technique that employs low temperature, negligible nutrient destruction and also least impact on the sensorial attributes of food products. Due to the increased consciousness of consumers towards health and well-being, use of chemical preservatives and high temperature and pressure methods of preservation are being frowned on that has paved a way for the use of non-thermal processes, such as: PEF, CP technology, etc.

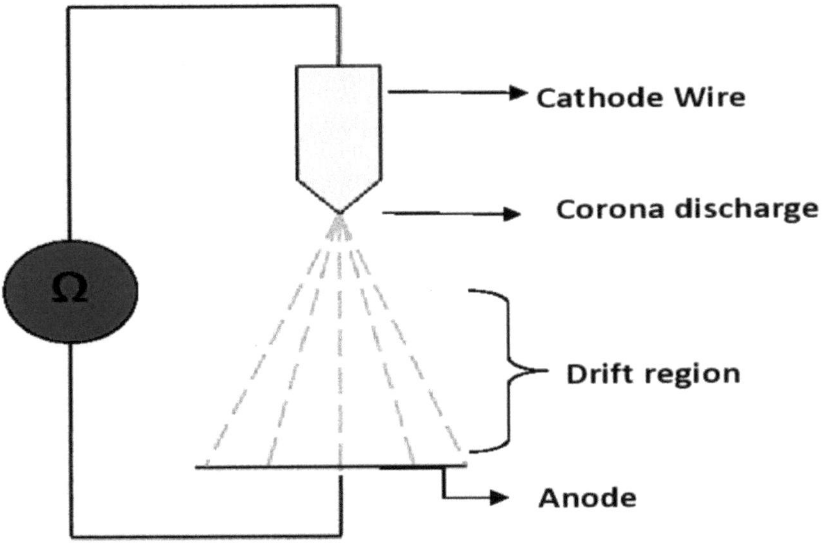

FIGURE 3.5 Schematic diagram of a corona discharge.
Source: Adapted from Ref. [53]

The use of CP technology for the destruction of microorganisms has been established for several food items (Table 3.1). In mandarin, CP technology has been effectively seen to inhibit growth of microorganisms [66]. The effectiveness of CP technology was dependent on plasma generation power and treatment time. It was found that CP technology is a potential method for the postharvest preservation of mandarin and antioxidant quality of the mandarin peel was increased without changing or destroying the biological and esthetic properties of the product.

The impact of surface topography on the effectiveness of cold atmospheric plasma treatment was observed using SEM (scanning electron microscope) that showed the clear images of microstructure of the inoculated filter, strawberry, lettuce, and surfaces of potato tissue [13]. Ultimately, cold atmospheric plasma can decontaminate the food produce thereby proving to

TABLE 3.1 Recent Applications of Plasma Technology for Microbial Decontamination in Different Food Materials

Food Item	Plasma Treatment	Results	References
Chicken eggs	HVACP (V = 85 kV, f = 60 Hz, ET = 15 min)	Attacked in *Salmonella enteridis*; and up to 5.53 log CFU/egg reduction.	[65]
Corn	HVACP (V = 90 kV, f = 50 Hz, RH = 40%, ET = 10 min)	Aflatoxin reduction up to 82%.	[54]
Dried sprout shreds	Corona discharge (V = 20 kV, f = 58 kHz, d = 25 mm, ET = 3 min)	Aerobic bacteria, marine bacteria and *S. Aureus* were inactivated by 2.0, 1.6, and 0.9 log units, respectively.	[9]
Groundnuts	Radiofrequency plasma (P = 60 W, RH = 45.3%, f = 13.56 MHz)	97.9% and 99.3% reduction in *Aspergillus parasiticus,* and *Aspergillus flavus,* respectively.	[11]
Lamb meat	DBD (V = 80 kV, f = 50 Hz, ET = 5 min)	2 log cycle reduction in *Brochothrix thermosphacta.*	[45]
Mandarin	Microwave discharge (P = 900 W, PL = 0.7 kPa, d = 24 cm, ET = 10 min)	Attacked in *Penicillium italicum* and 84% reduction in disease incidence.	[66]
Radish sprouts	Microwave discharge (P = 900 W, PL = 667 pa, ET = 20 min)	Up to 2.6 ± 0.4 log CFU/g reduction in *Salmonella typhimurium.*	[47]
Romaine lettuce	DBD (V = 42.6 kV, RH = 22%, d = 5 cm, ET = 10 min)	0.4–0.8 log CFU/g decrease in the number of *E. coli.*	[36]
Vacuum packaged beef loin	DBD (F = 9 kHz, d = 2 cm, P = 29.9 W)	≥2 log reduction in *S. aureus*, L. Monocytogenes and *E. coli.*	[5]

Legend: P: Power; V: Voltage; RH: Relative humidity; d: distance; f: frequency; and ET: Exposure time.

be a successful technique that can replace traditional thermal technologies. Although, CP has been established as remarkable non-thermal processing and decontamination technology in food industries, yet proper selection of gasses and electric power is crucial for obtaining desired results [14].

3.3.2 ENZYME INACTIVATION

Enzymes are the primary precursors of any chemical reaction in the biological systems. Enzyme activity tends to be an important parameter for any chemical activity; however, some chemical reactions have proven to be detrimental for the quality and esthetic appeal of the food products. For example, the enzymes, such as, peroxidase (PO), and polyphenol oxidase are found in fresh fruits and vegetables (FAV), which play a vital role in enzymatic browning resulting in the significant postharvest losses. POs are also concomitant with the enzymatic browning of vegetables at the cost of H_2O_2 leading to negative flavor changes during storage.

To prevent postharvest losses due to enzymatic browning, use of thermal techniques, anti-browning compounds, reducing agents, chelating agents and inorganic salts, etc., has been employed; however, the effects of ascorbic acid (AA), sodium bisulfate and other reducing agents on PPO have been debatable over the years [21]. The use of thermal processes for the inactivation of the enzymes is widely in use particularly in fruit juices and canned FAV; however, due to the use of high temperatures of about 80°C, there is sensorial degradation and loss of esthetic properties of these products resulting in the reduction of their marketability [63]. To avoid the quality degradation of fruits, vegetables, and their processed products due to high heat treatments, several non-thermal methods (such as: PEF, HHP, CPT) have been employed to tackle these issues.

CP treatment is capable of reducing the activity of the quality determining enzymes without any negative implications on the sensorial attributes of the food products thereby proving to be a significant alternative to the traditional thermal methodologies [58]. In a research study on the effects of atmospheric cold plasma (ACP) treatment on the deactivation of the wheat germ enzymes [62], it was found that the treatment for 25 min resulted in reduction of lipase and lipoxygenase (LOX) activity of wheat germ to 25.03% and 49.98% of initial extent, respectively; although, CP treatment could not inactivate the wheat germ enzymes permanently. Also, this treatment resulted in a significant increase in the storage shelf-life of wheat germ and the products.

3.3.3 FUNCTIONAL MODIFICATION OF FOOD PRODUCTS

The starch (a mixture of amylase and amylo-pectin molecules) is one of the most common constituents in most foods and is widely used in other industries. Certain enzymatic, physical, and chemical approaches are employed to improve the usefulness of native starch due to its limitations (poor solubility in cold temperatures, high viscosity, and retrogradation control). According to some research, plasma processing can cause beneficial alterations in the native constituents [61]. Increased surface energy, the inclusion of functional groups, cross linking, depolymerization, and a change in hydrophilic nature are all examples of these alterations [59]. Modification is induced by CP amylases through molecular breakdown and granular corrosion/etching. In a 3-minute time frame, reactive species generated by corona electrical discharge were used to change the techno-functionalities of banana starch [67]. Furthermore, protein-rich isolates from green peas could be modified using CP technique [6].

In the protein isolates, the investigators discovered increases in solubility and water uptake ability of 191% and 113%, respectively. Fluorescence emission intensity, which fluctuated from 328 nm to 355 nm with increasing treatment duration, was used to confirm conformational modification of tryptophan-containing proteins. In addition, the pH of the pea protein decreased by 3.4 over time. This may be due to an acidification because of generation of nitric and nitrous acids from NO via NO_2 [51]. In addition, the experiment revealed a significant increase in tensile strength to a maximum of 25.80 ± 1.94 MPa; while contact angle for all the experimented samples was lower than the control; this implied that zein hydrophilicity was improved to a greater extent. It was concluded that active species interacted with the surface of zein, leading to increase in polarity [13].

3.3.3.1 MODIFICATIONS OF GRAINS AND DERIVATIVE PRODUCTS

Non-equilibrium plasma can minimize cooking time, improve cooking qualities, and improve grain surface features in addition to its surface modification effect on starch and other food components. Cooking time was reduced from 30.25 to 20.45 minutes, and hardness was reduced from 22.50 to 12.36 N, according to the effect of plasma flow pressure on the physiochemical characteristics of black gram [52]. Furthermore, due to the synthesis of simple sugars (such as glucose, dextrin, maltose, and others), the water absorption capacity rose with plasma dose after 4 hours of soaking, while the contact angle decreased from 85.4° to 59.7°. Brown rice displayed alpha-amylase activity (up to 1.21 folds) and water absorption after plasma etching and surface depolymerization [32].

Plasma-assisted processing of grain byproducts, such as flour, pellets, and other products, would also be noteworthy. For example, increasing the exposure time and plasma power enhanced the water retention capacity of basmati rice flour [59]. In addition, the researchers discovered a significant rise ($p = 0.05$) in insolubility and swelling power, as well as a decrease in pasting viscosity from 692 to 591 BU, which matched a drop in peak temperatures. After plasma treatment, the amylase level of parboiled rice flour was reduced from 30.34 to 27.89%, with improved flour hydration characteristics and phenolic content liberation [59]. Micenthalpies and crystallinity percentages of the flour were also lowered, from 7.90 to 7.47 and 28.25 to 24.36, respectively. In these experiments, there were minor changes in grain proximate composition but a large rise in lightness values, implying that CP treatment had no significant effect on the sensory and nutritional profile of food items. However, there may be a need to fully explore the sensory impact and perception of consumers, as well as the viability in real-world food systems, such as incorporating plasma-treated grains or their derivatives into existing or novel foods.

3.3.3.2 COLD PLASMA (CP) IN PACKAGING MATERIALS

The food packaging minimizes the risk of damage in food materials due to the outside environment that are caused during transportation, handling, and distribution, but on the other hand it can also be a source of chemical food contamination [44]. The packaging material, which is in direct contact with the food is called the primary packaging. Various different types of primary packaging materials are commonly used, each with varying properties. However, to replace traditional packaging materials (such as: metals, glasses), use of paper and polymers has been increasing day by day. The switching of the packaging material from traditional to polymers is due to the flexibility, transparency, inertness, and low-cost.

However, there are certain specific properties, which are not met by these polymeric materials and are specific for certain packaging applications. To make these materials ready for the specific use, various surface treatments are being employed. The surface treatments are usually given to improve the wettability, sealabliity, printability, dye-uptake, etc., without having any negative impact on the desirable properties of the material. To have a safe packaging material for food products, several methods are being employed for surface decontamination. There are various chemical and physical methods that are applied to change the properties of the polymers; however, chemical methods are less desired as compared to the physical techniques. The most required physicochemical treatments employed for modification of polymeric materials include: corona treatment, low-pressure plasma, UV light, ion-beam techniques, gamma ray, and laser treatment [2].

The plasma treatment of the packaging materials leads to selective adhesion and anti-adhesion properties because of desirability, printability, sealability, and resistance to mechanical failure. The use of plasma technology can improve the barrier quality of the packaging material against the chemical solvents and gases. Use of plasma technology is also employed in the surface decontamination of the packaging materials and also sterilization of plastic bottles, caps, without negatively impacting the bulk properties of the material or leaving any residual effect [42].

3.3.3.3 IN-PACKAGE PLASMA TREATMENT

In recent research studies, DBDs have been documented in recent research studies as being utilized in the formation of plasma inside sealed-packages containing bacteria in strawberries. At 15–20 W, ACP was demonstrated to diminish background microflora on strawberries without inflicting any physiological or esthetic stress [37]. Microbial counts in fresh fruit, fish, and meat [7] can be effectively reduced by applying in-package plasma technology.

Furthermore, in the event of continuous processing in the food business, this is a simple strategy that can reduce post-packaging deterioration [39].

3.3.3.4 DEGRADATION OF PESTICIDE RESIDUES

DBD-generated CP had a significant effect in degrading imidacloprid and boscalid in blueberries by 75.62% and 80.18%, respectively, after being treated at 80 kV for 5 min; and it was observed that there was no visual change in the product, however, the firmness of fruits was found to be decreased due to collapsed internal structure [52]. In another investigation [38, 52], the use of high voltage ACP treatment in water samples had notably degraded the pesticide residues of endosulfan, malathion, and dichlorvos to 57.71%, 69.62%, 78.98%, respectively after treatment for 8 min at 80 kV. And the efficacy of He/O_2 induced plasma on malathion, and dichlorvos coated on glass sides was reduced on exposure to an inductively coupled plasma source [70]. Other studies have established that CP is a promising technique for reduction of pesticide residues including 17β-estradiol [19], endosulfan [48] and dichlorovas [70].

3.3.3.5 WASTEWATER TREATMENT

A viable innovation of the oxidation process for wastewater treatment seems a big task. Since the oxidation is needed for wastewater treatment and decontamination is difficult to complete using ultrasonication, photocatalysis, and UV/Ozone [49]. Plasma generates physicochemical effects for the formation of oxidizing species [25]. The most efficient method for treatment of wastewater could perform through the diffusion of gaseous phase into liquid phase [25].

Current studies on NTP have been analyzed extensively [25] on organic dyes [20], in water and phenols [30], modification in pharmaceutical compounds [29], pesticide residues [37] and pollutants mineralization by applying various catalysts [25]. Immense innovation for commercialization of NTP principle in wastewater treatment should be applied in Engineering and Applied Sciences. The mechanism of plasma oxidation for pollution minimization and other attributes need to be further studied [43].

3.4 FUTURE TRENDS

3.4.1 HYDROGENATION OF TRANS-FAT-FREE FATS

Hydrogenation is a chemical process involving the change of fat from liquid form to its solid state. A typical process of hydrogenation involves the use of catalyst (generally nickel). This process is used for the hardening of animal fats or vegetable oils and solidifying them at room temperature, ultimately it improves their stability. In the presence of a nickel catalyst, Hydrogen is reacted with carbon-carbon double bonds of fat or oil molecule in unsaturated fatty acid portion. In the past, hydrogenation of edible oils was used to improve the oxidative stability of oils, which contained polyunsaturated fatty acids, e.g., fish oil. The high sensitivity of oils to air oxidation can make them unusable in many food applications, particularly, in baking and frying. Moreover, oxidation decreases the storability of oil, as it prompts the off-flavor formation. Therefore, hydrogenation increases shelf-life of liquid oils as compared to store the oils a conventional method.

The deduction of unsaturation of the oil also results in an increase in melting point, ultimately increasing the transformation of oil into fat. This promotes the processing of natural oils into customized products with specific desired attributes. A complete hydrogenation facilitates the production of brittle and high-melting fats, which can be used as coatings, spreads, etc. [22]. However, the major drawback related to the hydrogenation process is the presence of trans fatty acids, due to chemical reaction that results in the formation of positional trans isomers [12]. As the use of fats and oils are an integral part of foods, it is necessary to explore the areas and technologies that could help in processing of the oils and fats in much healthier form. In a recent study, reaction of hydrogenation could increase 3% saturated fatty acids and 40% increase in trans fats; however, after high voltage atmospheric cold plasma (HVACP) treatment with exposure time of 12 h, there was 32.3% increase in saturated fatty acids of oils and no trans fats formation [68]. This study implies that HVACP process is more efficient than the conventional process. In spite of this, there are some hurdles still exist in commercialization of this process. The HVACP technique produces a by-product that could not be identified by the scientists yet. The usefulness of HVACP over traditional hydrogenation process stays in its operational conditions, i.e., ambient temperature, atmospheric pressure and lack of catalyst.

3.4.2 REDUCTION OF FOOD ALLERGENS

At varying power levels and plasma exposure periods, an attempt was made to investigate the influence of CP on germination rate in mung-bean seeds. The study [50] found that CP increased radical root, seed conductivity, and germination rate by 36.2, 20, and 102%, respectively, as well as the activity of hydrolytic enzymes (i.e., amylase, protease, and phytase), confirming the unique ability of CP to significantly speed up moisture uptake and support seed survival under extreme stress conditions. It's also probable that intrinsic species in plasma gas, such as hydroxyl radicals, disrupt peptide bonds, oxidize side chains, and cause fragmentation [58].

Disulfide bonds within peptides may be broken as a result of the presence of a hydroxyl radical, yielding RSH and RSO at the point of breakage. Furthermore, ROS, atomic oxygen, or hydroxyl radicals can target oxidation-prone amino acids, destroying antibody binding sites [35]. In both direct and remote CP Sodium dodecyl sulfate-poly acrylamide gel electrophoresis (SDS-PAGE) analysis of two major soy allergens (i.e., β-conglycinin (Glym5) and glycinin (Glym6), protein bands corresponding to these two proteins were less visible after plasma treatment, followed by formation of new bands at 50 kDa. In addition, for direct and reserved CP, immunological reactivity to soy protein changed by up to 100% and 89%, respectively. Milk allergens (casein, β-lactoglobulin (BLG), and α-lactalbumin) exposed to a remote plasma source, on the other hand, showed no discernible changes in IgE (immunoglobulins E) binding values or gel band intensities. Furthermore, it is well-known that food materials, particularly those derived from plants, contain naturally-occurring anti-nutritional factors (ANFs) that can hinder nutrient utilization by generating complex molecules, resulting in decreased nutrient absorption in the digestive system.

The digestive system is influenced by most anti-nutritional causes. Tannins, for example, can reduce protein digestibility by creating compounds with proteins, whereas lectins have been shown to impede breakdown and transit within intestinal absorptive cells. Anti-nutritional components in food matrices could be virtually or totally inactivated using non-thermal approaches (such as non-equilibrium plasma). The activities of soybean trypsin inhibitor contained in soymilk, for example, have been shown to decrease over time. In contrast to bovine trypsin, oxidative modifications and structural changes generated by the active species resulted in significant inactivation (96.1%) of soybean Kunitz-type tryps in inhibitor (SKTI). These studies established CP technology as a viable alternative to thermal approaches for allergy management and ANF inactivation in foods. As a result, broad conclusions cannot be drawn at this time until more research is done to uncover specific pathways and assess the nutritional and organoleptic profile after therapy.

3.4.3 MODIFICATION OF SEED GERMINATION PERFORMANCE

The drought, water logging, increased salt, and severe temperatures all have an unfavorable effect on seed germination, which can reduce agricultural productivity. Farmers must also contend with reduced seed viability due to microbial contamination and longer germination times. CP has been advocated by certain studies as a way to improve seed viability with different benefits such as uniform treatment and no seed destruction [34, 47, 57]. A DBD-PAW was also utilized to control the germination of some seedlings. In comparison to control samples, stem elongation was increased by 60% for tomatoes and pepper after 15 minutes of exposure to PAW, while germination rate increased by up to 80% for radish [55].

3.4.4 MANAGEMENT OF EFFLUENTS AND WASTEWATER

Food processing industry wastewater carries significant organic loads that biological approaches and classical technologies cannot entirely oxidize or disinfect. As a result, it is critical that sufficient technologies for treating wastewater released by the food industry be developed. When dispersed in liquid waste, oxidizing species of ROS have been shown to cause breakdown or degradation of contaminants. The use of fullness of CP for the treatment of effluents from fruit and vegetable processing plants has been recorded [41]. In a tomato processing plant, a plasma jet operating at 25 kV indicated that efficient decontamination of effluent was accomplished after 120 seconds.

After 180 seconds of exposure to blackberry and beetroot wastewater, total bacterial count reduced by 0.41 and 2.24 logs, respectively. The most vulnerable bacterial species was Escherichia coli, which had up to 90.22% of endotoxins eliminated [41]. Diffusion of hydroxyl radicals or other reactive species, photolytic activity of UV light during the destruction of organic compounds, and the introduction of shockwave that causes pyrolytic and chemical reactions in an indirect manner through electrohydraulic cavitation have all been mechanical pathways for CP activity [25].

3.5 CHALLENGES AND SAFETY ASPECTS OF COLD PLASMA (CP)

A study of the oral toxicity of edible films exposed to CP was recently published [23]. In this study, male and female Sprague Dawley (SD) rats were given edible film treated with CP at doses of 5,000 mg/kg or 1,000 mg/kg/day for 14 days and showed no evidence of chronic toxicity. Although there were minor changes in hemoglobin, hematocrit, bilirubin, creatinine, and aspartate aminotransferase levels in the blood. Furthermore, macroscopic examinations indicated no obvious alterations in the appearance or texture of the rats' representative liver sections [23]. As a result, the CP treatment of food products did not result in the creation of hazardous chemicals. CP's usefulness for treating fatty foods and other sensitive foods, as well as the possibility of producing chemical changes, has been investigated. For example, nuts exposed to CP for a long time and at high power showed a sharp increase in peroxide value [60].

The oxidation of fats leads to the formation of short chain fatty acids, and other unwanted compounds that cause off-flavors and foul odors. Furthermore, other food constituents, such as antioxidants might influence the decontamination efficiency of CP that calls for the optimization studies of plasma treatment in specific products. Other constraints of plasma treatment include the reduction in the firmness of FAV. Besides, the efficiency of plasma technique is influenced through the surface properties of the sample [15]. Future research should be focused on thorough detailing of CP interactions with food constituents, thereby, the overall effect on food quality.

3.6 SUMMARY

Several experimental investigations have been conducted so far on already established applications, while some other researchers have studied new pathways for exploration of NTP for food processing technologies. This chapter assembles the conclusions of recent studies on the applications and implications of CP in the food industry, such as microbial decontamination of food, modification, and treatment of packaging materials, modification of food components, improvement in seed germination performance and destruction of agrochemical residues. This review also focuses on other routes for the use of CP technology in the food industry with a foothold on future scope in several areas, such as hydrogenation of oils without production of trans fats, alleviation of food allergens and ANFs and wastewater management and effluent treatment. Although some several challenges and limitations are anticipated in the use of CP in the food industry and its implications, yet it is still safe to embrace CP technology as an impending significance for the food industry.

KEYWORDS

- **atmospheric pressure plasma jet**
- **cold plasma**
- **corona discharge**
- **dielectric barrier discharges**
- **food industry**
- **high voltage atmospheric cold plasma**
- **plasma jet**

REFERENCES

1. Abramzon, N., Joaquin, J. C., Bray, J., & Brelles-Marino, G., (2006). Biofilm destruction by RF high-pressure cold plasma jet. *IEEE Transactions on Plasma Sciences, 34*(4), 1304–1309.
2. Adler, H. J., Fischer, P., Heller, A., Jansen, I., Kuckling, D., Komber, H., Lehmann, D., et al., (1999). Trends in polymer chemistry. *Acta Polymerica, 50*(7), 232–239.
3. Bjiang, B., Zheng, J., Qiu, S., Wu, M., Zhang, Q., Yan, Z., & Xue, Q., (2014). Review on electrical discharge plasma technology for wastewater remediation. *Chemical Engineering Journal, 236*, 348–368.
4. Barekzi, N., & Laroussi, M., (2012). Dose-dependent killing of leukemia cells by low-temperature plasma. *Journal of Physics D: Applied Physics, 45*, 6. doi: 10.1088/0022-3727/45/42/422002.
5. Bauer, A., Ni, Y., Bauer, S., Paulsen, P., Modic, M., Walsh, J. L., & Smulders, F. J. M., (2017). The effects of atmospheric pressure cold plasma treatment on microbiological, physicochemical and sensory characteristics of vacuum packaged beef loin. *Meat Science, 128*, 77–87.
6. Bubler, S., Steins, V., Ehlbeck, J., & Schluter, O., (2015). Impact of thermal treatment versus cold atmospheric plasma processing on the techno-functional protein properties from *Pisum Sativum* var. Salamanca. *Journal of Food Engineering, 167*, 166–174.
7. Chiper, A. S., Chen, W., Mejlholm, O., Dalgaard, P., & Stamate, E., (2011). Atmospheric pressure plasma produced inside a closed package by dielectric barrier discharge in Ar/CO_2 for bacterial inactivation of biological samples. *Plasma Sources Science and Technology, 20*(2), E-article 025008; [Online]. https://doi.org/10.1088/0963-0252/20/2/025008 (accessed on 09 February 2023).
8. Chirokov, A., Gutsol, A., & Fridman, A., (2005). Atmospheric pressure plasma of dielectric barrier discharge. *Pure and Applied Chemistry, 77*(2), 487–495.
9. Choi, S., Puligundla, P., & Mok, C., (2017). Effect of corona discharge plasma on microbial decontamination of dried squid shreds including physicochemical and sensory evaluation. *LWT–Food Science and Technology, 75*, 323–328.
10. D'Agostino, R., Favia, P., Oehr, C., & Wertheimerv, M. R., (2005). Low-temperature plasma processing of materials: Past, present, and future. *Plasma Processes and Polymers, 2*, 7–15.
11. Devi, Y., Thirumdas, R., Sarangapani, C., Deshmukh, R. R., & Annapure, U. S., (2017). Influence of cold plasma on fungal growth and aflatoxins production on groundnuts. *Food Control, 77*, 187–191.
12. Dijkstra, A. J., (2006). Revisiting the formation of trans-isomers during partial hydrogenation of triacylglycerol oils. *European Journal of Lipid Science and Technology, 108*(3), 249–264.
13. Dong, S., Gao, A., Xu, H., & Chen, Y., (2016). Effects of dielectric barrier discharges (DBD) cold plasma treatment on physicochemical and structural properties of zein powders. *Food and Bioprocess Technology, 10*(3), 434–444.
14. Ekezie, F. G. C., Sun, D. W., & Cheng, J. H., (2017). A review on recent advances in cold plasma technology for the food industry: Current applications and future trends. *Trends in Food Science and Technology, 69*(part A), 46–58.
15. Fernandez, A., Noriega, E., & Thompson, A., (2013). Inactivation of *Salmonella enterica Serovar typhimurium* on fresh produce by cold atmospheric gas plasma technology. *Food Microbiology, 33*(1), 24–29.

16. Foster, J. E., Brandon, W., Gillman, E., & Yee, B., (2010). Underwater operation of a DBD plasma jet. *Plasma Sources Science and Technology, 2010*, E-article 025001; [Online]. doi: 10.1088/0963-0252/19/2/025001.
17. Fricke, K., Koban, I., Tresp, H., Jablonowski, L., Schroder, K., Kramer, A., Weltmann, K. D., et al., (2012). Atmospheric pressure plasma: A high-performance tool for the efficient removal of biofilms. *PLoS One, 7*(8), E-article 0042539; [Online]. doi: https://doi.org/10.1371/journal.pone.0042539.
18. Fridman, G., Shereshevsky, A., Jost, M. M., Brooks, A. D., Fridman, A., Gutsol, A., & Vasilets, V., (2007). Floating electrode dielectric barrier discharge plasma in air promoting apoptotic behavior in melanoma skin cancer cell lines. *Plasma Chemistry and Plasma Processing, 27*, 163–176.
19. Gao, L., Sun, L., Wan, S., Yu, Z., & Li, M., (2013). Degradation kinetics and mechanism of emerging contaminants in water by dielectric barrier discharge non-thermal: The case of 17β-estradiol. *Chemical Engineering Journal, 228*, 790–798.
20. Ghezzar, M. R., Abdelmalek, F., Belhadj, M., Benderdouche, N., & Addou, A., (2007). Gliding arc plasma assisted photo catalytic degradation of anthraquinonic acid green 25 in solution with TiO_2. *Applied Catalysis B: Environment, 72*(3, 4), 304–313.
21. Golan-Goldhirs, A., & Whitaker, J. R., (1984). Effect of ascorbic acid, sodium bisulfite, and thiol compounds on mushroom polyphenol oxidase. *Journal of Agriculture and Food Chemistry, 32*(5), 1003–1009.
22. Hamm, W., Hamilton, R. J., & Calliauw, G., (2013). *Edible Oil Processing* (2nd edn., p. 322). London: John Wiley and Sons, Ltd.
23. Han, S. H., Suh, H. J., Hong, K. B., Su, Y. K., & Min, S. C., (2016). Oral toxicity of cold plasma-treated edible films for food coating. *Journal of Food Science, 81*, T3052–T3057.
24. Herrmann, H. W., Henins, I., Park, J., & Selwyn, G. S., (1999). Decontamination of chemical and biological welfare (CBW) agents using an atmospheric pressure plasma jet (APPJ). *Physics of Plasma, 6*, E-article 2284. doi: https://doi.org/10.1063/1.873480.
25. Jiang, B., Zheng, J., Wu, S. Q. M., Zhang, Q., Yan, Z., & Xue, Q., (2014). Review on electrical discharge plasma technology for wastewater remediation. *Chemical Engineering Journal, 236*, 348–368.
26. Kalghatgi, S. U., Fridman, G., Fridman, A., Friedman, G., & Clyne, A. M., (2008). Non-thermal dielectric barrier discharge plasma treatment of endothelial cells. *Conference Proceeding IEEE Engineering Medical Biological Society, 2008*, 3578–3581.
27. Kieft, I. E., Laan, E. P., Der, V., & Stoffels, E., (2004). Electrical and optical characterization of the plasma needle. *New Journal of Physics, 6*, 1–14.
28. Kim, J. E., Oh, Y. J., Won, M. Y., Lee, K. S., & Min, S. C., (2017). Microbial decontamination of onion powder using microwave-assisted cold plasma treatments. *Food Microbiology, 62*, 112–123.
29. Krause, H., Schweiger, B., Schuhmacher, J., Scholl, S., & Steinfeld, U., (2009). Degradation of the endocrine disrupting chemicals (EDCs) carbamazepine, clofibric acid, and iopromide by corona discharge over water. *Chemosphere, 75*(2), 163–168.
30. Krishna, S., Ceriani, E., Marotta, E., Giardina, A., Spatenka, P., & Paradisi, C., (2016). Products and mechanism of verapamil removal in water by air non-thermal plasma treatment. *Chemical Engineering Journal, 292*, 35–41.
31. Laroussi, M., Tendero, C., Lu, X., Alla, S., & Hynes, W. L., (2006). Inactivation of bacteria by the plasma pencil. *Plasma Process and Polymer, 3*, 470–473.
32. Lee, H., Kim, J. E., Chung, M. S., & Min, S. C., (2015). Cold plasma treatment for the microbiological safety of cabbage, lettuce, and dried figs. *Food Microbiology, 51*, 74–80.

33. Lee, K. H., Kim, H. J., Woo, K. S., Jo, C., Kim, J. K., Kim, S. H., Park, H. Y., et al., (2016). Evaluation of cold plasma treatments for improved microbial and physicochemical qualities of brown rice. *LWT – Food Science and Technology, 73*, 442-447.
34. Ling, L., Jiafeng, J., Jiangang, L., Minchong, S., Xin, H., Hanliang, S., & Yuanhua, D., (2014). Effects of cold plasma treatment on seed germination and seedling growth of soybean. *Scientific Reports, 4*(1), E-article 5859; [Online]. doi: https://doi.org/10.1038/srep05859.
35. Meinlschmidt, P., Ueberham, E., Lehmann, J., Reineke, K., Schluter, O., Schweiggert-Weisz, U., & Eisner, P., (2016). The effects of pulsed ultraviolet light, cold atmospheric pressure plasma, and gamma-irradiation on the immune reactivity of soy protein isolate. *Innovative Food Science and Emerging Technologies, 38*(Part B), 374–383.
36. Min, S. C., Roh, S. H., Niemira, B. A., Boyd, G., Sites, J. E., Uknalis, J., & Fan, X., (2017). In-package inhibition of *E. coli* o157:h7 on bulk romaine lettuce using cold plasma. *Food Microbiology, 65*, 1–6.
37. Misra, N. N., Keener, K. M., Bourke, P., Mosnier, J. P., & Cullen, P. J., (2014). In-package atmospheric cold plasma treatment of cherry tomatoes. *Bioscience and Bioengineering, 118*, 177–182.
38. Misra, N. N., Koubaa, M., Roohinejad, S., Juliano, P., Alpas, H., Insico, R. S., Saraiva, J. A., & Barba, F. J., (2017). Landmarks in the historical development of twenty first century food processing technologies. *Food Research International, 97*, 318–339.
39. Misra, N. N., Tiwari, B. K., Raghavarao, K. S. M. S., & Cullen, P. J., (2011). Non-thermal plasma inactivation of food-borne pathogens. *Food Engineering Reviews, 3*, 159–170.
40. Mitra, A., Li, Y. F., Klampfl, T. G., Shimizu, T., Jeon, J., Morfill, G. E., & Zimmermann, J. L., (2014). Inactivation of surface-borne microorganisms and increased germination of seed specimen by cold atmospheric plasma. *Food and Bioprocess Technology, 7*(3), 645–653.
41. Mohamed, A. H., Shariff, S. M. A., Ouf, S. A., & Benghanem, M., (2016). Atmospheric pressure plasma jet for bacterial decontamination and property improvement of fruit and vegetable processing wastewater. *Journal of Physics D: Applied Physics, 49*(19), E-article 195401; [Online]. doi: 10.1088/0022-3727/49/19/195401.
42. Muranyi, P., Wunderlich, J., & Heise, M., (2017). Sterilization efficiency of a cascaded dielectric barrier discharge. *Journal of Applied Microbiology, 103*, 1535–1544.
43. Ognier, S., Iya-sou, D., Fourmond, C., & Cavadias, S., (2009). Analysis of mechanisms at the plasma–liquid interface in a gas–liquid discharge reactor used for treatment of polluted water. *Plasma Chemistry and Plasma Processing, 29*(4), 261–273.
44. Pankaj, S. K., & Keener, K. M., (2017). Cold plasma: Background applications and current trends. *Current Opinion in Food Science, 16*, 49–52.
45. Patange, A., Boehm, D., Bueno-Ferrer, C., Cullen, P. J., & Bourke, P., (2017). Controlling *Brochothrix thermosphacta* as a spoilage risk using in-package atmospheric cold plasma. *Food Microbiology, 66*, 48–54.
46. Phan, K. T. K., Phan, H. T., Brennan, C. S., & Phimolsiripol, Y., (2017). Non-thermal plasma for pesticide and microbial elimination on fruits and vegetables: An overview. *International Journal of Food Science and Technology, 52*, 2127–2137.
47. Puligundla, P., Kim, J. W., & Mok, C., (2017). Effects of non-thermal plasma treatment on decontamination and sprouting of radish (*Raphanus sativus* L.) seeds. *Food and Bioprocess Technology, 10*(6), 1093–1102.
48. Reddy, P. M. K., Mahammadunnisa, S. K., & Subrahmanyam, C., (2014). Catalytic non-thermal plasma reactor for mineralization of endosulfan in aqueous medium: A

green approach for the treatment of pesticide contaminated water. *Chemical Engineering Journal, 238*, 157–163.
49. Reddy, P. M. K., & Subrahmanyam, C., (2012). Green approach for wastewater treatment degradation and mineralization of aqueous organic pollutants by discharge plasma. *Industrial and Engineering Chemistry Research, 51*(34), 11097–11103.
50. Sadhu, S., Thirumdas, R., Deshmukh, R. R., & Annapure, U. S., (2017). Influence of cold plasma on the enzymatic activity in germinating mung beans (*Vigna radiate*). *LWT-food Science and Technology, 78*, 97–104.
51. Sakiyama, Y., & Graves, D. B., (2009). Neutral gas flow and ring-shaped emission profile in non-thermal rf-excited plasma needle discharge at atmospheric pressure. *Plasma Sources Science and Technology, 18*(2), E-article 025022; [Online]. https://doi.org/10.1088/0963-0252/18/2/025022.
52. Sarangapani, C., O'Toole, G., Cullen, P. J., & Bourke, P., (2017). Atmospheric cold plasma dissipation efficiency of agrochemicals on blueberries. *Innovative Food Science & Emerging Technologies, 44*, 235–241.
53. Schutze, A., Jeong, J. Y., Babayan, S. E., Park, J., Selwyn, G. S., & Hicks, R. F., (1998). The atmospheric pressure plasma jet: A review and comparison to other plasma sources. *IEEE Transactions on Plasma Science, 26*(6), 1685–1694.
54. Shi, H., Ileleji, K., Stroshine, R. L., Keener, K., & Jensen, J. L., (2017). Reduction of aflatoxin in corn by high voltage atmospheric cold plasma. *Food and Bioprocess Technology, 10*(6), 1042–1052.
55. Sivachandiran, L., & Khacef, A., (2017). Enhanced seed germination and plant growth by atmospheric pressure cold air plasma: Combined effect of seed and water treatment. *RSC Advances, 7*(4), 1822–1832.
56. Stoffels, E., Flikweert, A. J., Stoffetls, W. W., & Kroesen, G. M., (2002). Plasma needle: A non-destructive atmospheric plasma source for fine surface treatment of bio materials. *Plasma Sources Sciences and Technology, 11*(4), 383–388.
57. Su, L., Lan, Q., Pritchard, H. W., Xue, H., & Wang, X., (2016). Reactive oxygen species induced by cold stratification promote germination of *Hedysarum scoparium* seeds. *Plant Physiology and Biochemistry, 109*, 406–415.
58. Surowsky, B., Fischer, A., Schlueter, O., & Knorr, D., (2013). Cold plasma effects on enzyme activity in a model food system. *Innovative Food Science & Emerging Technologies, 19*, 146–152
59. Thirumdas, R., Deshmukh, R. R., & Annapure, U. S., (2016). Effect of low temperature plasma on the functional properties of basmati rice flour. *Journal of Food Science and Technology, 53*(6), 2742–2751.
60. Thirumdas, R., Sarangapani, C., & Annapure, U. S., (2014). Cold plasma: A novel non-thermal technology for food processing. *Food Biophysics, 10*(1), 1–11.
61. Thirumdas, R., Sarangpani, C., Ajinkya, M. T., Deshmukh, R. R., & Annapurre, U. S., (2016). Influence of low-pressure cold plasma on cooking and textural properties of brown rice. *Innovative Food Science & Emerging Technologies, 37*, 53–60.
62. Tolouie, H., Hashemi, M., Mohammadifar, M. A., & Ghomi, H., (2018). Cold atmospheric plasma manipulation of proteins in food systems. *Critical Reviews in Food Science and Nutrition, 58*(15), 2583–2597.
63. Vamos-Vigyazo, L., (1995). Prevention of enzymatic browning in fruits and vegetables: A review of principle and practice; Chapter 4. In: Le, C. Y., & Whitaker, J. R., (eds.), *Enzymatic Browning and its Preventions* (Vol. 600, pp. 49–62). Philadelphia – USA: American Chemical Society.

64. Walsh, J. L., Shi, J. J., & Kong, M. G., (2006). Contrasting characteristics of pulsed and sinusoidal cold atmospheric plasma jets. *Applied Physics Letter, 88*, E-article 171501. doi: https://doi.org/10.1063/1.2198100.
65. Wan, Z., Chen, Y., Pankaj, S. K., & Keener, K. M., (2017). High voltage atmospheric cold plasma treatment of refrigerated chicken eggs for control of *Salmonella enteritidis* contamination on eggshell. *LWT – Food Science and Technology, 76*, 124–130.
66. Won, M. Y., Lee, S. J., & Min, S. C., (2017). Mandarin preservation by microwave-powered cold plasma treatment. *Innovative Food Science and Emerging Technologies, 39*, 25–32.
67. Wu, T. Y., Sun, N. N., & Chau, C. F., (2017). Application of corona electrical discharge plasma on modifying the physicochemical properties of banana starch indigenous to Taiwan. *Journal of Food and Drug Analysis, 26*(1), 244–251.
68. Yepez, X. V., & Keener, K. M., (2016). High-voltage atmospheric cold plasma (HVACP) hydrogenation of soybean oil without trans-fatty acids. *Innovative Food Science and Emerging Technologies, 38*, 169–174.
69. Zhang, J., Sun, J., Wang, D., & Wang, X., (2006). A novel cold plasma-jet generated by atmospheric dielectric barrier capillary discharge. *Thin Solid Films, 506*, 404–408.
70. Zhu, W. C., Wang, B. R., Xi, H. L., & Pu, Y. K., (2010). Decontamination of surrogate malathion by atmospheric pressure radio-frequency plasma jet. *Plasma Chemistry and Plasma Processing, 30*(3), 381–389.
71. Grzegorzewski, F. (2011). Influence of Non-Thermal Plasma Species on the Structure and Functionality of Isolated and Plant-based 1,4-Benzopyrone Derivatives and Phenolic Acids. Thesis. Inst. Food Technology and Food Chemistry. file:///Users/Ash/Downloads/Influence_of_Non-Thermal_Plasma_Species_on_the_Str.pdf
72. Liao, X., Xiang, Q., Liu, D., Chen, S., Ye, X., & Ding, T. (2017). Lethal and sublethal effect of a dielectric barrier discharge atmospheric cold plasma on Staphylococcus aureus. Journal of food protection, 80(6), 928-932.
73. Pawłat, J., Starek, A., Sujak, A., Terebun, P., Kwiatkowski, M., Budzeń, M., & Andrejko, D. (2018). Effects of atmospheric pressure plasma jet operating with DBD on Lavatera thuringiaca L. seeds' germination. PloS one, 13(4), e0194349.

CHAPTER 4

POTENTIAL OF PLASMA ENGINEERING FOR DECONTAMINATION IN THE FOOD INDUSTRY

T. SUPRIYA and B. B. BORSE

ABSTRACT

Plasma systems and plasma engineering are emerging areas in the food industry. The necessity to embellish food health in microbial consistency, nutritional, and olfactory facets of food has built up a growing significance of this innovative method in the food industries. Non-thermal plasma (NTP) processing has a bigger effect on developing processes of production and packaging of food products.

4.1 INTRODUCTION

Plasma engineering is known for the application of partially ionized gases in the food industry [63]. A plasma state is accomplished when ample energy (such as heat or electricity) is applied to a gas [28]. The energy used ionizes the gas by dissociating molecular bonds [10]. Currently, consumers stipulate safe foods that should meet other quality criteria. Plasma may be produced by any energy source capable of ionizing gasses, such as electrical, thermal, optical (UV light), radioactive (gamma radiation), and electromagnetic X-ray radiation [45].

The foods may be tainted by different micro-organisms or may be subject to degradation due to natural food enzymes. Because of their major public health threats and economic effects, the pathogenic and spoilage micro-organisms are troublesome in the food industry. These micro-organisms are

Advanced Research Methods in Food Processing Technologies. Junaid Ahmad Malik, Megh R. Goyal, Preeti Birwal, & Ritesh B. Watharkar (Eds.)
© 2024 Apple Academic Press, Inc. Co-published with CRC Press (Taylor & Francis)

eradicated by many methods: Thermal technologies (such as sterilization, ohmic heating, pasteurization), other non-thermal technologies (such as pulsed electrical fields, high hydrostatic pressure (HHP)), and high voltage arc discharge [48, 61].

Treatment of foodstuff by cold plasma (CP) discharge leads to a decline in micro-organisms (bacteria, fungi) count on its surface. Plasma could also be used for the disintegration of copious undesirable chemical compounds, which are present in foodstuffs [3]. Some techniques are capable of inducing atmospheric pressure plasma at a macroscopic temperature close to the surface, and this occurrence of plasmas during a potential differential is formed between the electrodes at atmospheric pressure planted in a gas [39].

Plasma is assembled beneath the limitation that the normal kinetic energy of an electron extensively exceeds the average Coulomb energy desired for an ion to attach the electron [43]. The existence of free ions, utilizing plasma for ion origin is certain. Plasma is a relevant form of low-pressure gas discharge [65] as shown in Figure 4.1. Plasma can be factitiously spawned by the condition of a neutral gas or inflicting it to a strong electromagnetic field to indicate where an ionized gaseous substance metamorphosis to increasingly electrically conductive. The culminate charged ions and electrons evolve and persuade by long-range electromagnetic fields, making the plasma dynamics additional precise to these fields than a neutral gas [40]. Because of its divergent disposition, plasma has achieved application in numerous fields (Figure 4.2), such as surface treatment, controlled thermonuclear fusion, medicine, biomedical devices, electronics, transport, and propulsion to particle, particle mechanics, plasma solid-state, gas lasers, etc., [1]. Plasma properties include ion and electron densities, ion, and electron temperatures, diverse exciting species densities, total particle density, and gas temperature [33].

This chapter highlights the potential of plasma engineering in the food industry for the decontamination and quality improvement of foods.

4.2 INTRODUCTION TO PLASMA ENGINEERING

In 1898, Sir William Crookes invented the term ionization from neutral to electron atom and strong ion to define the disintegration process in a system. In 1928, Irving Langmuir proposed the term plasma to describe an essentially electro-neutral set of electrons and ions, which can or cannot contain a neutral gas and which is ready to react to electrical and magnetic fields [7, 25]. Dielectric discharge barriers and microwave (MW) and plasma jet discharges are typically used for spawning CP [24]. The electrical plasma emissions were defined by a voltage test in the time domain at the electrodes, using a high-stress probe [49]. The plasmas are efficient in melting or vaporizing abundant materials and benefit the industry for plasma flame spraying, welding, etc., [51].

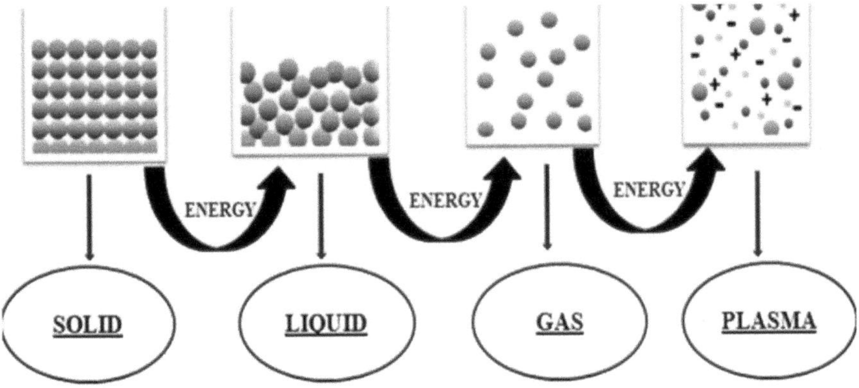

FIGURE 4.1 Schematic view of the four related states of a matter [27].

Source: Adapted from Ref. [27]

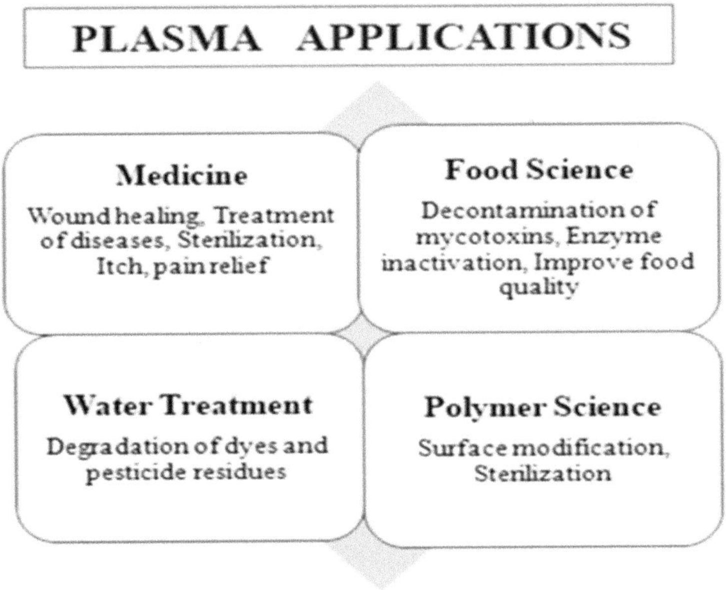

FIGURE 4.2 Various applications of plasma technology [15, 54].

Source: Adapted from Ref. [15] and Ref. [54]

Plasma in medicine is a dominating field merging in plasma physics and engineering, medicine, and bioengineering to utilize plasmas in the remedial treatments [22]. Plasma engineering is consistently decisive, reproducible, relatively, economical, and appropriate to diverse geometries of samples and distant components, such as metals, polymers, ceramics, and composites [6].

4.3 SCOPE OF PLASMA TECHNOLOGY IN FOOD INDUSTRY

Plasma is a persuasive, cost-effective, atmospheric secure manner of critical cleaning [12]. It is used to determine the biocidal effects of microbes in the liquid inactivation process [32]. The food packaging industry insists on reasonable packaging with high obstruction adjacent to gas, flavor, and moisture [38]. Plasma treatment is primarily practiced for the cleaning, surface etching of the films, and sterilization in food packaging [26].

The chart-topping technique for convening this criterion applies to biaxial-driven vacuum-coated polypropylene (BOPP) films [38]. Components, including proteins, polysaccharides, and lipids are the most commonly used raw materials for the development of packaging food films [50]. The Biaxial orientation (BO) strengthens the film's delicate structure and inevitably decreases turbidity, embellishing the obstructive properties and flexibility at a low temperature. Biaxial-focused polypropylene (BOPP) film is frequently glazed with an additional polymer layer to improve its gaseous and humidity mechanical properties [64]. Plasma treatment improved the hydrophilicity of the BOPP films on the surface and improved their resistance to the coating [30].

Plasma application was limited to impervious materials with heat and vacuum [25]. Plasma can work at low atmospheric pressure and temperature for processing [17]. The polymerized plasma components may be used typified by IR, NMR, and other analytical approaches. The polymer has been introduced to be highly cross-linked, amorphous, and also contains functional groups not found in the monomer. A specific technique called plasma polymerization can improve the surface properties of food contact materials [7].

The application of non-thermal ionized gasses, often known as cold gas plasmas, has materialized as another decontamination method for fresh fruit and vegetables [56]. Pulsed discharge of the plasma gas is an exceptionally qualified method for handling suspensions of polluted liquid material. This alleviates micro-organisms by applying pulses of high voltage (20–30 KV) to a liquid test, which is being sparged [13].

4.4 PLASMA TECHNOLOGIES IN THE FOOD DECONTAMINATION

4.4.1 NON-THERMAL PLASMA (NTP) TECHNOLOGY

For the last 50 years, in the content manufacturing field non-thermal plasmas (NTPs) had erect implementations. One important advantage of NTP is that for reaction with divergent surfaces it consists of bountiful chemically convincing material. Plasma technology can deliver special content handling possibilities at low temperatures [10]. NTP is set up electrically and contains extremely susceptible species introduced by gas molecules, charged particles, Electrons, free radicals, and electromagnetic radiation concentrations at close-space temperatures. The platform is NTP, a materialized technology with possible sterilization applications within the food sector [37].

The traditional technology can be discretionary to NTP. Throughout the generation of NTP, the bulk of electrical energy goes moderately into energy-matter output because of gas heating. The NTP energy is used to dissociate and ionize the electron-impact gas to yield NTP species [9, 52]. NTP has recently been investigated for the inactivation of micro-organisms, ready-to-eat food preparation, etc., and this continues to be a rising anti-microbial technique mainly used to decontaminate polluted surfaces. For healthcare, cancer cells were treated with plasma and apoptosis, prion, and another biomolecule silencing, infection control, or contaminated wound therapy was initiated. Among the novel ways of inactivating the inadmissible microflora, NTP contributes to an ample clout escort by minimizing the mangle of biomaterials released, thermo-sensitive labeling products, such as plant and animal tissues, and packaged foods [55].

The advantages of NTP are related to its potential to operate inside low-temperature levels except for the need for inducing fascinating altered to heat-sensitive food materials diminishing nutrient and sensory property degradation. NTP is a green technology that is uncluttered and eco-friendly, because of its qualification to boost the safety and shelf-life of food items. Many commercial plasma systems require products processing to change or handle bottles, ink, textiles, and appliances, water treatment and exhaust gases, sterilization of medical instruments, water management, substance dehydration and synthesis, air gases, deposition, and synthesis, and recently, in three-dimensional tissue printing to Micron-size application to imitate natural tissue microstructure [41].

NTP has countless promising food production uses include dry disinfection of crop surfaces (such as dairy, flesh, and fowl), Particulate granules and seed matter (cereals, plants, and spices), and sprouting seeds. In conjunction with the decontamination, plasma technology offers diversified assistance in the cereal industry. This is an assertive non-thermal technique used for various crop types in the cereals industry. Because of its indigent cooking and eating qualities, the brown rice is less irresistible. Nonetheless, due to its nutritionally good brown rice is selected. Processing plasma causes the properties of brown rice to change. The product of cooking CP low pressure and parboiled rice consistency, protected by several treatment times and power [34, 35, 53].

4.4.1.1 COLD PLASMA (CP)

Cold plasma (CP) treatments for non-thermal aspects demonstrated no or negligible effects on distinct products physicochemical, sensory, and nutritional attributes. CP is a special technology used for the non-thermal handling of food strong, for inactivation of reactive gases for meat, poultry, fruit, and vegetables contaminating microbes [42]. CP technology has proved its effectiveness and different methods for food decontamination and extension of the shelf-life. The use of CP in the food sector has been demonstrated for agricultural decontamination, inactivation of enzymes and toxins, modifications to food packaging, and wastewater treatment [45]. The competence to introduce cold atmospheric gas plasmas for the treatment of whole fruit surfaces or rinds before cutting and peeling helps in the commercial preparation of distinct fruit products [47].

CP is a fairly cutting-edge technical innovation to ensure food quality and health. CP technology makes use of electrical power that is applied to carrier gasses like hydrogen, oxygen, nitrogen, or helium with the help of 11 electrodes [31]. Dry processing techniques such as CP may be an ideal method for disinfecting food processing equipment and the atmosphere of food products manufacturers. A reduced need for the use of chemicals and water may also enable manufacturers to save costs in addition to potentially making it safe and chemical-free processes. A dry process also ensures that the system can be used during food production to handle plants, distribution lines, or equipment with hazardous areas to maintain low bacterial levels. This will reduce the risk of cross-contamination and introducing and growing bacteria into biofilms [14].

CP processing studied for multi-stage requirements of primary and secondary food production, storage of raw materials, intermediate or finished products, facilities for packaging equipment, and the environment [5]. CP is under demanding analysis as a novel decontamination technology [46].

4.4.2 THERMAL PLASMA (TP) TECHNOLOGY

Thermal plasma (TP) may be used to remove or induce in solid, liquid, and gaseous toxic halogenated and reactive compounds, thermal barriers, anti-corrosion, anti-wear coatings, etc. Recent manufacturing technologies focused on the use of TPs show enormous destruction efficiency due to the achievement of additional exalted temperatures and also high process flexibility reactivity in substantial volumes. TP guides creating ceramic or metallic thick films with high adhesion and strong thickness control on various metal substrates [4].

The ambient temperature innovations appeared in search of traditional heat treatment possibilities [36]. Regulation of kinetic release parameters by plasma treatment may be a valuable tool in the food industry to customize specific food items easily and increase their shelf life under restricted circumstances [44]. Thermal methods (such as the processing and pasteurization of effectively inactivated micro-organisms and enzymes) and their inimical effects on food color, flavor, and nutritional content have not always fulfilled consumer requirements, while TP may offer solutions to these stated drawbacks and may emerge as a major clean and green technology as a relief to mother earth and humanity. TP is produced at high pressures and needs high-power inputs [36].

4.5 PLASMA IN IMPROVING FOOD QUALITY

Plasma therapy has a varied impact on the sowing process and morphological characteristics of seeds because of complicated interactions with organic materials and living cells [57]. Food treated with plasma is disclosed to the responsive factor as in the contaminating micro-organisms. Accordingly, the goal is to carry out the maximum imaginable microbe count depreciation with the minimum attainable outcome on the quality of food. This refers to the effect of plasma therapy on future improvements in the composition of the food, in particular, where potentially dangerous [60].

4.5.1 MICROBIAL CONTAMINATION

Microbial contamination of cereals is obtained from different sources, Sources, including dirt, sand, water, plants, insects, birds, and animal feces, and also during the planting, processing, drying, and storing of the crops [29]. *Salmonella* and *E. coli* are the most widely linked human pathogens of leafy green crops. The effectiveness of CP therapies on microbial inactivation of vegetables and the efficacy of these therapies for micro-organism decontamination on vegetables have been accomplished [59]. The microbial result of liquid plasma therapy may be decorated with the intense ion bombardment.

4.5.2 PHYSICAL CHARACTERISTICS

The related non-thermal methods for starch modifications are process technology with high pressure, pulsed electrical field, ultrasonics, MW treatment, irradiation, and CP technology [2]. Plasma treatment is a persuasive procedure to boost the hydrophilicity of the biopolymer surfaces [62].

4.5.3 CHEMICAL CHARACTERISTICS

Modification of chemicals affects the molecular weight, gelatinization rates, pH, rheological, pasting, and other useful characteristics. The synergy of plasma species with starch will change in three distinct attainable mechanisms, such as cross-linking, depolymerization, and etching of plasma [66].

4.5.4 SENSORY CHARACTERISTICS

The sensory characteristics are essential criteria for the consistency and survival of all foods including dairy products and their derivatives. The ions in CP have an antimicrobial effect and possible negative effects on the product's physicochemical and sensory characteristics, particularly those with high nutritional value [8].

4.6 ROLE OF PLASMA TECHNOLOGY IN THE PROCESSING AND SAFETY OF FOODS

Today the safeguarding of food and quality food has become an important activity than in some significant time. New threats are facing due to the growth characteristics of suitable micro-organisms, improvements in processing methodologies, diversified climate and the climate, and the growth of global food trading [37]. A tremendous health standard may be set attained by precisely heat-sterilizing of food, hereby devastating great flavor and nutritious benefit, wherein now novel plasma food processing has a greater role to play to ward off all the evils of the processes which are not full-proof. Irradiation is another form of simple microbial risk management; however, it is not recommended for the general public in many countries due to costly procedures [19].

CP technology to improve the microbiological protection of food by using non-thermal food processing [59]. Plasma treatment applied to fresh foods does not increase the degree of discoloration, dehydration, or increase the sensory and quality characteristics of the food products in any way, therefore, it may become an appropriate technology for the food sector. Since CP treatment is a contactless, waterless, and chemical-less process therefore it can lower microbial growth [58]. Applications of CP in livestock and food industries are indicated in Figure 4.3.

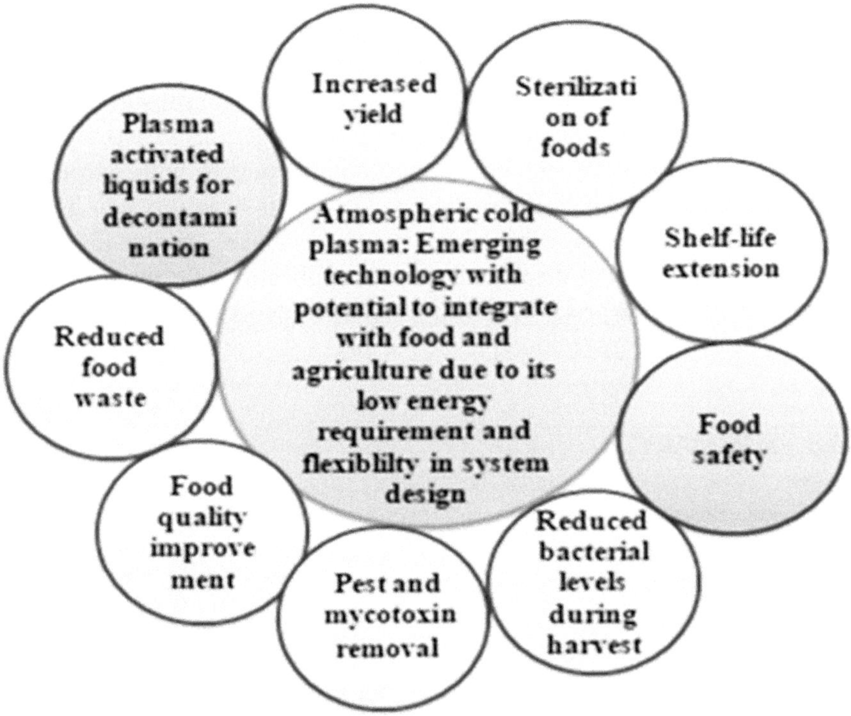

FIGURE 4.3 Applying cold plasma technology to the livestock and food industries [12].

Source: Adapted from Fig. [12]

CP food therapy has become an advanced and active methods are need to be identified to carry out regular procedures on a wide scale. Significant amounts of efforts are needed in terms of R&D at a bench, pilot, and industry level to make CP the choice for food processing. Treatment of plasma was conducted by researchers and has established an atmospheric plasma pressure under vacuum conditions that result in low price, improved pace of care, and applicability to food industry [23].

4.7 RECENT TRENDS IN PLASMA ENGINEERING

The CP technology proved to accomplish and protein-rich isolates. CP technology may be used to treat the surface of packaging material to enhance performance, etching, or cleaning, and deposition [21]. The oleo chemical process is hydrogenation, which includes the transformation of changing chemical structures from liquid oil to solid fat [23]. Atmospheric CP is a flexible approach that explains productiveness for handling many circumstances over these sectors, and it can administer transformative and sustainable technologies [12]. CP is one of the latest innovative nutritional management systems because of plasma's ability to control surface properties without deterring the quality of food [54].

Fresh raw products are turned into food products with added value continuing a sequence of unit activities [48]. High volatility can be found in the plasma configurations both advantageous and restrictive [57]. Novel methods for processing displayed possible applications for the reduction of food allergens in the processing of hypoallergenic foods for individuals with a history of an allergy to food [5, 11, 18, 20].

4.8 SUMMARY

The generation of atmospheric pressure plasma admits progressive processing, which is a benefit for applications of food processing. For lab applications, the plasma systems are illustrated. The use of cold air plasma pressure is a positive factor and pleasant technique for dry food decontamination. Owing to the dynamics of the non-thermal ambient plasma pressure and the breathing cell, the basic process for interplaying plasma with microorganisms can be a little complicated.

KEYWORDS

- **biaxial-driven vacuum-coated polypropylene**
- **biaxial orientation**
- **cold plasma**
- **non-thermal plasma**
- **plasma engineering**
- **thermal plasma**

REFERENCES

1. Adhikari, B. R., & Khanal, R., (2013). Introduction to the plasma state of matter. *Himalayan Physics, 4*, 60–64.
2. Ashogbon, A. O., & Akintayo, E. T., (2014). Recent trend in the physical and chemical modification of starches from different botanical sources: A review. *Starch-Stärke, 66*(1, 2), 41–57.
3. Bartos, P., Kriz, P., Havelka, Z., Bohata, A., Olsan, P., Spatenka, P., & Dientsbier, M., (2017). Plasma technology in food industry: Review. *Kvasny. Prumysl., 63*(3), 134–138.
4. Bonizzoni, G., & Vassallo, E., (2002). Plasma physics and technology, industrial applications. *Vacuum, 64*(3, 4), 327–336.
5. Bourke, P., Ziuzina, D., Boehm, D., Cullen, P. J., & Keener, K., (2018). The potential of cold plasma for safe and sustainable food production. *Trends in Biotechnology, 36*(6), 615–626.
6. Chu, P. K., Chen, J. Y., Wang, L. P., & Huang, N., (2002). Plasma-surface modification of biomaterials. *Materials Science and Engineering: R: Reports, 36*(5, 6), 143–206.
7. Çökeliler, D., Erkut, S., Zemek, J., Biederman, H., & Mutlu, M., (2007). Modification of glass fibers to improve reinforcement: A plasma polymerization technique. *Dental Materials, 23*(3), 335–342.
8. Coutinho, N. M., Silveira, M. R., Rocha, R. S., Moraes, J., Ferreira, M. V. S., Pimentel, T. C., & Borges, F. O., (2018). Cold plasma processing of milk and dairy products. *Trends in Food Science & Technology, 74*, 56–68.
9. Deng, S., Ruan, R., Mok, C. K., Huang, G., Lin, X., & Chen, P., (2007). Inactivation of *Escherichia coli* on almonds using non-thermal plasma. *Journal of Food Science, 72*(2), M62–M66.
10. Dey, A., Chroneos, A., Braithwaite, N. S. J., Gandhiraman, R. P., & Krishnamurthy, S., (2016). Plasma engineering of graphene. *Applied Physics Reviews, 3*(2), 021301.
11. Dong, X., Wang, J., & Raghavan, V., (2020). Critical reviews and recent advances of novel non-thermal processing techniques on the modification of food allergens. *Critical Reviews in Food Science and Nutrition, 61*(2), 196–210.

12. Ekezie, F. G. C., Sun, D. W., & Cheng, J. H., (2017). Review on recent advances in cold plasma technology for the food industry: Current applications and future trends. *Trends in Food Science & Technology, 69*, 46–58.
13. Fernandez, A., Noriega, E., & Thompson, A., (2013). Inactivation of *Salmonella enterica Serovar typhimurium* on fresh produce by cold atmospheric gas plasma technology. *Food Microbiology, 33*(1), 24–29.
14. Fridman, G., Brooks, A. D., Balasubramanian, M., Fridman, A., Gutsol, A., Vasilets, V. N., & Friedman, G., (2007). Comparison of direct and indirect effects of non-thermal atmospheric pressure plasma on bacteria. *Plasma Processes and Polymers, 4*(4), 370–375.
15. Gavahian, M., & Khaneghah, A. M., (2020). Cold plasma as a tool for the elimination of food contaminants: Recent advances and future trends. *Critical Reviews in Food Science and Nutrition, 60*(9), 1581–1592.
16. Gómez-López, V. M., Orsolani, L., Martínez-Yépez, A., & Tapia, M. S., (2010). Microbiological and sensory quality of sonicated calcium-added orange juice. *LWT-Food Science and Technology, 43*(5), 808–813.
17. Güleç, H. A., Sarıoglu, K., & Mutlu, M., (2006). Modification of food contacting surfaces by plasma polymerization technique, part I: Determination of hydrophilicity, hydrophobicity and surface free energy by contact angle method. *Journal of Food Engineering, 75*(2), 187–195.
18. Guo, J., Huang, K., & Wang, J., (2015). Bactericidal effect of various non-thermal plasma agents and the influence of experimental conditions in microbial inactivation: A review. *Food Control, 50*, 482–490.
19. Havelaar, A. H., Brul, S., De Jong, A., De Jonge, R., Zwietering, M. H., & Ter Kuile, B. H., (2010). Future challenges to microbial food safety. *International Journal of Food Microbiology, 139*, S79–S94.
20. Hertwig, C., Meneses, N., & Mathys, A., (2018). Cold atmospheric pressure plasma and low energy electron beam as alternative non-thermal decontamination technologies for dry food surfaces: A review. *Trends in Food Science & Technology, 77*, 131–142.
21. Keener, K. M., & Misra, N. N., (2016). Future of cold plasma in food processing. In: *Cold Plasma in Food and Agriculture* (pp. 343–360). Boston – MA: Academic Press.
22. Keidar, M., & Beilis, I., (2013). *Plasma Engineering: Applications from Aerospace to Bio-and Nanotechnology* (p. 424). Boston – MA: Academic Press.
23. Kim, B., Yun, H., Jung, S., Jung, Y., Jung, H., Choe, W., & Jo, C., (2011). Effect of atmospheric pressure plasma on inactivation of pathogens inoculated onto bacon using two different gas compositions. *Food Microbiology, 28*(1), 9–13.
24. Kim, J. E., Oh, Y. J., Won, M. Y., Lee, K. S., & Min, S. C., (2017). Microbial decontamination of onion powder using microwave-powered cold plasma treatments. *Food Microbiology, 62*, 112–123.
25. Knorr, D., Froehling, A., Jaeger, H., Reineke, K., Schlueter, O., & Schoessler, K., (2011). Emerging technologies in food processing. *Annual Review of Food Science and Technology, 2*, 203–235.
26. Lei, J., Yang, L., Zhan, Y., Wang, Y., Ye, T., Li, Y., & Li, B., (2014). Plasma treated polyethylene terephthalate/polypropylene films assembled with chitosan and various preservatives for antimicrobial food packaging. *Colloids and Surfaces – B: Biointerfaces, 114*, 60–66.
27. Liao, X., Liu, D., Xiang, Q., Ahn, J., Chen, S., Ye, X., & Ding, T., (2017). Inactivation mechanisms of non-thermal plasma on microbes: A review. *Food Control, 75*, 83–91.

28. Lopez, M., Calvo, T., Prieto, M., Mugica-Vidal, R., Muro-Fraguas, I., Alba-Elias, F., & Alvarez-Ordonez, A., (2019). Review on non-thermal atmospheric plasma for food preservation: Mode of action, determinants of effectiveness and applications. *Frontiers in Microbiology, 10,* 622–628.
29. Los, A., Ziuzina, D., Akkermans, S., Boehm, D., Cullen, P. J., Van, I. J., & Bourke, P., (2018). Improving microbiological safety and quality characteristics of wheat and barley by high voltage atmospheric cold plasma closed processing. *Food Research International, 106,* 509–521.
30. Lu, P., Guo, M., Xu, Z., & Wu, M., (2018). Application of nano fibrillated cellulose on BOPP/LDPE film as oxygen barrier and antimicrobial coating based on cold plasma treatment. *Coatings, 8*(6), 207–211.
31. Mandal, R., Singh, A., & Singh, A. P., (2018). Recent developments in cold plasma decontamination technology in the food industry. *Trends in Food Science & Technology, 80,* 93–103.
32. Marsili, L., Espie, S., Anderson, J. G., & MacGregor, S. J., (2002). Plasma inactivation of food-related micro-organisms in liquids. *Radiation Physics and Chemistry, 65*(4, 5), 507–513.
33. Mattox, D. M., (2010). *Handbook of Physical Vapor Deposition (PVD) Processing* (p. 946). New York: William Andrew.
34. Mir, S. A., Bosco, S. J. D., Shah, M. A., Mir, M. M., & Sunooj, K. V., (2016). Variety difference in quality characteristics, antioxidant properties and mineral composition of brown rice. *Journal of Food Measurement and Characterization, 10*(1), 177–184.
35. Mir, S. A., Shah, M. A., & Mir, M. M., (2016). Understanding the role of plasma technology in food industry. *Food and Bioprocess Technology, 9*(5), 734–750.
36. Misra, N. N., Schlüter, O., & Cullen, P. J., (2016). Plasma in food and agriculture. In: *Cold Plasma in Food and Agriculture* (pp. 1–16). Academic Press.
37. Misra, N. N., Tiwari, B. K., Raghavarao, K. S. M. S., & Cullen, P. J., (2011). Non-thermal plasma inactivation of food-borne pathogens. *Food Engineering Reviews, 3*(3, 4), 159–170.
38. Moosheimer, U., & Bichler, C., (1999). Plasma pretreatment of polymer films as a key issue for high barrier food packaging. *Surface and Coatings Technology, 116,* 812–819.
39. Moreau, M., Orange, N., & Feuilloley, M. G. J., (2008). Non-thermal plasma technologies: New tools for bio-decontamination. *Biotechnology Advances, 26*(6), 610–617.
40. Morozov, A. I., (2012). *Introduction to Plasma Dynamics* (1st edn., p. 840) Boca Raton–FL: CRC Press.
41. Muhammad, A. I., Xiang, Q., Liao, X., Liu, D., & Ding, T., (2018). Understanding the impact of non-thermal plasma on food constituents and microstructure—A review. *Food and Bioprocess Technology, 11*(3), 463–486.
42. Niemira, B. A., (2012). Cold plasma decontamination of foods. *Annual Review of Food Science and Technology, 3,* 125–142.
43. Nishikawa, K., & Wakatani, M., (2000). Basic properties of plasma. In: *Plasma Physics* (Vol. 8, pp. 6–13). Springer Series on Atoms & Plasma. Berlin – Heidelberg: Springer-Verlag.
44. Pankaj, S. K., Bueno-Ferrer, C., Misra, N. N., O'Neill, L., Jiménez, A., Bourke, P., & Cullen, P. J., (2014). Surface, thermal and antimicrobial release properties of plasma-treated zein films. *Journal of Renewable Materials, 2*(1), 77–84.
45. Pankaj, S. K., Wan, Z., & Keener, K. M., (2018). Effects of cold plasma on food quality: A review. *Foods, 7*(1), 4–10.

46. Patange, A., Boehm, D., Giltrap, M., Lu, P., Cullen, P. J., & Bourke, P., (2018). Assessment of the disinfection capacity and eco-toxicological impact of atmospheric cold plasma for treatment of food industry effluents. *Science of the Total Environment, 631*, 298–307.
47. Perni, S., Liu, D. W., Shama, G., & Kong, M. G., (2008). Cold atmospheric plasma decontamination of the pericarps of fruit. *Journal of Food Protection, 71*(2), 302–308.
48. Priyadarshini, A., Rajauria, G., O'Donnell, C. P., & Tiwari, B. K., (2019). Emerging food processing technologies and factors impacting their industrial adoption. *Critical Reviews in Food Science and Nutrition, 59*(19), 3082–3101.
49. Ragni, L., Berardinelli, A., Vannini, L., Montanari, C., Sirri, F., Guerzoni, M. E., & Guarnieri, A., (2010). Non-thermal atmospheric gas plasma device for surface decontamination of shell eggs. *Journal of Food Engineering, 100*(1), 125–132.
50. Romani, V. P., Olsen, B., Collares, M. P., Oliveira, J. R. M., Prentice-Hernández, C., & Martins, V. G., (2019). Improvement of fish protein films properties for food packaging through grow discharge plasma application. *Food Hydrocolloids, 87*, 970–976.
51. Roth, J. R., (2001). Industrial plasma engineering. *Book Series: Applications to Non-thermal Plasma Processing* (Vol. 2, p. 658) Boca Raton – FL: CRC Press.
52. Ruan, R., Han, W., Ning, A., Deng, S., Chen, P. L., & Goodrich, P., (1999). Effects of design parameters of planar, silent discharge, plasma reactors on gaseous ammonia reduction. *Transactions of the ASAE, 42*(6), 1841–1850.
53. Sarangapani, C., Devi, Y., Thirundas, R., Annapure, U. S., & Deshmukh, R. R., (2015). Effects of low-pressure plasma on physicochemical properties of parboiled rice. *LWT-Food Science and Technology, 63*(1), 452–460.
54. Schlüter, O., Ehlbeck, J., Hertel, C., Habermeyer, M., Roth, A., Engel, K. H., & Eisenbrand, G., (2013). Opinion on the use of plasma processes for treatment of foods. *Molecular Nutrition & Food Research, 57*(5), 920–927.
55. Scholtz, V., Pazlarova, J., Souskova, H., Khun, J., & Julak, J., (2015). Non-thermal plasma: A tool for decontamination and disinfection. *Biotechnology Advances, 33*(6), 1108–1119.
56. Shen, M., & Bell, A. T., (1979). Review of recent advances in plasma polymerization. In: *Encyclopedia of Plasma Technology* (pp. 1–36). Washington – DC: American Chemical Society.
57. Šimončicová, J. L., Kryštofová, S., Medvecká, V., Ďurišová, K., & Kaliňáková, B., (2019). Technical applications of plasma treatments: Current state and perspectives. *Applied Microbiology and Biotechnology, 103*(13), 5117–5129.
58. Smeu, I., & Nicolau, A. I., (2014). Enhancement of food safety–antimicrobial effectiveness of cold plasma treatments. *The Annals of the University Dunarea de Jos of Galati, Fascicle VI – Food Technology, 38*(1), 9–20.
59. Song, A. Y., Oh, Y. J., Kim, J. E., Song, K. B., Oh, D. H., & Min, S. C., (2015). Cold plasma treatment for microbial safety and preservation of fresh lettuce. *Food Science and Biotechnology, 24*(5), 1717–1724.
60. Sookwong, P., Yodpitak, S., Doungkaew, J., Jurithayo, J., Boonyawan, D., & Mahatheeranont, S., (2014). Application of oxygen-argon plasma as a potential approach of improving the nutrition value of pre-germinated brown rice. *Journal of Food and Nutrition Research, 2*(12), 946–951.
61. Stoica, M., Alexe, P., & Mihalcea, L., (2014). Atmospheric cold plasma as new strategy for foods processing: Overview. *Innovative Romanian Food Biotechnology, 15*, 1–8.

62. Thirumdas, R., Kadam, D., & Annapure, U. S., (2017). Cold plasma: An alternative technology for the starch modification. *Food Biophysics, 12*(1), 129–139.
63. Thring, M. W., (1966). Plasma engineering. In: *Les Proprietes et les Applications des Plasmas (Properties and Applications of Low-Temperature Plasma)* (pp. 329–343). Boston – MA: Springer.
64. Vukušić, T., Vesel, A., Holc, M., Ščetar, M., Jambrak, A. R., & Mozetič, M., (2018). Modification of physicochemical properties of acryl-coated polypropylene foils for food packaging by reactive particles from oxygen plasma. *Materials, 11*(3), 372–378.
65. Wiesemann, K., (2014). *Short Introduction to Plasma Physics* (p. 38). Senec, Slovakia: The CAS-CERN Accelerator School: Ion Sources.
66. Xu, L., Garner, A. L., Tao, B., & Keener, K. M., (2017). Microbial inactivation and quality changes in orange juice treated by high voltage atmospheric cold plasma. *Food and Bioprocess Technology, 10*(10), 1778–1791.

CHAPTER 5

IMPACT OF COLD PLASMA ON PHYSICAL SURFACE MODIFICATIONS OF FOODS AND FOOD PRODUCTS

M. DHARINI, RATISH K. RAMANAN, E. J. RIFNA, C. K. SUNIL, N. VENKATACHALAPATHY, and R. MAHENDRAN

ABSTRACT

The fourth state of matter [also known as non-thermal plasma (NTP)] is produced by supplying energy to gas molecules, and it consists of highly reactive species and ions at ambient temperature. The cold plasma (CP) has been applied to the food products for decontamination of foods and surface modification. This chapter discusses the impact of CP on the beneficial effects of physical modification on foods. The physical modification is one of the productive approaches as a pretreatment, which alters the biological interactions and composition to enhance the properties of food material for further processing. Further research on CP-assisted physical modification in food processing will undoubtedly augment its applications in the food industry.

5.1 INTRODUCTION

The surge in demand for food with reference to safety, novelty, and nutrition motivates the food industry to adopt newer technologies, which are efficient, effective, and eco-friendly. The physical modification alters the biological interactions and food composition to improve the performance of food material for further processing, enhance the mechanical strength, increase bulk

Advanced Research Methods in Food Processing Technologies. Junaid Ahmad Malik, Megh R. Goyal, Preeti Birwal, & Ritesh B. Watharkar (Eds.)
© 2024 Apple Academic Press, Inc. Co-published with CRC Press (Taylor & Francis)

properties, and elevate the bio-integrity of certain preferential biocompounds [13]. Thermal pretreatment methods have caused some undesirable effects on foods [6]. Therefore, the focus has completely turned to non-thermal methods of pretreatment and food preservation.

The effect of numerous non-thermal methods on surface modification has been studied around the globe. In recent days, the interest in the application of cold plasma (CP) technology on physical modification has significantly increased. CP is a promising novel non-thermal and non-residual technology that is being adopted for effective decontamination of foods from foodborne pathogens, pest management in stored products and surface modification. Further, plasma processing can effectively modify the food materials by alleviating certain physical attributes while maintaining the nutritional components of food and reducing the microbial load [34].

This chapter presents the physical changes that plasma imparts on food materials and their potential applications in food processing. The chapter also includes a discussion on utilization of CP in cooking quality, modification, drying, elevation of powder quality, seed germination and enhancement of extraction yield.

5.2 DEFINITION OF COLD PLASMA (CP)

Plasma is the ionized gas (4th state of matter) that is either wholly or partially ionized gas (Figure 5.1). It comprises of numerous reactive species, such as: positive and negative ions, electrons, atoms, and molecules in the excited state, free radicals, and radiation (such as: UV photons and visible light) [32]. In the beginning, the name plasma was used only for the liquid obtained when corpuscles were removed from the blood. Later Irving Langmuir termed plasma for ionized gases in the year 1928 [29], and he explained that the plasma carrying positive and negative ions evoked him of the plasma in blood that carries white and red blood corpuscles.

Plasma, in general, is classified into thermal and non-thermal plasma (NTP). In thermal plasma (TP), the heavy particles and electrons are in the thermodynamic equilibrium having higher temperature; whereas there exists a non-thermodynamic equilibrium in CP [52]. As there is no thermal equilibrium between electrons and species, it leads to the chilling of ions. Further, more efficient energy transfer is obtained from electrons than the reactive species, thus the gas can be maintained at an atmospheric temperature [35]. The major studies on the impact of plasma are related to the changes in the microbial inactivation, physical property modification and pest management of the food products [12]. Table 5.1 indicates several studies on the impact of CP on applications in food products.

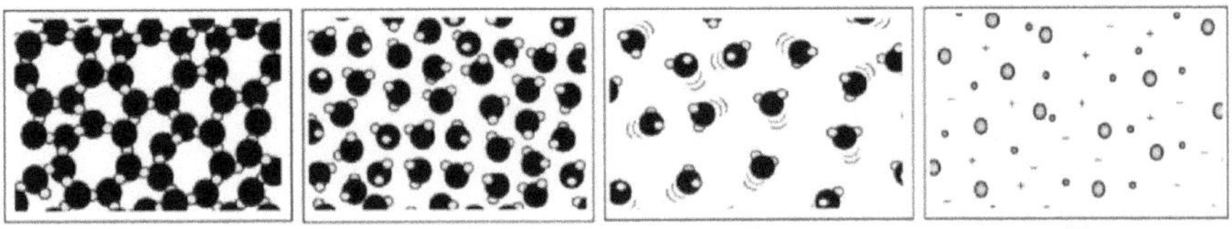

FIGURE 5.1 The schematic representation of four states of matter.

Apart from studies in Table 5.1, more studies have been documented to affirm the favorable effects of CP on food products. As plasma is composed of several reactive species, therefore the mechanism of microbial reduction or property modification is different. Physical modification is one of the mechanisms applicable for plasma processing: it is the only type of modification, which can be visible to the naked eye in some cases. This chapter discusses various applications of the physical surface modification that plasma imparts on the food material.

5.3 IMPACT OF COLD PLASMA (CP) ON MODIFICATION OF PHYSICAL PROPERTY

When food is subjected to NTP, numerous modifications (such as: physical, chemical, and nutritional characteristics) occur [2]. CP is effective in retaining the texture, sensory, and functional properties of the food system [17]. Plasma is known for its application for increasing microbial and enzymatic safety in foods. Apart from which, it has also been widely researched for its application on surface modification of foods. The major physical changes observed during CP processing are etching and erosion of the surface layer of the target material. These surface modifications of plasma are highly effective techniques for elevating the functional and textural properties of a food. The application of physical modifications of foods includes improvement in cooking time of rice, increase in functional properties, seed germination, increase in extraction, and increase in the texture characteristics of powdered foods (Figure 5.2).

TABLE 5.1 Selected Research Studies on Cold Plasma for Different Food Commodities

Commodity	Key Findings	References
	I. Effect of Cold Plasma on the Surface Characteristics	
Basmati rice	The plasma treatment reduced the cooking time and increased the water uptake with an increase in power.	[53]
Kiwi fruit	Atmospheric plasma treatment increased the quality parameters of kiwi; it improved the fruit color and reduced the browning that occurs during storage.	[44]
Leafy vegetables	Atmospheric plasma is effective in retaining the quality of the food material in terms of antioxidants, key components and minerals.	[41]
Pea flour	Plasma treatment increased the solubility of pea protein by 191% thereby increasing the water holding and binding capacity of pea flour.	[9]
Pomegranate juice	The ellagic acid in pomegranate juice is elevated after plasma treatment, due to hydrolysis and depolymerization of ellagitannin to ellagic acid in plasma treatment.	[24]
Rice flour	Plasma treatment on rice increased the flour and gel-hydration properties in flour. Depolymerization of starch during plasma treatment resulted in a reduction of amylose content.	[48]
	II. Effect of Cold Plasma on Microbial Inactivation	
Black pepper	RF-assisted jet plasma and microwave generated plasma were examined for reducing the microbial load in whole black pepper. Microwave remote plasma achieved a maximum reduction in microbes compared to direct jet plasma.	[25]
Corn salad	The Shiga toxin producing *E. coli* bacterial strain was alleviated by 3.3 log after 2 min exposure of argon based atmospheric plasma jet treatment.	[3]
Food containers	Atmospheric plasma jet reduced the biofilm on packaging materials. 3 to 4 log reduction of microorganisms was achieved at 10 min plasma exposure.	[27]
Lettuce, carrot, tomato	The highest power level and the increase in exposure time helped in reducing the pathogenic strain of *E. coli* (1.6 log reduction).	[7]
Queso fresco cheese	5 min plasma exposure (dry air) reduced the microbial load of *Listeria innocua* by 1.4 log CFU (colony forming unit) and *E. coli* by 3.5 log CFU.	[54]

TABLE 5.1 *(Continued)*

Commodity	**Key Findings**	**References**
Raw chicken	The combination of oxygen and nitrogen atmospheric pressure plasma with 20 mm of distance and 10 min exposure time resulted in highest deactivation of *E. coli* in raw chicken breast.	[57]
Red chicory radicchio vegetable leaves	Radicchio leaves treated with atmospheric pressure DBD plasma achieved a reduction of 2.2 log *E. coli* after 30 min of exposure time, without causing any changes in other quality parameters.	[41]
Sour cherry juice	6 log reduction of *E. coli* was observed (DBD, atmospheric pressure plasma, 20 kV).	[26]
	III. Effect of Cold Plasma on Pest Management	
Red flour beetle	Optimum parameters of 2,500 V at an exposure time of 15 min with 3.5 cm electrode distance resulted in 100% of all stages of *Tribolium castaneum*	[46]
Vinegar fruit fly	The first sign was the effect of atmospheric plasma on the larval stage. The ROS produced irreversible damages in the cells.	[22]
Yellow mealworms (edible insect flour)	Cold atmospheric treatment of insect flour aided in decreasing the microbial load and led to the future of protein-rich insect product development.	[8]
	IV. Effect of Cold Plasma on Modification of Biomolecules	
Apple juice and cubes	The enzyme activity was decreased due to the amino acid side-chain modification by reactive oxygen species in the plasma.	[21]
King prawn	The plasma species acts on the proteins to form fragmentation and crosslinking. The Vander Waal bond disruption occurred with α-helix structure reduction and an increase in β-sheets.	[20]
Vegetable oil	The reactive species act on lipids and oxidize them readily, which is followed by the series of lipid oxidation reactions.	[38]
Xanthan gum	An increase in crosslinking of protein networks is the cause for the rise in viscosity and emulsifying properties.	[36]

FIGURE 5.2 Applications of physical modification imparted by cold plasma on food products.

5.3.1 APPLICATIONS OF PLASMA-ASSISTED PHYSICAL MODIFICATION OF FOODS

5.3.1.1 IMPACT OF PLASMA ON COOKING OF RICE

Rice is one of the most important cereal crops, and it is composed of protein and starch. There are numerous nutritious traditional rice varieties. Although these traditional varieties provide more nutrition compared to other rice varieties, yet the prevalence of using traditional varieties is low due to the increased cooking time required for the traditional varieties. The etching property of CP was used to improve the rice cooking process [11]. The investigation of surface topography showed that the starch structure of brown rice was influenced by CP treatment and the enthalpy and crystallinity was found to decrease.

A similar study was done to analyze the cooking quality of basmati rice on treatment with low-temperature plasma [53]. Plasma treatment reduced the cooking time to 20s with a substantial rise in water uptake. The decrease in contact angle, hardness, and stickiness was notable; and the bombardment of species makes the surface more hydrophilic thus rice absorbs more water and ultimately it decreases the cooking time. In Figure 5.3, the SEM image depicts the morphology of basmati rice before and after plasma application for various treatments.

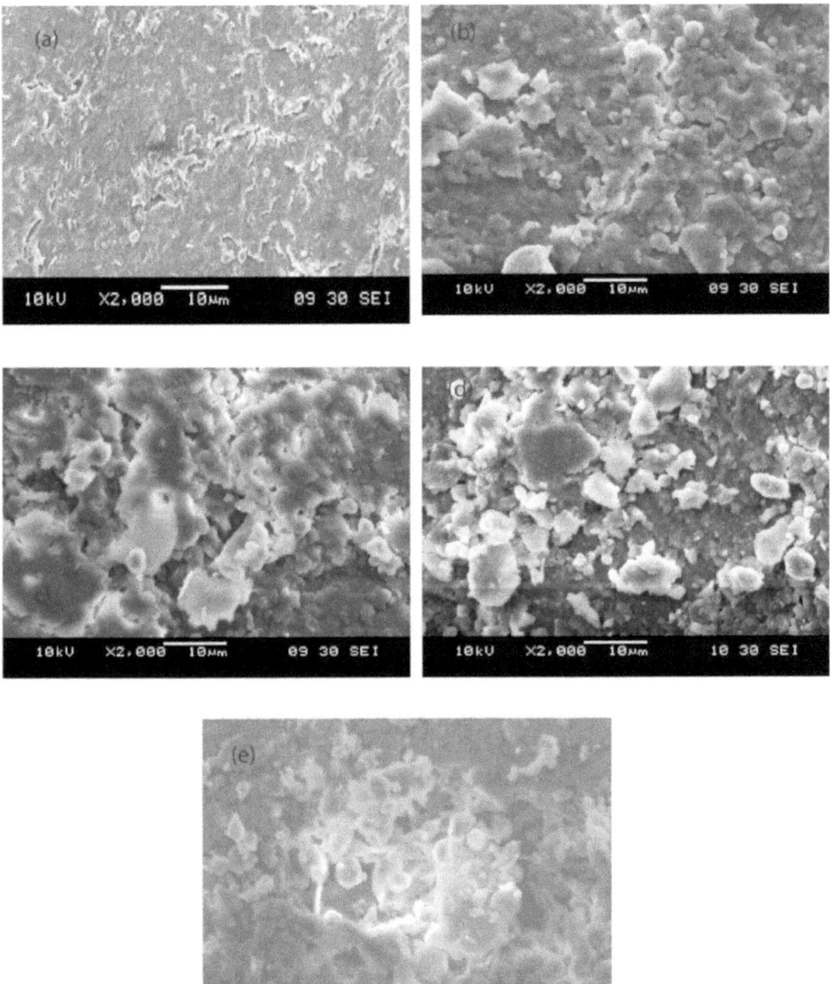

FIGURE 5.3 Surface topography image of basmati rice before and after the plasma treatment: (a) control; (b) plasma 30 W and 5 min; (c) plasma 30 W and 10 min; (d) plasma 40 W and 5 min; and (e) plasma 40 W and 10 min.

Source: Reprinted with permission from Ref. [53]; @ Copyright 2020. Elsevier.

Bamboo rice has higher cooking time and it was subjected to plasma. The study indicated the reduction in the cooking time from 50 min to 12 min [42]. The SEM analysis of the rice surface revealed the presence of fissures, crakes, and erosion on the rice bran, which was the reason of increasing the water absorption.

Similarly, Liu et al. [31] used CP to enhance the properties of Chinese milled rice. They succeeded in decreasing the cooking time and hardness of rice; also, they observed the increase in surface rupture with the increase in the applied plasma power. They reported that the plasma application gave a chalky appearance on the surface, which in turn increased the water absorption rate and gelatinization with reduced cooking time. Apart from decreasing the cooking time of rice, CP was also found effective in stabilizing the color of brown rice during storage. Additionally, it was also observed to maintain the volatile biocompounds of stored rice [31]. Thus, plasma can be effective for modifying the texture of rice to aid in easy absorption of water and reduces cooking time.

5.3.1.2 PLASMA-ASSISTED DRYING OF FOODS

Drying is the process of removal of moisture from food surface to enhance drying and the quality of food, and certain pretreatments may be necessary before subjecting the food to drying. Additionally, the presence of waxy outer epidermal layers of certain food materials makes it hard to dry [55]. In general, chemical treatments are done to increase the drying rate; however, the presence of chemical residues on the dried food is a threat to health.

CP can be used as a pretreatment before drying due to the physical modification on the food surface. CP pretreatment could modify the epicuticular waxy layer of chili pepper through the surface etching property of reactive species. It was reported that the 30s (atmospheric plasma 750 W) of pretreatment was enough to increase the hot air assisted drying rate of chili peppers [59].

When shiitake mushroom was subjected to plasma treatment, the surface of the mushroom was highly altered with the presence of visible cavities, pores, intracellular spaces on the surface [51]. These alterations on mushroom surface were reported to be the reason for the rise in drying rate and increase in the mass transfer of moisture from the surface. Similarly, jujube slices subjected to plasma increases the process of drying due to the surface etching property of plasma [4]. Similar to the cooking process, the etching on the surface of the food material makes it easy for evaporation of moisture from the inside of the food products.

5.3.1.3 PLASMA-ASSISTED ENHANCEMENT OF FLOUR PROPERTIES

The properties of the powdered food materials can be altered by subjecting these to plasma treatment. The study on the impact of low-pressure plasma on the properties of parboiled rice flour had been executed [48] and the results depicted that plasma treatment improves the flour and gel hydration property of flour due to surface etching action. The plasma treatment also reduced the amylose to amylopectin ratio, due to fragmentation of starch that occurs at the time of plasma exposure. The result also explains that the antioxidant and total phenolic content were increased for plasma exposure time of <15 min. The reaction of total antioxidants with free radicals above 15 minutes forms an intermediate compound, which degrades the phenolic compounds and decreases the antioxidant activity. Similarly, the research on wheat flour using CP also revealed that the surface area of flour was increased after plasma treatment, which in turn raised the hydration properties of wheat flour [10].

The effect of plasma on zein powder and its surface modification during film formation has been studied [18]. Zein is a by-product of maize used as a biomaterial in packaging, pharmaceutical, and also as a food ingredient. The results showed that an increase of voltage fortified the surface hydrophilicity and tensile strength of Zein films; and explained that the plasma had great control on physicochemical properties of Zein powders with better quality of film formation. In Figure 5.4, the SEM analysis revealed that the topography of the surface roughness was increased with the increase in plasma power level.

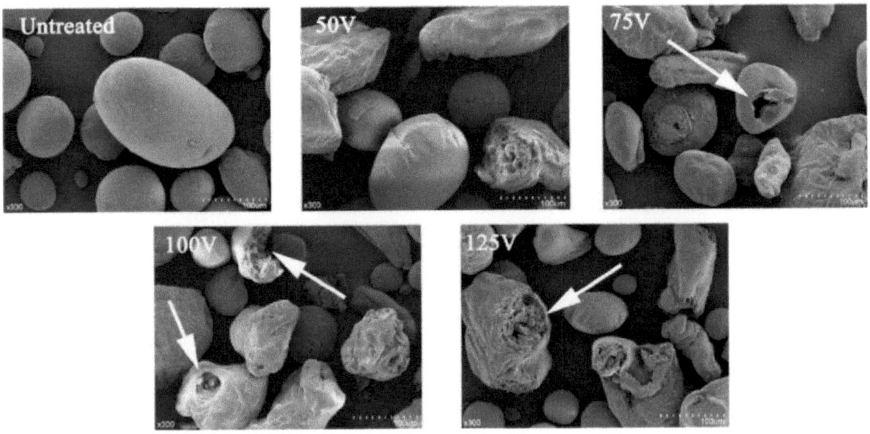

FIGURE 5.4 SEM image of plasma-treated zein powder for various power levels.

Source: Reprinted with permission from: Ref. [18]; Copyright 2020. Elsevier.

Eggshell, which is a biowaste with less solubility, was treated in the form of powder with CP. The results revealed that the plasma treatment reduced the bulk density of the powder due to the surface etching physical modification by plasma [1]. The treated eggshell powder was then added to coffee decoction to increase the calcium content; further, the study reported that the solubility of eggshell powder was increased after plasma application. Therefore, plasma can be used for increasing the properties of the powdered foods without affecting the quality of the powders.

5.3.1.4 IMPACT OF COLD PLASMA (CP) ON SEED GERMINATION

CP has high potential to improve the seed germination by modifying the seed surface. Seed germination, in general, is a process in which the embryo breaks the dormant stage, which involves the break of the seed coat. Application of plasma or plasma-activated water can enhance this process and will improve the growth rate, total germination, root, and stem elongation by increasing soaking and absorption properties of nutrients [23].

The study on SEM image of seeds of mung bean treated with plasma activated water (PAW) depicted that there were eroded surfaces on the seeds without ridges [60] due to the reaction of reactive oxygen species (ROS) induced by oxygen on the seed surface. Further, the study on seed growth showed improved seed germination in the seeds treated in plasma-activated water proving the surface etching property of plasma on seeds. CP treatment of cumin seeds revealed that the low-pressure plasma (118 W) application increased the germination rate by 43.24%; the contact analysis unraveled that the plasma treatment increased the wettability, water absorption of seeds after treatment [39]. Figure 5.5 represents the contact angle of plasma-treated cumin seeds for exposure time of 1, 2, 3, and 4 min.

CP was used on radish seeds to measure the ability of plasma on seed germination enhancement in terms of root and stem length and the weight dried plant [33]. The length of root and stem were measured after seven days; the seeds treated with 140 W of plasma showed the difference in growth rate for both stem and root. The length of roots was 4.5 and 7.5 cm for control and plasma-treated seeds, respectively. Further, the study on dried weight measurement gave a view on the power and exposure time of plasma on seed germination. The dry weight was higher for seeds treated at 140 W compared to the one treated at 90 W. Additionally, the increase in exposure time at 90 W power level increased the dry weight. Whereas the exposure time at 140

W power level reduced the growth due to damage of seed material. Therefore, a proper study on power level and exposure time can enhance the seed germination process with better results.

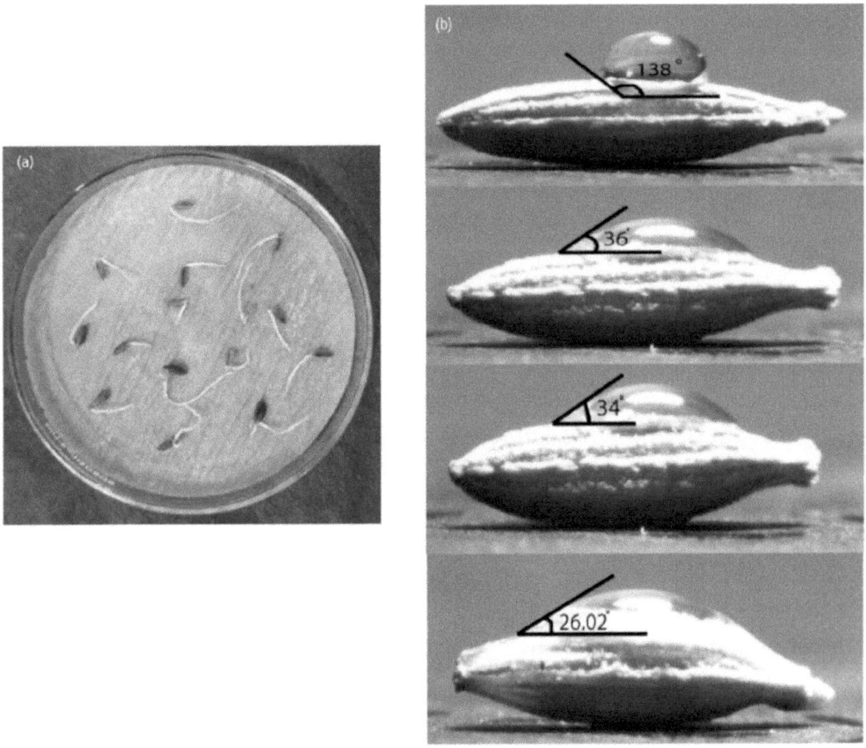

FIGURE 5.5 Contact angle analysis of cumin seeds treated with plasma.

Source: Reprinted with permission from: Ref. [50]; Copyright 2017. Elsevier

Apart from increasing the germination rate, plasma application was also observed to increase the availability of bioactive compounds in rice [56]. Germinated *indica* paddy rice was subjected to plasma treatment and it was found that the plasma increased the water uptake, germination, and gamma aminobutyric acid (GABA: a free amino acid, which aids in preventing non-communicable diseases in human) content in the paddy rice [58]. Therefore, apart from increasing the germination rate of seeds, CP was found efficient in increasing the key components in seeds during germination.

5.3.1.5 ROLE OF COLD PLASMA (CP) ON ENHANCEMENT OF EXTRACTION

CP is known for its ability to modify the surface properties of polymers. Surface morphology of biological materials can also be modified by CP to enhance the extraction of a desirable component from the subsequent layers. This assists the extraction yield of juice or other essential components from different parts of plants. The possible reason for this property of CP is due to the etching of the surface layer by CP [15]. In a study, CP was utilized as a pretreatment to extract phenolic compounds in the grape pomace [5]. Plasma was reported to disrupt the cell structure of pomace and thus aided in easy drying, and easy extraction of phenolic extraction; the extraction yield of the treated sample was increased from 10.9 to 22.8%.

Similarly, fenugreek seeds treated with high voltage plasma increased the extraction yield of galactomannan. Surface structural disruptions on the seed surface was observed after treatment; additionally, the increase in the acidity of the surrounding saline solution also proved the increase in reactive species after plasma application [45]. Apart from the extraction of principal components, CP can also be used for increasing the extraction of essential oils from biological materials.

The lemon peels were subjected to O_2, N_2, Ar, and air plasma treatments followed by measurement of surface properties [28]. It was observed that the plasma treatment for 1 min was efficient in increasing the oil extraction yield. While studying the microscopic structure of lemon peels, surface modifications were visible in plasma-treated peel.

The oil extraction process was enhanced due to the diffusion of essential oil from oil glands after surface physical modifications by CP. When cumin seeds were subjected to plasma in the form of bubbling of inside water, the extraction yield of cumin oil by steam distillation method yielded higher quantities of essential oil in comparison with the sample without bubbling. In this study, bubbling of plasma was reported to increase the oil yield due to the hydrodynamic cavitation of bubbles, which in turn increased the reaction rate and distribution of reactive species [49].

A study on plasma application of plasma on fenugreek increased the extraction of key components. The scanning electron microscope (SEM) analysis of the surface of the seed indicated the formation of cracks and fissures on the surface of fenugreek (Figure 5.6), which in turn increases the extraction yield. Similarly, a study on oil extraction of lemon peel was carried out using a combination of MW and CP [43]. Apart from the physical interaction of plasma with food materials, edible packaging materials can also be treated to increase the property of the packaging films.

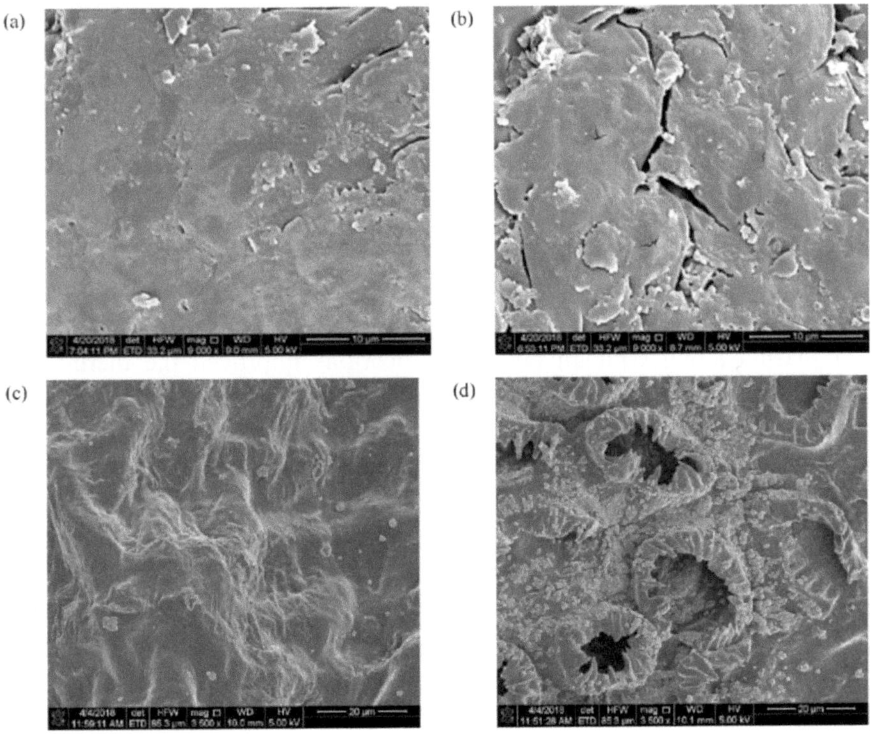

FIGURE 5.6 SEM image comparison of dry and moist fenugreek seeds: (a) dry fenugreek seed; (b) dry fenugreek after plasma treatment; (c) moist fenugreek seed; and (d) moist fenugreek after plasma exposure.

Source: Reprinted from: Ref. [45]; Copyright 2015. Elsevier

5.3.2 IMPACT OF PLASMA-ASSISTED PHYSICAL MODIFICATION ON FOOD PACKAGING

Biodegradable food packaging materials form the major trend due to the global concerns on excess use of plastics and pollution caused by the use of plastic-packaging materials [37]. Plant, crustacean, and microbial based biodegradable packaging materials have come into existence; however, the functional and barrier properties of these biodegradable materials must be improved to imitate the plastic packaging materials.

Plant-based packaging materials are mainly made of proteins, which are renewable, low-cost packaging material. To enhance the properties of these films, many thermal, chemical, and enzymatic treatments are being carried out. CP is one of these processing methods that can efficiently alter the properties of the packaging material; there are two or more mechanisms involved in the film modifications, such as functional group formation, physical modification and protein structure modification through reactive species chemical interactions [40].

The impact of the torpedo of reactive species on the film surface aids in the formation of fissures, capillary pores and cracks, which in turn increase the water absorption capacity of the film. When zein film was exposed to atmospheric pressure plasma unit at 65 V, it was able to increase the surface roughness and functional groups on the film surface. Surface etching property of plasma along with all other modifications improved the thermal and mechanical properties, hydrophilic nature, contact angle, and adhesiveness; it also reduced the water vapor transmission rate of the film [19].

Another interesting physical phenomenon that has been observed other than etching is the cleaning process, where the reactive species remove the surface low molecular weight compounds that are weakly bonded to the surface [16]. This physical modification such as cleaning and etching along with other changes were reported to decrease the water vapor transition and solubility of the fish protein film [47]. CP was also used for modifying the surface of encapsulated active packaging material for sustained release of active ingredient for a prolonged duration, and surface roughening effect of plasma plays a significant part in the release of active ingredients [14].

The major physical modification of plasma, the surface etching property was found efficient in numerous applications. The fissures, cracks, and erosion that occur on the surface helps in easy mass transfer into and out of the food, which aids in cooking, drying, extraction, and germination. The plasma-treated edible packaging material also forms an important application for improving the flexibility and transmissibility of the film, which in turn maintains the quality of the food.

5.4 SUMMARY

Thermal processing of foods for surface modification has been carried out for years, but the application of severe heat has resulted in detrimental effects on physical properties of a food product, such as: negative modification in color, texture, structure, and degradation of nutrients. From this scenario, the productive approach of CP technology is an alternative tool for food physical property modification, due to its versatile, non-thermal, and eco-friendly nature. CP treatments have revealed more positive impacts on the physical structure, chemical properties and nutritional characteristics of various food products. The polymer-processing, electronics, and textile sectors are so far enjoying the fruitfulness of plasma applications for structural modification. The initiative of CP-assisted change in food structure and thereby its effect on functional, structural, and behavioral characteristics are very recent to the field of food processing. This chapter on physical modifications assisted by CP provided the understanding of the applications of the physical changes that plasma imparts to food surfaces. The surface etching property of plasma forms the major reason for different applications in food processing. Improvement in cooking, functional properties of food powders, extraction, seed germination and drying are major applications of plasma on food. This exhibits the importance of realizing the effect of CP on physical property modification when applied to food products and potential industrial applications.

KEYWORDS

- **cold plasma**
- **etching**
- **extraction**
- **food packaging**
- **gamma-aminobutyric acid**
- **germination**
- **scanning electron microscopy**
- **surface modification**
- **thermal plasma**

REFERENCES

1. Aditya, S., Gnanasekaran, S., Stephen, J., & Radhakrishnan, M., (2020). Enhancing the properties of eggshell powder by cold plasma for improved calcium fortification in black coffee. *Journal of Food Process Engineering, 43*(8), E-article 13450; [Online]. doi: 10.1111/jfpe.13450.

2. Afshari, R., & Hosseini, H., (2014). Non-thermal plasma as a new food preservation method, its present and future prospect. *Journal of Paramedical Sciences, 5*(1), 116–120.
3. Baier, M., Janßen, T., Wieler, L. H., Ehlbeck, J., Knorr, D., & Schlüter, O., (2015). Inactivation of Shiga toxin-producing *Escherichia Coli* O104: H4 using cold atmospheric pressure plasma. *Journal of Bioscience and Bioengineering, 120*(3), 275–279.
4. Bao, T., Hao, X., Shishir, M. R. I., Karim, N., & Chen, W., (2021). Cold plasma: Emerging pretreatment technology for the drying of jujube slices. *Food Chemistry, 337*(866), E-article 127783. doi: 10.1016/j.foodchem.2020.127783.
5. Bao, Y., Reddivari, L., & Huang, J. Y., (2020). Enhancement of phenolic compounds extraction from grape pomace by high voltage atmospheric cold plasma. *LWT, 133*, E-article 109970. doi: 10.1016/j.lwt.2020.109970.
6. BeMiller, J. N., (2018). *Physical Modification of Starch.* New York: Elsevier Ltd. E-book 100868. doi: 10.1016/B978-0-08-100868-3.00005-6.
7. Bermúdez-Aguirre, D., Wemlinger, E., Pedrow, P., Barbosa-Cánovas, G., & Garcia-Perez, M., (2013). Effect of atmospheric pressure cold plasma (APCP) on the inactivation of *Escherichia Coli* in fresh produce. *Food Control, 34*(1), 149–157.
8. Bußler, S., Rumpold, B. A., Fröhling, A., Jander, E., Rawel, H. M., & Schlüter, O. K., (2016). Cold atmospheric pressure plasma processing of insect flour from Tenebrio Molitor: Impact on microbial load and quality attributes in comparison to dry heat treatment. *Innovative Food Science and Emerging Technologies, 36*, 277–286.
9. Bußler, S., Steins, V., Ehlbeck, J., & Schlüter, O., (2015). Impact of thermal treatment versus cold atmospheric plasma processing on the techno-functional protein properties from *Pisum sativum* "Salamanca." *Journal of Food Engineering, 167*, 166–174.
10. Chaple, S., Sarangapani, C., Jones, J., Carey, E., Causeret, L., Genson, A., Duffy, B., & Bourke, P., (2020). Effect of atmospheric cold plasma on the functional properties of whole wheat (*Triticum aestivum*) grain and wheat flour. *Innovative Food Science and Emerging Technologies, 66*, E-article 102529. doi: 10.1016/j.ifset.2020.102529.
11. Chen, H. H., (2014). Investigation of properties of long-grain brown rice treated by low-pressure plasma. *Food and Bioprocess Technology, 7*(9), 2484–2491.
12. Chizoba, E. F. G., Sun, D. W., & Cheng, J. H., (2017). Review on recent advances in cold plasma technology for the food industry: Current applications and future trends. *Trends in Food Science and Technology, 69*, 46–58.
13. Chu, P. K., Chen, J. Y., Wang, L. P., & Huang, N., (2002). Plasma-surface modification of biomaterials. *Materials Science and Engineering: R: Reports, 36*, 143–206.
14. Cui, H., Surendhiran, D., Li, C., & Lin, L., (2020). Biodegradable zein active film containing chitosan nanoparticle encapsulated with pomegranate peel extract for food packaging. *Food Packaging and Shelf Life, 24*, E-article 100511. doi: 10.1016/j.fpsl.2020.100511.
15. Dakshayani, R., Paul, A., Mahendran, R., & Preparation, A. S., (2020). Cold plasma-induced effects on bioactive constituents and antioxidant potential of lotus petal powder. *International Journal of Food Engineering, 2020*, 1–6.
16. Dawood, N., (2021). Surface modification of date palm leaves by cold plasma treatment. *Journal of King Saud University–Science, 33*(5), E-article 101465. doi: 10.1016/j.jksus.2021.101465.
17. Dey, A., Rasane, P., Choudhury, A., Singh, J., Maisnam, D., & Rasane, P., (2016). Cold plasma processing: A review. *Journal of Chemical and Pharmaceutical Sciences, 9*(4), 2980–2984.

18. Dong, S., Gao, A., Zhao, Y., Li, Y. T., & Chen, Y., (2017). Characterization of physicochemical and structural properties of atmospheric cold plasma (ACP) modified zein. *Food and Bioproducts Processing, 106*, 65–74.
19. Dong, S., Guo, P., Chen, Y., Ji, H., Ran, Y. E., Li, S. H., & Chen, Y. E., (2018). Surface modification via atmospheric cold plasma (ACP): Improved functional properties and characterization of zein film. *Industrial Crops and Products, 115*, 124–133.
20. Ekezie, F. G. C., Sun, D. W., & Cheng, J. H., (2019). Altering the IgE binding capacity of king prawn (*Litopenaeus Vannamei*) Tropomyosin through conformational changes induced by cold argon-plasma jet. *Food Chemistry, 300*, E-article 125143. doi: 10.1016/j.foodchem.2019.125143.
21. Farias, T. R. B., Rodrigues, S., & Fernandes, F. A. N., (2020). Effect of dielectric barrier discharge plasma excitation frequency on the enzymatic activity, antioxidant capacity and phenolic content of apple cubes and apple juice. *Food Research International, 136*, E-article 109617. doi: 10.1016/j.foodres.2020.109617.
22. Ferreira, M. I., Gomes, J. G. L., Benilov, M. S., & Khadem, M., (2016). Effects of non-thermal atmospheric-pressure plasma on drosophila development. *Plasma Medicine, 6*(2), 115–124.
23. Groot, G. J. J. B., Hundt, A., Murphy, A. B., Bange, M. P., & Mai-Prochnow, A., (2018). Cold plasma treatment for cotton seed germination improvement. *Scientific Reports, 8*(1), 1–10.
24. Herceg, Z., Kovačević, D. B., Kljusurić, J. G., Jambrak, A. R., Zorić, Z., & Dragović-Uzelac, V., (2016). Gas phase plasma impact on phenolic compounds in pomegranate juice. *Food Chemistry, 190*, 665–672.
25. Hertwig, C., Reineke, K., Ehlbeck, J., Knorr, D., & Schlüter, O., (2015). Decontamination of whole black pepper using different cold atmospheric pressure plasma applications. *Food Control, 55*, 221–229.
26. Hosseini, S. M., Rostami, S., Hosseinzadeh, S. B., & Lorigooini, Z., (2020). The effect of atmospheric pressure cold plasma on the inactivation of *Escherichia coli* in sour cherry juice and its qualitative properties. *Food Science and Nutrition, 8*(2), 870–883.
27. Kim, H. J., Jayasena, D. D., Yong, H. I., Alahakoon, A. U., Park, S., Park, J., Choe, W., & Jo, C., (2015). Effect of atmospheric pressure plasma jet on the foodborne pathogens attached to commercial food containers. *Journal of Food Science and Technology, 52*(12), 8410–8415.
28. Kodama, S., Thawatchaipracha, B., & Sekiguchi, H., (2014). Enhancement of essential oil extraction for steam distillation by DBD surface treatment. *Plasma Processes and Polymers, 11*(2), 126–132.
29. Laroussi, M., (2008). The biomedical applications of plasma: Brief history of the development of a new field of research. *IEEE Transactions on Plasma Science, 36*(4, part-3), 1612–1614.
30. Liu, J., Wang, R., Chen, Z., & Li, X., (2021). Effect of cold plasma treatment on cooking, thermomechanical and surface structural properties of Chinese milled rice. *Food and Bioprocess Technology, 14*(5), 866–886.
31. Liu, Q., Wu, H., Luo, J., Liu, J., Zhao, S., Hu, Q., & Ding, C., (2021). Effect of dielectric barrier discharge cold plasma treatments on flavor fingerprints of brown rice. *Food Chemistry, 352*(3), E-article 129402; [online]. doi: 10.1016/j.foodchem.2021.129402.
32. Mahendran, R., Abirami, C. V. K., & Alagusundaram, K., (2017). Cold plasma technology: An emerging non-thermal processing of foods – A review. In: *Engineering Interventions in Agricultural Processing* (pp. 33–55). Boca Raton – FL: CRC Press.

33. Matra, K., (2016). Non-thermal plasma for germination enhancement of radish seeds. In: *Procedia Computer Science, 86*, 132–135.
34. Mir, S. A., Shah, M. A., & Mir, M. M., (2016). Understanding the role of plasma technology in the food industry. *Food and Bioprocess Technology, 9*(5), 734–750.
35. Misra, N. N., Schlüter, O., & Cullen, P. J., (2016). Plasma in food and agriculture. In: *Cold Plasma in Food and Agriculture: Fundamentals and Applications* (pp. 1–13). Boston – MA: Academic Press.
36. Misra, N. N., Yong, H. I., Phalak, R., & Jo, C., (2018). Atmospheric pressure cold plasma improves viscosifying and emulsion stabilizing properties of xanthan gum. *Food Hydrocolloids, 82*, 29–33.
37. Moosavi, M. H., Khani, M. R., Shokri, B., Hosseini, S. M., Shojaee-Aliabadi, S., & Mirmoghtadaie, L., (2020). Modifications of protein-based films using cold plasma. *International Journal of Biological Macromolecules, 142*, 769–777.
38. Na, H. S., Mok, C. K., & Lee, J. H., (2020). Effects of plasma treatment on the oxidative stability of vegetable oil containing antioxidants. *Food Chemistry, 302*, E-article 125306;
39. Paatre, S. S., Ramireddy, L., & Radhakrishnan, M., (2020). Stimulation of the germination and seedling growth of *Cuminum cyminum* seeds by cold plasma. *Journal of Applied Research on Medicinal and Aromatic Plants, 18*, E-article 100259. doi: 10.1016/j.jarmap.2020.100259.
40. Pankaj, S. K., Bueno-Ferrer, C., & Misra, N. N., (2014). Applications of cold plasma technology in food packaging. *Trends in Food Science and Technology, 35*(1), 5–17.
41. Pasquali, F., Stratakos, A. C., Koidis, A., et al., (2016). Atmospheric cold plasma process for vegetable leaf decontamination: A feasibility study on radicchio (red chicory, *Cichorium Intybus*). *Food Control, 60*, 552–559.
42. Potluri, S., Sangeetha, K., Santhosh, R., Nivas, G., & Mahendran, R., (2018). Effect of low-pressure plasma on bamboo rice and its flour. *Journal of Food Processing and Preservation, 42*(12), 1–7.
43. Pragna, C. H., Ranjitha, G. T. K., Mahendran, R., & Anandharamakrishnan, C., (2019). Effects of microwave and cold plasma assisted hydrodistillation on lemon peel oil extraction. *International Journal of Food Engineering, 2019*, 1–10.
44. Ramazzina, I., Berardinelli, A., Rizzi, F., et al., (2015). Postharvest biology and technology effect of cold plasma treatment on physico-chemical parameters and antioxidant activity of minimally processed kiwifruit. *Postharvest Biology and Technology, 107*, 55–65.
45. Rashid, F., Bao, Y., Ahmed, Z., & Huang, J. Y., (2020). Effect of high voltage atmospheric cold plasma on extraction of fenugreek galactomannan and its physicochemical properties. *Food Research International, 138*(PA), E-article 109776. doi: 10.1016/j.foodres.2020.109776.
46. Ratish, R. K., Sarumathi, R., & Mahendran, R., (2018). Influence of cold plasma on mortality rate of different life stages of *Tribolium castaneum* on refined wheat flour. *Journal of Stored Products Research, 77*, 126–134.
47. Romani, V. P., Olsen, B., Pinto, C. M., Meireles, O. J. R., Prentice, C., & Guimarães, M. V., (2019). Plasma technology as a tool to decrease the sensitivity to water of fish protein films for food packaging. *Food Hydrocolloids, 94*, 210–216.
48. Sarangapani, C., Thirumdas, R., Devi, Y., Trimukhe, A., Deshmukh, R. R., & Annapure, U. S., (2016). Effect of low-pressure plasma on physicochemical and functional properties of parboiled rice flour. *LWT – Food Science and Technology, 69*, 482–489.

49. Sharanyakanth, P. S., Lokeswari, R., & Mahendran, R., (2021). Plasma bubbling effect on essential oil yield, extraction efficiency, and flavor compound of *Cuminum cyminum* seeds. *Journal of Food Process Engineering, 2021*, E-article 13730.
50. Shashikanthalu, S. P., Ramireddy, L., & Radhakrishnan, M., (2020). Stimulation of the germination and seedling growth of *Cuminum cyminum* L. seeds by cold plasma. *Journal of Applied Research on Medicinal and Aromatic Plants*: *Journal of Food Process Engineering, 2020*, E-article 100259. doi: 10.1016/j.jarmap.2020.100259.
51. Shishir, M. R. I., Karim, N., Bao, T., et al., (2020). Cold plasma pretreatment: Novel approach to improve the hot air-drying characteristics, kinetic parameters, and nutritional attributes of shiitake mushroom. *Drying Technology, 38*(16), 2134–2150.
52. Tendero, C., Tixier, C., Tristant, P., Desmaison, J., & Leprince, P., (2006). Atmospheric pressure plasmas: A review. *Spectrochimica Acta – Part B: Atomic Spectroscopy, 61*(1), 2–30.
53. Thirumdas, R., Deshmukh, R. R., & Annapure, U. S., (2015). Effect of low temperature plasma processing on physicochemical properties and cooking quality of basmati rice. *Innovative Food Science and Emerging Technologies, 31*, 83–90.
54. Wan, Z., Misra, N. N., Li, G., & Keener, K. M., (2021). High voltage atmospheric cold plasma treatment of *Listeria innocua* and *Escherichia coli* K-12 on queso fresco (fresh cheese). *LWT – Food Science and Technology, 146*, E-article 111406. doi: 10.1016/j.lwt.2021.111406.
55. Wu, W., Jin, X., Liu, Z., Xu, Y., Zhu, F., & Zhang, J., (2021). Effect of low temperature plasma pretreatment on drying process of vegetables with waxy layer. *Journal of Food Process Engineering, 44*(12), E-article 13911.
56. Yodpitak, S., Mahatheeranont, S., Boonyawan, D., Sookwong, P., Roytrakul, S., & Norkaew, O., (2019). Cold plasma treatment to improve germination and enhance the bioactive phytochemical content of germinated brown rice. *Food Chemistry, 289*, 328–339.
57. Yong, H. I., Kim, H. J., Park, S., Choe, W., Oh, M. W., & Jo, C., (2014). Evaluation of the treatment of both sides of raw chicken breasts with an atmospheric pressure plasma jet for the inactivation of *Escherichia coli*. *Foodborne Pathogens and Disease, 11*(8), 652–657.
58. Zargarchi, S., & Saremnezhad, S., (2019). Gamma-aminobutyric acid, phenolics and antioxidant capacity of germinated indica paddy rice as affected by low-pressure plasma treatment. *LWT – Food Science and Technology, 2019*, [online]. doi: 10.1016/j.lwt.2018.12.014.
59. Zhang, X. L., Zhong, C. S., & Mujumdar, A. S., (2019). Cold plasma pretreatment enhances drying kinetics and quality attributes of chili pepper (*Capsicum annuum*). *Journal of Food Engineering, 241*, 51–57.
60. Zhou, R., Li, J., Zhou, R., Zhang, X., & Yang, S., (2019). Atmospheric pressure plasma treated water for seed germination, seedling growth of mung bean and its sterilization effect on mung bean sprouts. *Innovative Food Science and Emerging Technologies, 53*, 36–44.

CHAPTER 6

ULTRASOUND PROCESSING OF FOODS AND FOOD PRODUCTS: AN OVERVIEW

GITANJALI GAUTAM, MAINAO ALINA GAYARY,
NIKHIL KUMAR MAHNOT, and CHARU LATA MAHANTA

ABSTRACT

Enhancement of food quality and shelf-life of foods has been a major concern for processed food manufacturers. Ultrasonication is considered to be an effective eco-friendly technology of high quality and productivity with minimum physical and chemical hazards. Various studies for maximizing cavitation activity in sample matrix with different experimental parameters include duration, temperature, frequency, power, etc. This chapter provides an insight into the mechanism of operation and effect of ultrasonication on food systems and highlights its various applications in the food systems, such as food preservation, filtration, drying, cooking, processing of fruits, vegetables, and meat, and enzyme technology. The use of ultrasonication in biomolecule extraction, food emulsion stability, and microbial inactivation in conjunction with other unit operations is also explained.

6.1 INTRODUCTION

Food is a complex mixture of carbohydrates, proteins, fats, vitamins, minerals, and other non-nutritive components (such as water, pigments, and bioactive compounds). Conventionally, food processing and preservation methods involve thermal treatments (such as pasteurization, sterilization,

Advanced Research Methods in Food Processing Technologies. Junaid Ahmad Malik, Megh R. Goyal, Preeti Birwal, & Ritesh B. Watharkar (Eds.)
© 2024 Apple Academic Press, Inc. Co-published with CRC Press (Taylor & Francis)

evaporation, blanching, etc.) and treatments with various natural and chemical preservatives. Such methods ensure food safety and confer longer shelf-life but can negatively alter the nutritional and organoleptic characteristics due to physical and chemical changes. Hence, there is a constant tussle to balance the nutritive and sensory characteristics along with food safety during processing. On the other hand, with increasing consumer awareness towards a healthy diet and lifestyle, the importance of food products retaining maximum nutritive value or natural-like characteristics is on the rise.

With the rising demand and strict monitoring of compliance with food and environmental regulations, traditional food processing techniques are gradually being replaced by non-thermal technologies [61]. Ultrasonication is one such emerging green technology and has received great acceptance due to strategic advantages and multifaceted benefits in food processing and preservation [8, 18]. High energy ultrasonic waves are now-a-days used in operations like cleaning, microbial, and enzyme inactivation, bio-component separation, emulsification, heat, and mass transfer enhancement, cutting, crystallization enhancement and extraction of bioactive components from foods and plants [35].

The chapter covers mechanisms of ultrasound technology along with types of sonication and ultrasonic systems. Possible applications of ultrasonic energy in the food technology sector are described. The chapter also introduces the novel application of ultrasonication as a hurdle technology in combination with other technologies.

6.2 OVERVIEW OF ULTRASOUND METHOD

6.2.1 *BASICS OF ULTRASOUND ENERGY*

Ultrasound energy refers to a vibration frequency (>16 kHz) higher than the audible limit of humans. It is transmitted through various substances (such as solid, liquid, or gas) that have elastic properties. The upper limit of ultrasonic frequency is not specifically defined but is usually 5 MHz in gas systems and 500 MHz for solid and liquid systems [59].

During the ultrasonication process, a sound source is transferred to the molecules in a sample medium, which transmits their motion to the conjoined particles and returns back to the original position. This continuous movement of molecules produces two types of actions, such as: (i) the molecules are compressed together; and (ii) there is a deficiency of molecules that creates a void space. This gives rise to the alternate compression and rarefaction regions of sound wave. The particle oscillation occurs in the direction of the propagation and thereby produces longitudinal waves in liquids and gases. Since the solids possess shear elasticity, transverse waves are produced, where the particle movement takes place perpendicular to the direction of propagation of wave [59].

The ultrasound waves can be categorized into two intensity ranges: (i) low amplitude and high frequency (0.1–1 W/cm^2, 100 kHz–10 MHz) propagation (signal ultrasound); and (ii) high energy and low frequency (10–1,000 W/cm^2, 20–100 kHz) propagation (power ultrasound). Signal ultrasound is basically applied in the food quality analysis, medical imaging, and non-destructive inspection, whereas power ultrasound is considered for cleaning, inactivation of microbes, denaturation of enzymes, etc. [35] and has importance in the food processing sector. Therefore, power ultrasound applications will only be elaborated in this chapter.

6.2.2 BASIC PRINCIPLE OF OPERATION OF ULTRASOUND METHOD

The basic working principle of ultrasound is cavitation, which refers to the formation of bubbles, their growth and subsequent collapse [25] as presented in Figure 6.1 [78, 85]. The cavitation process takes place under the influence of sound waves in a liquid medium and is regarded as acoustic cavitation. When ultrasonic waves pass through a liquid medium, all molecules in the test sample experience variations in pressure. If the temporary reduction of pressure falls below the lower limit of tensile strength of the liquid, a rupture in the liquid continuum occurs, i.e., there is formation of a number of small cavitation bubbles.

During liquid compression and rarefaction, the cavitation bubbles behave in two ways, i.e., either stable or transient [59, 67]:

1. **Stable Bubbles or Cavities:** These containing gas and some vapors are formed at fairly low ultrasonic intensities (1–3 W/cm^2) that oscillate non-linearly around some equilibrium size and are relatively long-lived and continue to oscillate for many acoustic cycles. It is important to note that stable bubbles are capable of transformation into transient bubbles and undergo collapse.
2. **Transient Bubbles or Cavities:** These are voids or vapor-filled bubbles produced at higher ultrasonic intensities (in excess of 10 W/cm^2) and exist for few acoustic cycles during which they expand to

at least double or multiple times their original size. The transient bubbles subsequently collapse violently and can disintegrate into a mass of smaller bubbles, which may act as nuclei for more bubbles or can simply dissolve into the solution due to surface tension if the radius is sufficiently small.

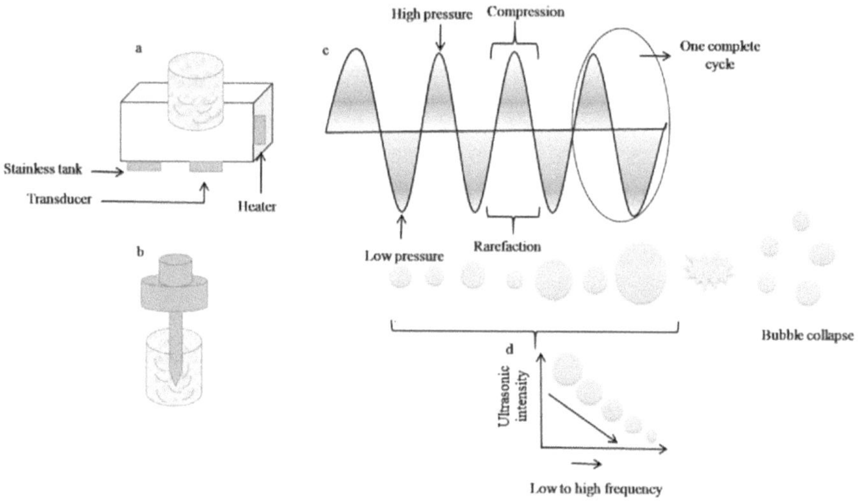

FIGURE 6.1 Schematic presentation of the action of ultrasound: (a) bath ultrasound; (b) probe ultrasound; (c) mechanism of ultrasonic cavitation; and (d) bubble collapse with increase in frequency and intensity.

Source: (b) Adapted from Ref. [19]; (c) & (d) Adapted from Refs. [85] and [49], respectively.

Cavitation during ultrasound results in mechanical, thermal, and chemical effects [4]. For example, an aqueous system is subjected to 20 kHz sound frequency, each cavitation bubble after collapsing acts as a localized hot spot generating approximately 5,000 K temperature and pressures of around 1,000 atm in an adiabatic process. Such localized hot spots lead to thermal effects, such as microbial inactivation. The thermal effects provide energy for chemical effects like bond fission of the vapors present in the bubbles. Under such drastic conditions, water molecules break into free radicals and also generate active molecules, which have antimicrobial effects and can cause chemical changes as they are highly reactive in nature. Also, such conditions assist in extraction of volatile components into the bubbles, which can be effectively utilized. In a solid-liquid system, a rapid collapse of bubbles causes a sudden influx of surrounding liquid medium at high velocities and induces huge, localized pressure drops.

This phenomenon is referred to as micro-jetting. Micro-jetting leads to mechanical effects, such as surface erosion (cleaning), wetting of particles, and particle size reduction owing to the generation of high shear force [49]. This broad mechanism involved in ultrasound treatment generates physical and chemical effects that provide scope for several potential applications in the food industry [85].

6.2.3 ULTRASOUND SYSTEMS

Ultrasound is applied in two ways: (i) by directly applying to the sample; and (ii) by indirectly applying through the walls of the container, where the sample is placed. The principal equipment used for direct and indirect application of ultrasound is of probe-type and bath-type, respectively.

Ultrasonic probe device consists of a generator, a transducer, fixed horns and detachable horns (probes) [74, 77]. The generator converts electrical energy into required high frequency alternating current (AC) to drive the transducer assembly. Transducer transforms the AC into mechanical vibrations of fixed or variable frequencies. Piezoelectric-type transducers are commonly used due to their scalability. The fixed horns are used to increase the amplitude of vibrations while detachable horns/probes resonate at the same frequency as that of the fixed horns and transmit ultrasonic energy directly into the sample. The probes are variously designed and their selection is based on the amount of power to be applied to the sample, sample type, sample size or volume.

A bath-type sonicator is a simple setup and consists of a stainless-steel tank with thin walls fitted with a transducer underside of the tank. Several factors for bath-type application include frequency of the ultrasound wave, sonication time, particle size of the sample, sample position in the bath, ultrasound transmission property of the medium (basically a liquid) in which the sample is placed, and temperature of the medium [77]. Ultrasonic bath device differs from probe sonicator in two ways: (a) probes remain in direct contact with the sample inside the container; and (b) ultrasound intensity in probe type configuration is much higher (up to 100 times more) than that of a bath sonicator [18].

In ultrasound treatment, an important consideration is the generation of heat, which might limit its applicability to heat sensitive materials. Increase of temperature also leads to low effectiveness and reproducibility of sonication treatment. However, this can be overcome by using temperature control systems or use of pulse modes to avoid temperature build-up.

Ultrasound also causes erosion of the probes or bath housing; therefore, cautious use is required. Modern-day ultrasonic systems are designed accordingly to generate uniform cavitation and have variable power and frequency settings, electronic timers, and thermostatically controlled heating [49].

6.2.4 TYPES OF SONICATION

1. **Ultrasonication (US):** It involves the application of power at low temperature. Widespread application of ultrasonication is targeted as a non-thermal technique for heat sensitive food products as it retains sensory, nutritional, and functional characteristics, and provides microbial safety and enhanced shelf-life [61]. However, this process requires longer exposure time to achieve significant microbial inactivation that increases energy consumption. Moreover, the generation of heat during cavitation needs to be controlled for better process output.
2. **Manosonication (MS):** It involves the combined application of ultrasound waves and moderate pressure (100–300 kPa) at low temperature. The mechanism of generation of cavitation in MS depends on raising the hydrostatic pressure above the saturated vapor pressure. The increased hydrostatic pressure causes a decrease in the vapor pressure inside the bubbles leading to more violent collapse, which disintegrates microbial cells compared to US [4].
3. **Thermosonication (TS):** It uses ultrasound and mild heat simultaneously. The better efficacy of TS against microbes and enzymes is due to the additive effect created by the combination of heat and cavitation. However, at times the occurrence of additive effect is hindered when the temperature of the liquid rises above 55°C. Depending on the temperature range, TS may be divided into sub-lethal (<45°C) and lethal (>45°C) applications. TS has shown rapid and reliable results to improve the nutritional and sensory quality and shelf-life of juices [6]. This method has the same potential of inactivation as that of conventional heat treatments.
4. **Manothermosonication (MTS):** It combines the simultaneous benefits of ultrasound, mild temperature, and moderate and static pressure. This increases the lethality of heat treatment (approx. 10–30 folds) resulting in a drastic reduction in thermal processing

time. During MTS, cell structures are exposed to intense cavitation stress causing strong shaking up of molecules or breaking of bonds. Vegetative cells lose the integrity of cell wall while cortex degradation of bacterial spores occur with the release of dipicolinic acid and some low molecular weight polypeptides, which finally lead to rehydration of protoplast resulting in loss of heat resistance [75].

6.3 APPLICATIONS OF POWER ULTRASOUND IN FOOD SYSTEMS

6.3.1 APPLICATIONS IN FOOD PRESERVATION

Ultrasound processing has emerged as an innovative method for preservation of foods, which refers to the extension of its shelf-life owing to microbial and enzyme inactivation for delaying of food decay. Present-day consumers demand safe foods having natural-like quality. Ultrasound has proved to be an alternative technology to inactivate microbes at room temperature, unlike the conventional thermal method, thus minimal quality changes occur. The antimicrobial action of ultrasonic treatment is attributed to the cavitation phenomena leading to physical, thermal, and chemical effects [30]. Microbial spores and enzymes at times being somewhat resistant to ultrasound.

The ultrasonication has also been used in combination with other strategies, such as temperature (thermosonication), pressure (manosonication) or both (manothermosonication). The efficacy of microbial inactivation of ultrasound is dependent on factors, such as: intensity, temperature, time, volume, food composition, type of microbes and initial microbial load [82]. Ultrasound has been effective to cause reduction of *Salmonella* sp., *Listeria* spp., *Escherichia coli*, *Clostridium* spp., *Campylobacter jejuni* and *Saccharomyces cerevisiae*, and other yeasts and molds found in fruits, vegetables, milk, and poultry [2, 33, 50, 72]. Ultrasonic cavitation initiates hot spot generation and free radical induced oxidation causing thermal denaturation of enzyme and enzymatic depolymerization, respectively to result in enzyme inactivation [30, 37]. Thermosonication (TS) methods have shown better results than the ultrasound or temperature alone in inactivating polyphenol oxidase (PPO), peroxidase (PO) and pectin methylesterase (PME); thus reducing enzymatic browning of food products [72], and increase cloud stability in various juices [30]. Thus, the ultrasonic effect on overall inactivation of microbes and enzymes helps in maintaining product quality for a longer period of time.

6.3.2 APPLICATIONS IN MIXING AND HOMOGENIZATION AND FILTRATION

Ultrasound has found applications in mixing of liquids, homogenization, and filtration. Ultrasound can reduce membrane fouling caused by clogging of the pores, which is a common problem associated with filtration systems resulting in reduced filtration rate, and regular cleaning is required to maintain the same flux. Ultrasound enhances the filtration flux as the vibrational forces break the concentration polarization and keep the particulate matter in suspension, thus preventing clogging of pores or cake formation [44]. These forces cause the filter or screen to vibrate creating a frictionless surface and allow smaller particles of interest to passthrough easily [72]. Moreover, ultrasound does not affect the intrinsic permeability of the filter membrane or screen. Although one must be careful about membrane erosion while applying ultrasound to filter membranes, yet membrane durability needs to be accounted [55]. Increased filtration flux by ultrasound application near the filtration membrane is commonly observed. Such ultrasound-assisted filtration is found effective to increase the permeation flux in the filtration of cow's milk, grape pomace extracts, and cherry concentrate [24].

Power ultrasound application has been studied for mixing of foods with the aerated texture of cakes. Tan et al. [87] observed that ultrasound-assisted mixing of sponge cake batter resulted in higher air content in the baked bread compared with conventional mixing. Such batters had lower batter density, high over-run and viscosity compared with normal mixing. Similar results were obtained on mixing bread dough in case of low densities and high loaf volume [69]. Ultrasound-assisted physical mixing of napin with lactalbumin protein was able to increase *in vitro* protein digestibility of napin due to changes in the disulfide bond and protein structure [71]. Thus, this opens up further possibility of ultrasound application in the field of protein digestibility.

Homogenization unit operation is carried out to achieve particle size reduction (solid-liquid interaction) or to change the degree of dispersion (liquid-liquid interaction). For example, fat globule size reduction in milk is a solid-liquid interaction, whereas emulsification of two immiscible liquid is a liquid-liquid interaction. Ultrasound application in tomato juice processing has shown to improve the consistency in comparison with untreated juice owing to higher breakage of cell-structure and increase in suspended material [12]. Reduction in fat globule size has also been reported for yogurt and ice-cream processing [1] due to the effect of high shear rates on bubble collapse and cavitation caused by ultrasonication. Ultrasound has proved to be effective in increasing emulsion stability in low-fat emulsions, such as salad dressings, mayonnaise, and in reducing the dependency on emulsifiers [14, 43].

6.3.3 APPLICATION IN DRYING/DEHYDRATION

Drying/dehydration involves removal of water to aid in processing along with shelf-life enhancement of products. Conventional drying techniques have certain limitations, such as longer process time, energy intensive and overall quality loss of products due to extended heating time due to slow rate of mass transfer. Ultrasonic waves have proved to be beneficial in increasing the mass transfer rates and decreasing drying time resulting in better product quality. Ultrasound-assisted convective drying increases the drying rate in fruits and vegetables (FAV), meat, fish, etc. [88].

Ultrasound-assisted operation has been widely studied to enhance hot air drying, fluidized bed drying (FBD), osmotic dehydration (OD), vacuum drying or freeze-drying [10, 13, 79].

In FBD drying, use of ultrasound reduces the required minimum fluidization velocity and particle agglomeration to enhance the heat transfer coefficient. In OD application, ultrasound helps in easier water removal from the food tissues due to cavitation followed by a series of rapid contraction and expansion that expels the water. It is also postulated that cavitation in ultrasound treatment creates microscopic capillaries or channels in food tissues, thus reducing subsequent drying time for ultrasound pre-treated or assisted OD products by increasing osmotic solute influx into the food material [13]. The same micro-channels also improve diffusion and convective mass transfer rates and have shown to improve vacuum drying on ultrasound pre-treatment. In ultrasound-assisted freeze drying, the ice crystal formation was better controlled and decrease in sublimation drying time was observed. Ultrasonication improves freeze drying efficiency and lowers energy consumption [63, 92].

6.3.4 APPLICATION IN PROCESSING OF FRUITS AND VEGETABLES (FAV)

In processing of FAV, ultrasound technologies alone or as a hurdle have been effective in the disinfection, sterilization, enzyme inactivation, and dehydration. Ultrasound-assisted chemical disinfection methods have shown to reduce the dependence on harsh chemicals and achieve sufficient microbial reduction. Ultrasound-assisted commercial disinfectants have shown to reduce *S. enterica, L. monocytogenes, E. coli* and natural microbiota in cherry tomatoes, bell peppers, plum, strawberries, lettuce, and spinach [20]. Ultrasound treatment has been successful to inactivate both PPO and PO enzymes in various FAV (50–90% inactivation in fresh cut pineapple, guava, apples, and tomatoes) to delay the browning on storage [20]. PO is quite heat-resistant; therefore, a harsh heat treatment is required for inactivation, thus causing nutritive losses [20, 47].

Ultrasonic pre-treatment of FAV tend to reduce quality losses owing to reduced time required for subsequent drying methods with better rehydration properties of the final product [13, 26, 53, 92]. More recently, ultrasound treatment has also shown to be beneficial in fortification of iron in food matrices, such as, pumpkin, potatoes, and kidney beans [73]. Recent research studies have also indicated that ultrasound treatment alone or in combination with chemical and other strategies are effective in pesticide residue removal from FAV [9]. On the whole, the shelf-life of FAV is increased with ultrasound treatment without significantly affecting the overall sensory properties and quality characteristics.

6.3.5 APPLICATIONS IN COOKING

Cooking involves changes in texture, color, flavor, and aroma in foods making them more palatable and digestible. Application of ultrasound in the cooking of meat products has been strongly recommended. Generation of hydroxyl radicals and hydrogen peroxide (H_2O_2) on cavitation leads to oxidation of lipids and proteins. Oxidation products (such as: free fatty acids, aldehydes, ketones, and amino acids) are responsible for flavor and aroma in different types of meat. Spiced beef cooked using ultrasound (80–120 min, 20 kHz, 40–1,000 W) resulted in enhanced flavor, improved tenderness, and greater water holding capacity [93, 94].

For meat-emulsion product like mortadella, ultrasound treatment (25 kHz) caused a faster increase in the internal temperature and temperature homogeneity without affecting the flavor and improved gel-formation suggesting textural improvement [83]. In brown rice, ultrasound pre-treatment improved the absorption of water and greater expansion occurred on subsequent cooking with flavor in comparison with cooked white rice [91]. Ultrasound-assisted vacuum frying has shown to induce greater water loss and lower oil-uptake, to reduce frying time and to delay oil degradation [20]. In case of purple potato chips and mushroom, ultrasound-assisted vacuum microwave (MW) frying led the fried products to have a porous texture with 16–34% less oil retention [27, 86].

6.3.6 MODERN APPLICATIONS IN FOOD ENGINEERING

6.3.6.1 FOOD EMULSIONS

The stability and properties of food emulsions (protein-lipid or protein-protein or protein-water interactions) to a large extent depend on the intensity, time of exposure and the frequency of ultrasound treatment. The ultrasound treated (570 W) casein resulted in a stable food-grade Pickering high internal phase emulsion (HIPE), which will have probable application in 3D printing [11]. Application of dual frequency (20/28 kHz for 10 min) as compared to mono-frequency-based ultrasound treatment on whey emulsion gel has improved mechanical and gel properties [22]. The emulsion stability (O/W) of protein: polysaccharide (high methoxyl pectin and soy protein isolate in 1:1 ratio at pH of 3.5) complex with soybean oil is increased with ultrasonication [3]. Apart from protein, the micro-nano-cellulose crystals were also explored for Pickering-type emulsion formation (O/W) using flaxseed oil with ultrasound treatment (frequency of 20 kHz, power of 525 W and 4 min of time duration) [62]. Physical stability of O/W emulsion of lecithin was improved after sonication treatment [38].

Recently, Nishad et al. [68] developed grapefruit-peel-phenolic (GPP) nano-emulsion in mustard oil and they reported that 9.5 min treatment time using 30% amplitude (20 kHz) and Span-80 (sorbitan monooleate; 0.52%) resulted in 29.73±1.62 nm size stable GPP nano-emulsion. Combination of high-power ultrasound (HPU) with supercritical carbon dioxide can be used as an alternative to thermal-based pasteurization of lipid emulsions containing heat sensitive molecules [39]. The emulsifying functionality and stability of chicken myofibrillar protein (MP) was enhanced using intense ultrasound treatment (power of 450 W, frequency of 20 kHz and intensity of 30 W/cm^2) [58].

6.3.6.2 FOOD SAFETY WITH RESPECT TO MICROBIAL AND ENZYME INACTIVATION

The safety and shelf-life of a particular food depends on microbial, or enzyme related biochemical activities. The thermal treatments are generally employed to ascertain the food safety in terms of microbial load (pathogenic and spoilage micro-organisms) and inactivation of enzymes. The thermal treatments not only cause sensory changes but also degrade heat labile bio-molecules in the food matrices. Therefore, application of non-thermal technology like ultrasound in assuring food safety is gaining much interest for both industry and academia. Ultrasound at suitable intensity, temperature, and time destroys cell-wall and causes intracellular cavitation in micro-organism, which destroys intracellular components leading to lysis or death.

Ultrasound along with heat treatment (thermosonication) has been effective in the inactivation of heat-resistant *Saccharomyces cerevisiae* ascospores in beer. The researchers have suggested treatment time of 1.9 min and 26s for 50°C and 55°C, respectively to pasteurize beer in combination with ultrasound [64]. Also, combined effect of zinc oxide (ZnO, 40 mM) and low-frequency ultrasound (20 kHz, 120 μm amplitude, 43–45 W) effectively resulted in reducing >5 log CFU/mL of *Listeria innocua* population within 8 min [29]. Even, *Bacillus subtilis* spores subjected to combined thermal and ultrasound treatments of 80°C and 20 W/mL were reduced to 2.43 ± 0.08 log [34]. Recently, Li et al. [57] reported on the application of mild heat and ultrasound treatments (63°C, 400 W) to achieve 5-log population reduction of *B. subtilis* in Chinese bayberry juice in 9.59 min.

The cavitation effect of ultrasound generates shock waves that increase temperature; and formation of free radicals (H^+ and OH^-) from the sonolysis of water cause denaturation/inactivation of macromolecules, such as proteins and enzymes. The ultrasound application (20 kHz, 5/5s ON/OFF pulse, intensity of 452 W/cm^2, 12 min) was effective in the inactivation of polyphenol oxidase (PPO) and polyphenol peroxidase (POD) in bayberry juice [15]. Inactivation of enzymes phosphatase and lactoperoxidase (indicators of pasteurization efficiency) was also achieved in raw milk samples on application of high-intensity ultrasound (475 W, 19 kHz) using a 13-mm ultrasonic probe [81, 89]. Wang et al. [89] used combined treatment of ultrasound and ultraviolet (UV) treatment (600 W for 10 min) for inactivation of PME, polyphenol oxidation enzyme and peroxidase in mango juice.

6.3.6.3 EXTRACTION OF BIOMOLECULES

The application of ultrasound for extraction of various biomolecules (such as: phenolic compounds, polysaccharides, oil, proteins, etc.), can be done effectively maintaining the functionality and structure in the food matrix at industrial scale. UAE-based technique requires minimum consumption of extraction solvent and is very fast and efficient. The extraction solvent, frequency, temperature, power, and solvent to substrate ratio are various parameters for implementing an UAE technique. Table 6.1 enlists the applications of ultrasound in extraction/recovery of various biomolecules from different sources:

1. **Polyphenol Extraction:** Polyphenols are functional phytocomponents widely present in various plant sources. The ultrasound interventions are effective in increasing the yield and maintaining their stability. The effect of various ultrasonic parameters on phenolic extraction/recovery as reported by various research groups has been summarized in Table 6.1.
2. **Oil Extraction:** The ultrasonic treatment, because of the acoustic cavitation phenomenon, causes cell-wall disruption, which helps in effective diffusion of solvent and release of oil. Thus, application of this bio-refining technology can increase oil extraction efficiency, preserve sensory and nutritional properties and slow down the oil oxidation rate. The ultrasound-assisted pre-treatment requires less solvent, less energy consumption, less time and low temperature for extraction. The efficiency can be increased in combination with other advanced extraction techniques, such as enzyme assisted, MW, pulse-field, supercritical, etc. Table 6.1 summarizes the oil extraction from different sources with ultrasonic interventions.
3. **Protein Isolate Extraction:** Protein extracted from different sources is an essential ingredient in various food industries for improving viscosity, emulsion stability and gelation properties. The ultrasound treatment can affect to a large extent the functional properties of proteins (such as gelling ability, emulsifying ability, water-holding capacity, solubility, and foaming property). Thus, selection of proper parameters in ultrasound application (such as frequency, time, solvent to be used, pH of extraction solvent, ionic strength, solvent polarity, etc.) is critical for retaining the functionality of a particular protein without altering its structure. Table 6.1 summarizes the functionality and yield of the protein as reported by different research groups.

TABLE 6.1 Applications of Ultrasound-Assisted Extraction (UAE) of Biomolecules from Different Sources

Major Components	Source	Operating Conditions	Major Findings	References
Oil	Olive	Operating frequencies: 0.4 and 2 MHz; Power: 280 W at 100% output; Water: paste ratio – 1:3; Time: 5 min	Oil extractability: 91.4%; Malaxation duration: 10 min (reduced); 0.4 MHz megasonic treatment (better than 2 MHz); Specific energy: 252 kJ kg^{-1}	[5]
Oil	Olive pomace	Frequency: 60.0 kHz; Power: 280 W; Solvent: hexane	Solid/liquid ratio and temperature combination: • β-sitosterol: 1.399 mg g^{-1} at 1:12 g mL^{-1} and 60°C. • Campestral: 0.220 mg g^{-1} at 1:10 g mL^{-1} and 55°C. • Stigmasterol: 0.105 mg g^{-1} at 1:12 g mL^{-1} and 50°C. • Squalene: 4.575 mg g^{-1} at 1:12 g mL^{-1} and 60°C.	[16]
Oil	Umbu (*Spondias tuberosa*) seed	Frequency: 20 kHz; Power: 500 W; Solid/solvent ratio of 1:30 (w/v, g/mL); Time: 4 min; Solvent: ethanol/water	Palmitic acid (C16:0): 18.5%; Stearic acid (C18:0): 8.5±0.1%; Oleic acid (C18:1): 33.2±0.1%; Linoleic acid (C18:2): 37.0±0.1%; Linolenic acid (C18:3): 2.7±0.3%; MUFA: 33.3±0.1% FFA content; PUFA: 39.7±0.2% FFA content	[28]
Phenolic compounds	Baobab (*Adansonia digitata*) seeds	900 W power probe sonicator; methanol as solvent. RSM conditions for extraction: Time: 20 min; Temperature: 60°C; Amplitude: 30%; Solvent to solid ratio: 30 mL g^{-1}	TFC: 1633.84±10.75 mg; RE 100 g^{-1} d.w.; Catechin: 87.18±14.57 μg g^{-1}	[48]

TABLE 6.1 *(Continued)*

Major Components	Source	Operating Conditions	Major Findings	References
Phenolic compounds	Brazilian pomegranate (of the molar variety) peel	Power: 180 W; 70% ethanol (v/v); Frequency: 37 kHz (normal or pulse modes); Temperatures: 50–60°C for better TPC and yield. Frequency: 80 kHz (continuous mode); Temperature: 70°C for better antioxidant capacity.	TPC: 274.65 ± 3.17 mg GAE g^{-1} of dry sample; Total extract yield: 59.02±1.06%; Antioxidant capacity (AC): 2796.14±6.22 µmol Teq g^{-1} of dry sample	[60]
Polyphenolic compounds	*Ilex guayusa* Loes. leaves	Frequency: 40 kHz; Power: 110 W; RSM conditions: 60°C; 29.9 min; and 76.8/23.2 ratio of ethanol/water	TPC: 3.46 ±0.17 g GAE 100^{-1} g d.w.; FRAP: 0.080 mmol TROLOX equivalents 100 g^{-1} d.w.; ABTS: 40.71 µmol TROLOX equivalents 100 g^{-1} d.w.	[7]
Polyphenols	Wheat (*Triticum aestivum*) waste biomass (bran)	Combining acidic deep eutectic solvent (glycerol/citric acid/glycine); ultrasound-assisted pretreatment (Frequency: 100 Hz; Power: 120 W; and Acoustic energy density: 120 W L^{-1}; and thermal treatment	Total polyphenols: 94.62 mg FAE g^{-1} of d.w.	[23]
Polyphenols	Thyme (*Thymus vulgaris*) and rosemary (*Rosmarinus officinalis*) plants	Ethanol: water (50:50) and water; mass-to-solvent ratio of 1:50 (w: v); Temperature: 40°C; Power: 100% (400 W); Ultrasonic emitter: 2.2 cm of diameter	TFC: 16.29–24.99 mg CE g^{-1} d.w. for rosemary; 10.42–15.22 mg CE g^{-1} d.w. for thyme; Total carotenoid content: 0.123 mg carotenoid g^{-1} d.w. (for rosemary); 0.131 mg carotenoid g^{-1} d.w (for thyme). Trolox equivalent antioxidant capacity (TEAC): 5.15–8.35 mmol TE (extracted in water); 5.55–10.60 mmol TE (extracted in ethanol)	[66]
Protein	*Eurycoma longifolia* roots	Time: 5 min; Frequency: 20 kHz; Amplitude: 9 W; Duty cycle: 63%	Protein yield: 9.543±0.946%	[31]

TABLE 6.1 *(Continued)*

Major Components	Source	Operating Conditions	Major Findings	References
Protein isolate	Canola (*Brassica napus* L.)	Frequency: 40 kHz; Power: 130 W; Temperature: 25°C; Time: 30 min	Protein: 86.54±0.12 g 100 g^{-1}; *in vitro* protein digestibility: 4.22±1.37%; WAC: 2.52±0.02 g H_2O g^{-1} protein; OAC: 3.26±0.07 g oil g^{-1} protein	[36]
Protein isolate	Soybean	Ultrasound treatments with acid: 0.001 M HCl, pH 3.0); Power: 200 W; Frequency: 20 kHz; Temperature: 25±2°C; 2 s/2 s ON/OFF pulse	For 10 min ultrasound treatment protein solubility: 80.9%; Secondary protein structure: α-helix (8.9 ± 0.6%), β-sheet (29.9 ± 0.7%), β-turn (23.4 ± 0.1%), random coil (37.8 ± 1.3%).	[46]
Protein isolates	*Chenopodium album*	Frequency: 20 kHz; Power: 500 W; Amplitude: 25%; Temperature: 10°C	Increased solubility, surface hydrophobicity, foaming capacity and foam stability after 25 min treatment	[65]

Legend: CE: Catechin equivalent; d.w.: Dry weight; FAE: Ferulic acid equivalent; GAE: Gallic acid equivalent; MUFA: Monounsaturated fatty acids; OAC: Oil absorption capacity; PUFA: Polyunsaturated fatty acids; RE: Rutin equivalent; RSM: Response surface methodology; SFA: Saturated fatty acid; TFC: Total flavonoid content; TPC: Total polyphenol content; WAC: Water absorption capacity.

6.4 INTEGRATED NOVEL APPLICATIONS OF ULTRASOUND IN COMBINATION WITH OTHER TECHNOLOGIES

6.4.1 INTEGRATED WITH PULSED ELECTRIC FIELD (PEF)

Pulsed electric field (PEF) works on the principle of pulse-based electric field (electric voltage) using two electrodes, which result in electroporation and electro-permeability of the cell membrane due to an electromechanical force. It helps in destabilizing the membranes and inducing mass transfer out of the components from the food matrixes [84]. It is being used as a non-thermal process for increasing the shelf-life of food products.

Palgan et al. [70] used MTS followed by PEF (bipolar square-wave pulses through an electrode gap of 3 mm) for effective reduction of 5.6 log CFU mL^{-1} of *Listeria innocua* in milk-based smoothie, while 4.2 log CFU mL^{-1} was noted in reverse sequence of the treatment. TPC value of 2.6 g GAE 100 g^{-1} in fresh weight defatted canola seed cake was observed with the application of PFE (at fixed parameter: burst pulses of 900, pulse width of 20 μs, electrode gap of 80 mm and energy of 10 kJ) followed by MW and 200 W of ultrasound at 70°C for 20 min, in that order [88]. Recent study reported the inactivation of *Listeria monocytogenes* by pre-ultrasound treatment (500 W, 40 kHz and 20°C) followed by dielectric barrier discharge (DBD) cold plasma (50 V, 10 kHz, 1,000 W and 5 mm gap between sample surface and quartz plate) [71].

6.4.2 INTEGRATED WITH SUPERCRITICAL FLUID EXTRACTION (SFE)

The supercritical fluid extraction (SFE) process involves the use of temperature and pressure conditions above the critical point of the components (with different polarities) like supercritical CO_2 or water resulting in high mass transfer along with higher solubility [32]. The bioactive components (such as: tocotrienols, polyphenols, etc., present in very low amount and difficult to extract) can be effectively extracted with the integrated approaches of ultrasound interventions (at a specific frequency) followed by SFE from the solid phase to the solvent phase. The application of ultrasound creates cavitations that shear the tissues/cells and allow the solvent (above their critical point) to enter the cells efficiently, resulting in high mass transfer of the target compound.

Hu et al. [45] used adlay seeds (*Coix lachrymal-jobi* L.) for optimum extraction of oil and coixenolide using ultrasound-assisted supercritical fluid extraction (USFE) and reported an increase in the extraction yield by 14% using SFE conditions of 20 MPa pressure, 40°C temperature, 3.5 h extraction time and 3.0 L h^{-1} of CO_2 flow rate. Eight different isoflavones were extracted from sea algae (*Hypnea spinella*, *Sargassum muticum*, *Undaria pinnatifida*, *Sargassum vulgare*, *Porphyra* sp., *Halopytis incurvus* and *Chondrus crispus*) and cyanobacteria (*Scenedesmus*, *Spongiochloris spongiosa* and *Nostoc* 17), which were subjected to sonication followed by supercritical CO_2 at 40°C with 35 MPa pressure for 60 min using 3% (v/v) of MeOH/H_2O mixture (9:1, v/v) [51].

Reategui et al. [76] applied SFE-US (Supercritical CO_2 extraction assisted by ultrasound; 13 mm titanium end probe) technique on bagasse of blackberry (*Rubus* sp.) and the resulting extract was high in phenolic and anthocyanin content and exhibited better antioxidant properties. Wei et al. [90] explored ultrasound-assisted supercritical carbon dioxide (USC-CO_2) extraction on clove and recovered 23.31±0.77% (g/g clove, w/w%) of clove oil along with α-humulene 2.75±0.09 (g/g oil, w/w%), which was better than the conventional extraction techniques.

The microorganism inactivation can be effectively achieved with the combined effect of HPU and high-pressure carbon dioxide (supercritical fluid), as shown for the inactivation *of Listeria monocytogenes* from dry cured ham [52, 80]. The authors also recommended the use of 12 MPa pressure at 35°C with 10 W applied power for 5 min with minimal effect on total acidity, color, and texture profile of the dried cured ham.

6.4.3 INTEGRATED WITH ENZYME ASSISTED TECHNOLOGY

The specificity of a particular enzyme in respect to a particular substrate makes it ideal for the extraction of polyphenols, which are bound with hydrogen bonds or hydrophobic interactions with other compounds. The specificity, activities, and extraction efficiency of amylases, cellulase, pectinase, proteases, etc., depend largely on conditions (solid-liquid ratio, incubation time-temperature combinations and pH), solvent property and matrix of the substrate, respectively. The high cost of extraction using enzyme-based interventions has limited its application at industrial scale, which can be overcome by an integrated approach with ultrasound. Generation of cavitation due to ultrasound enhances the enzyme-substrate interaction, resulting in higher product formation and increased yield.

Ultrasound at 130 W, 20 kHz (Ti-Al-V probe of 13 mm) with Peclyve-V enzyme (concentration of 2% w/w; 55°C and 6:1 mL g^{-1} liquid-solid ratio) resulted in oil recovery of 95.8% in 10 min from pomegranate seeds [42]. Gorguç et al. [40] used ultrasound-assisted enzymatic system for recovering protein and total phenolic contents (TPC) from sesame bran. Alcalase enzyme with a concentration of 1.248 AU 100 g^{-1} at 9.8 pH at 43°C (for 98 min) and ultrasound power 836 W yielded 88.4% of protein and TPC of 6.03 mg GAE g^{-1}. Very recently, Gorguç et al. [41] used vacuum and ultrasound assisted enzymatic extraction simultaneously (optimum condition of 238 mm of Hg vacuum pressure for 8 min) with restoration time of 68 min along with alcalase at 1.82 AU 100 g^{-1} concentration and ultrasound power of 550 W (43°C); and they obtained protein yield and TPC of 65.9±3.04% and 5.31±0.0 mg GAE per g, respectively.

6.4.4 INTEGRATED WITH MICROWAVE (MW) TECHNOLOGY

In MW technology, the generated electromagnetic waves at certain frequencies cause ionic motions and dipole rotations resulting in the generation of high heat. Application of MW assisted extraction (MAE) along with ultrasound will reduce the amount of solvent required and the time of extraction/digestion and increase efficiency. The combined effects of ultrasound and MW result in cell-wall damage/rupture by ultrasound and penetration of solvent in the matrix by MW heating that results in higher recovery of target compounds in the solvent [19].

Chemat et al. [17] combined and illustrated a single mode MW system along with an ultrasound system of 2.45 GHz electromagnetic and 20 kHz mechanical for sample digestion/extraction purpose. Citric acid was extracted using sequential ultrasound-microwave assisted extraction (UMAE) from pomelo peel powder [56]. Authors found that the yield was increased to 38% of citric acid at optimum conditions of 27.5 min of sonication at pH 1.80 followed by 643 W of MW treatment for 6.4 min. The combined effect of ultrasound (at low power) along with MW resulted in increased drying rate of carrots [54].

Chen et al. [21] designed an integrated ultrasonic-microwave assisted extraction (UMAE) for extraction of green coffee oil (GCO) from coffee beans using ethanol as solvent. The researchers optimized the extraction parameters using response surface methodology (RSM) and they reported that the ultrasonic-microwave cooperative extractor (CW-2000 with 10 min of extraction time, 350 W MW power and 28:1 mLg^{-1} of liquid-solid ratio) was able to extract 10.58% of GCO.

6.5 SUMMARY

Ultrasound is considered as an emerging non-thermal technology being used in preservation, filtration, mixing, homogenization, and even in drying/dehydration of foods. Extraction of polyphenols, oil, protein, etc., from a complex matrix using ultrasound results in extracts in a concentrate form that are free from any residual solvents and contaminants in a shorter time as compared to the conventional extraction techniques. The power, frequency, temperature, duration, etc., have a great influence on FAV processing, formation of food emulsions, cooking, and inactivation of microbes and enzymes. The application of ultrasound ensures nutritionally superior food quality. Integration of other novel techniques (such as: PEF, MW technology, SFE and enzyme assisted technology) further enhances process efficiency and product quality. However, more future research on sustainability and validation of this technology is required for use on an industrial scale to produce high-quality and safe food products.

KEYWORDS

- **cavitation**
- **extraction**
- **shelf-life**
- **ultrasonic frequency**
- **ultrasonic intensity**
- **ultrasonication**

REFERENCES

1. Akdeniz, V., & Akalın, A. S., (2019). New approach for yoghurt and ice cream production: High-intensity ultrasound. *Trends in Food Science and Technology, 86*, 392–398.
2. Alenyoregea, E. A., Ma, H., Aheto, J. H., Chikaria, I. A. F., Osaea, R., & Zhou, C., (2020). Response surface methodology centered optimization of mono-frequency ultrasound

reduction of bacteria in fresh-cut Chinese cabbage and its effect on quality. *LWT – Food Science and Technology, 122*, E-article 108991.

3. Albano, K. M., Cavallieri, A. L. F., & Nicoletti, V. R., (2020). Electrostatic interaction between soy proteins and pectin in O/W emulsions stabilization by ultrasound application. *Food Biophysics, 15*, 297–312.
4. Al-Juboori, R., Aravinthan, V., & Yusaf, T., (2010). Review of common and alternative methods for disinfection of microorganisms in water. In: *2010 Southern Region Engineering Conference (SREC 2010)* (pp. 18–27). Toowoomba, Australia.
5. Amarillo, M., Perez, N., Blasina, F., Gambaro, A., Leone, A., Romaniello, R., Xu, X. Q., & Juliano, P., (2019). Impact of sound attenuation on ultrasound-driven yield improvements during olive oil extraction. *Ultrasonics Sonochemistry, 53*, 142–151.
6. Anaya-Esparza, L. M., Velazquez-Estrada, R. M., Roig, A. X., Garcia-Galindo, H. S., Sayago-Ayerdi, S. G., & Montalvo-Gonzalez, E., (2017). Thermosonication: An alternative processing for fruit and vegetable juices. *Trends in Food Science & Technology, 61*, 26–37.
7. Arteaga-Crespo, Y., Radice, M., Bravo-Sanchez, L. R., Garcia-Quintana, Y., & Scalvenzi, L., (2020). Optimization of ultrasound-assisted extraction of phenolic antioxidants from *Ilex guayusa* leaves using response surface methodology. *Heliyon, 6*, E-article 03043.
8. Arvanitoyannis, I. S., Kotsanopoulos, K. V., & Savva, A. G., (2017). Use of ultrasound in the food industry – methods and effects on quality, safety, and organoleptic characteristics of foods: A review. *Critical Reviews in Food Science and Nutrition, 57*, 109–128.
9. Azam, S. M. R., Ma, H., Xu, B., Devi, S., Siddique, M. A. B., Stanley, S. L., Bhandari, B., & Zhu, J., (2020). Efficacy of ultrasound treatment in the removal of pesticide residues from fresh vegetables: A review. *Trends in Food Science & Technology, 97*, 417–432.
10. Başlar, M., Toker, O. S., Karasu, S., Tekin, Z. H., & Yildirim, H. B., (2015). Ultrasonic applications for food dehydration. In: Ashokkumar, M., (ed.), *Handbook of Ultrasonics and Sonochemistry* (pp. 1247–1270). Singapore: Springer.
11. Bi, A. Q., Xu, X. B., Guo, Y., Du, M., Yu, C. P., & Wu, C., (2020). Ultrasound pre-fractured casein and in-situ formation of high internal phase emulsions. *Ultrasonics Sonochemistry, 64*, E-article 104916.
12. Bot, F., Calligaris, S., Cortella, G., Nocera, F., Peressini, D., & Anese, M., (2017). Effect of high-pressure homogenization and high-power ultrasound on some physical properties of tomato juices with different concentration levels. *Journal of Food Engineering, 213*, 10–17.
13. Bozkir, H., Ergün, A. R., Serdar, E., Metin, G., & Baysal, T., (2019). Influence of ultrasound and osmotic dehydration pretreatments on drying and quality properties of persimmon fruit. *Ultrasonics Sonochemistry, 54*, 135–141.
14. Cabrera-Trujillo, M. A., Filomena-Ambrosio, A., Quintanilla-Carvajal, M. X., & Sotelo-Diaz, L. I., (2018). Stability of low-fat O/W emulsions obtained by ultra turrax, rotor-stator and ultrasound homogenization methods. *International Journal of Gastronomy and Food Science, 13*, 68–64.
15. Cao, X., Cai, C., Wang, Y., & Zheng, X., (2018). Inactivation kinetics of polyphenol oxidase and peroxidase in bayberry juice during thermal and ultrasound treatments. *Innovative Food Science & Emerging Technologies, 45*, 169–178.
16. Chanioti, S., & Tzia, C., (2019). Evaluation of ultrasound assisted and conventional methods for production of olive pomace oil enriched in sterols and squalene. *LWT, 99*, 209–216.

17. Chemat, S., Lagha, A., & Amar, H. A., (2004). Ultrasound-assisted microwave digestion. *Ultrasonics Sonochemistry, 11*, 5–8.
18. Chemat, F., & Ashokkumar, M., (2017). Ultrasound in the processing of liquid foods, beverages and alcoholic drinks. *Ultrasonics Sonochemistry, 38*, 753–760.
19. Chemat, F., Rombaut, N., Sicaire, A. G., Meullemiestre, A., Fabiano-Tixier, A. S., & Abert-Vian, M., (2017). Ultrasound-assisted extraction of food and natural products. Mechanisms, techniques, combinations, protocols and applications: A review. *Ultrasonics Sonochemistry, 34*, 540–560.
20. Chen, F., Zhang, M., & Yang, C. H., (2020). Application of ultrasound technology in processing of ready-to-eat fresh food: A review. *Ultrasonics Sonochemistry, 63*, E-article 104953.
21. Chen, Q., Dong, W., Wei, C., Hu, R., & Long, Y., (2020). Combining integrated ultrasonic-microwave technique with ethanol to maximize extraction of green coffee oil from Arabica coffee beans. *Industrial Crops and Products, 151*, E-article 112405.
22. Cheng, Y., Donkor, P. O., Ren, X., Wu, J., Agyemang, K., Ayim, I., & Ma, H., (2019). Effect of ultrasound pretreatment with mono-frequency and simultaneous dual frequency on the mechanical properties and microstructure of whey protein emulsion gels. *Food Hydrocolloids, 89*, 434–442.
23. Cherif, M. M., Grigorakis, S., Halahlah, A., Loupassaki, S., & Makris, D. P., (2020). High-efficiency extraction of phenolics from wheat waste biomass (bran) by combining deep eutectic solvent, ultrasound-assisted pretreatment and thermal treatment. *Environmental Processes, 7*, 845–859.
24. Cordova, A., Astudillo-Castro, C., Ruby-Figueroa, R., Valencia, P., & Soto, C., (2020). Recent advances and perspectives of ultrasound-assisted membrane food processing. *Food Research International, 133*, E-article 109163.
25. Czechowska-Biskup, R., Rokita, B., Lotfy, S., Ulanski, P., & Rosiak, J. M., (2005). Degradation of chitosan and starch by 360KHz ultrasound. *Carbohydrate Polymers, 60*, 175–184.
26. Dehghannya, J., Gorbani, R., & Ghanbarzadeh, B., (2016). Shrinkage of Mirabelle plum during hot air drying as influenced by ultrasound-assisted osmotic dehydration. *International Journal of Food Properties, 19*, 1093–1103.
27. Devi, S., Zhang, M., & Law, C. L., (2018). Effect of ultrasound and microwave assisted vacuum frying of mushroom (*Agaricus Bisporus*) chips quality. *Food Bioscience, 25*, 111–117.
28. Dias, J. L., Mazzutti, S., De Souza, J. A., Ferreira, S. R., Soares, L. A., Stragevitch, L., & Danielski, L., (2019). Extraction of umbu (*Spondias Tuberosa*) seed oil using CO_2, ultrasound and conventional methods: Evaluations of composition profiles and antioxidant activities. *The Journal of Supercritical Fluids, 145*, 10–18.
29. Dolan, H. L., Bastarrachea, L. J., & Tikekar, R. V., (2018). Inactivation of *Listeria innocua* by combined treatment of low-frequency ultrasound and zinc oxide. *LWT, 88*, 146–151.
30. Dolas, R., Saravanan, C., & Kaur, B. P., (2019). Emergence and era of ultrasonics in fruit juice preservation: A review *Ultrasonics Sonochemistry, 58*, E-article 104609.
31. Elhag, H. E. E. A., Naila, A., Nour, A. H., Ajit, A., Sulaiman, A. Z., & Abd, A. B., (2019). Optimization of protein yields by ultrasound assisted extraction from *Eurycoma longifolia* roots and effect of agitation speed. *Journal of King Saud University–Science, 31*, 913–930.

32. Elizabeth, A. E. M., Moreira, S. A., Castro, L. M., Pintado, M., & Saraiva, J. A., (2018). Emerging technologies to extract high added value compounds from fruit residues: Sub/supercritical, ultrasound, and enzyme-assisted extractions. *Food Reviews International, 34*, 581–612.
33. Evelyn, & Silva, F. V. M., (2015). Use of power ultrasound to enhance the thermal inactivation of *clostridium perfringens* spores in beef slurry. *International Journal of Food Microbiology, 206*, 17–23.
34. Fan, L., Hou, F., Muhammad, A. I., Ruiling, L. V., Watharkar, R. B., Guo, M., Ding, T., & Liu, D., (2019). Synergistic inactivation and mechanism of thermal and ultrasound treatments against *Bacillus subtilis* spores. *Food Research International, 116*, 1094–1102.
35. Feng, H., Yang, W., & Hielscher, T., (2008). Power ultrasound. *Food Science and Technology International, 14*, 433–436.
36. Flores-Jimenez, N. T., Ulloa, J. A., Silvas, J. E. U., Ramirez, J. C. R., Ulloa, P. R., Rosales, P. U. B., Carrillo, Y. S., & Leyva, R. G., (2019). Effect of high-intensity ultrasound on the compositional, physicochemical, biochemical, functional and structural properties of canola (*Brassica Napus* L.) protein isolate. *Food Research International, 121*, 947–956.
37. Gallo, M., Ferrara, L., & Naviglio, D., (2018). Application of ultrasound in food science and technology: A perspective. *Foods, 7*, 164–172.
38. Gavahian, M., Chen, Y. M., Khaneghah, A. M., Barba, F. J., & Yang, B. B., (2018). In-pack sonication technique for edible emulsions: Understanding the impact of acacia gum and lecithin emulsifiers and ultrasound homogenization on salad dressing emulsions stability. *Food Hydrocolloids, 83*, 79–87.
39. Gomez-Gomez, A., Brito-De La, F. E., Gallegos, C., Garcia-Perez, J. V., & Benedicto, J., (2020). Non-thermal pasteurization of lipid emulsions by combined supercritical carbon dioxide and high-power ultrasound treatment. *Ultrasonics Sonochemistry, 67*, E-article 105138.
40. Gorguç, A., Bircan, C., & Yılmaz, F. M., (2019). Sesame bran as an unexploited byproduct: Effect of enzyme and ultrasound-assisted extraction on the recovery of protein and antioxidant compounds. *Food Chemistry, 283*, 637–645.
41. Gorguç, A., Ozer, P., & Yılmaz, F. M., (2020). Simultaneous effect of vacuum and ultrasound assisted enzymatic extraction on the recovery of plant protein and bioactive compounds from sesame bran. *Journal of Food Composition and Analysis, 87*, E-article 103424.
42. Goula, A. M., Papatheodorou, A., Karasavva, S., & Kaderides, K., (2018). Ultrasound-assisted aqueous enzymatic extraction of oil from pomegranate seeds. *Waste and Biomass Valorization, 9*, 1–11.
43. He, K., Zhang, X., Li, Y., Li, B., & Liu, S., (2020). Water-insoluble dietary-fibers from *Flammulina velutiper* used as edible stabilizers for oil-in-water Pickering emulsions. *Food Hydrocolloids, 101*, 105519.
44. Heloisa, W. D., Camara, H. D., & Lohi, L., (2020). *In-situ* ultrasound-assisted control of polymeric membrane fouling. *Ultrasonics, 108*, E-article 106206.
45. Hu, A. J., Zhao, S., Liang, H., Qiu, T. Q., & Chen, G., (2007). Ultrasound assisted supercritical fluid extraction of oil and coixenolide from adlay seeds. *Ultrasonics Sonochemistry, 14*, 219–224.
46. Huang, L., Ding, X., Li, Y., & Ma, H., (2019). The aggregation, structures and emulsifying properties of soybean protein isolate induced by ultrasound and acid. *Food Chemistry, 279*, 114–119.

47. Iqbal, A., Murtaza, A., Hu, W., Ahmad, I., & Xu, X., (2019). Activation and inactivation mechanisms of polyphenol oxidase during thermal and non-thermal methods of food processing. *Food and Bioproducts Processing, 117*, 170–182.
48. Ismail, B. B., Guo, M., Pu, Y., Wang, W., Ye, X., & Liu, D., (2019). Valorization of baobab (*Adansonia digitata*) seeds by ultrasound-assisted extraction of polyphenolics: Optimization and comparison with conventional methods. *Ultrasonics Sonochemistry, 52*, 257–265.
49. Jiang, Q., Zhang, M., & Xu, B., (2020). Application of ultrasonic technology in post-harvested fruits and vegetables storage: A review. *Ultrasonics Sonochemistry, 69*, E-article 105261.
50. Kassem, A., Meade, J., McGill, K., Walsh, C., Gibbons, J., Lyng, J., & Whyte, P., (2018). An investigation of high intensity ultrasonication and chemical immersion treatments on *Campylobacter Jejuni* and spoilage bacteria in chicken. *Innovative Food Science and Emerging Technologies, 45*, 298–305.
51. Klejdus, B., Lojkova, L., Plaza, M., Snoblova, M., & Sterbova, D., (2010). Hyphenated technique for the extraction and determination of isoflavones in algae: Ultrasound-assisted supercritical fluid extraction followed by fast chromatography with tandem mass spectrometry. *Journal of Chromatography A, 1217*, 7956–7965.
52. Koubaa, M., Mhemdi, H., & Fages, J., (2018). Recovery of valuable components and inactivating microorganisms in the agro-food industry with ultrasound-assisted supercritical fluid technology. *The Journal of Supercritical Fluids, 134*, 71–79.
53. Kowalski, S. J., Szadzinska, J., & Pawlowski, A., (2015). Ultrasonic-assisted osmotic dehydration of carrots followed by convective drying with continuous and intermittent heating. *Drying Technology, 33*, 1570–1580.
54. Kroehnke, J., Szadzinska, J., Stasiak, M., Radziejewska-Kubzdela, E., Bieganska-Marecik, R., & Musielak, G., (2018). Ultrasound– and microwave-assisted convective drying of carrots: Process kinetics and product's quality analysis. *Ultrasonics Sonochemistry, 48*, 249–258.
55. Kyllonen, H. M., Pirkonen, P., & Nystrom, M., (2005). Membrane filtration enhanced by ultrasound: A review. *Desalination, 181*(1–3), 319–335.
56. Liew, S. Q., Ngoh, G. C., Yusoff, R., & Teoh, W. H., (2016). Sequential ultrasound-microwave assisted acid extraction (UMAE) of pectin from pomelo peels. *International Journal of Biological Macromolecules, 93*, 426–435.
57. Li, J., Cheng, H., Liao, X., Liu, D., Xiang, Q., Wang, J., Chen, S., et al., (2019). Inactivation of *Bacillus subtilis* and quality assurance in Chinese bayberry (*Myrica rubra*) juice with ultrasound and mild heat. *LWT, 108*, 113–119.
58. Li, K., Fu, L., Zhao, Y. Y., Xue, S. W., Wang, P., Xu, X. L., & Bai, Y. H., (2020). Use of high-intensity ultrasound to improve emulsifying properties of chicken myofibrillar protein and enhance the rheological properties and stability of the emulsion. *Food Hydrocolloids, 98*, E-article 105275.
59. Lorimer, J. P., & Mason, T. J., (1987). Sonochemistry, part 1: Physical aspects. *Chemical Society Reviews, 16*, 239–274.
60. Machado, A. P. D. F., Sumere, B. R., Mekaru, C., Martinez, J., Bezerra, R. M. N., & Rostagno, M. A., (2019). Extraction of polyphenols and antioxidants from pomegranate peel using ultrasound: Influence of temperature, frequency and operation mode. *International Journal of Food Science & Technology, 54*, 2792–2801.

61. Majid, I., Nayik, G. A., & Nanda, V., (2015). Ultrasonication and food technology: A review. *Cogent Food & Agriculture, 1*, E-article 1071022.
62. Meirelles, A. A. D., Costa, A. L. R., & Cunha, R. L., (2020). The stabilizing effect of cellulose crystals in O/W emulsions obtained by ultrasound process. *Food Research International, 128*, E-article 108746.
63. Merone, D., Colucci, D., Fissore, D., Sanjuan, N., & Carcel, J. A., (2020). Energy and environmental analysis of ultrasound-assisted atmospheric freeze-drying of food. *Journal of Food Engineering, 283*, E-article 11003.
64. Milani, E. A., & Silva, F. V., (2017). Ultrasound-assisted thermal pasteurization of beers with different alcohol levels: Inactivation of *Saccharomyces Cerevisiae* ascospores. *Journal of Food Engineering, 198*, 45–53.
65. Mir, N. A., Riar, C. S., & Singh, S., (2019). Physicochemical, molecular and thermal properties of high-intensity ultrasound treated protein isolates from album (*Chenopodium Album*) seed. *Food Hydrocolloids, 96*, 433–441.
66. Munekata, P. E. S., Alcantara, C., Zugcic, T., Abdelkebir, R., Collado, M. C., Garcia-Perez, J. V., Jambrak, A. R., et al., (2020). Impact of ultrasound-assisted extraction and solvent composition on bioactive compounds and *in vitro* biological activities of thyme and rosemary. *Food Research International, 134*, E-article 109242.
67. Neppiras, E. A., (1980). Acoustic cavitation. *Physics Reports, 61*, 159–251.
68. Nishad, J., Dutta, A., Saha, S., Rudra, S. G., Varghese, E., Sharma, R. R., Tomar, M., et al., (2020). Ultrasound-assisted development of stable grapefruit peel polyphenolic nano-emulsion: Optimization and application in improving oxidative stability of mustard oil. *Food Chemistry, 334*, E-article 127561.
69. Pa, N. F. C., Chin, N. L., Yusof, Y. A., & Aziz, N. A., (2014). Power ultrasound-assisted mixing effects on bread physical properties. *Agriculture and Agricultural Science Procedia, 2*, 60–66.
70. Palgan, I., Munoz, A., Noci, F., Whyte, P., Morgan, D. J., Cronin, D. A., & Lyng, J. G., (2012). Effectiveness of combined pulsed electric field (PEF) and manothermosonication (MTS) for the control of *Listeria innocua* in a smoothie type beverage. *Food Control, 25*, 621–625.
71. Pan, M., Xu, F., Wu, Y., Yao, M., Xiao, X., Zhang, N., Ju, X., & Wang, L., (2020). Application of ultrasound-assisted physical mixing treatment improves *in vitro* protein digestibility of rapeseed napin. *Ultrasonics Sonochemistry, 67*, 105136.
72. Paniwnyk, L., (2014). Applications of ultrasound. In: Sun, D. W., (eds.), *Emerging Technologies for Food Processing* (2nd edn., pp. 271–291). New York – USA; Elsevier.
73. Pla, M. F. D. E., Carolina, S. K., & Genevois, C. E., (2020). Innovative strategies and nutritional perspectives for fortifying pumpkin tissue and other vegetable matrices with iron. *Food Science and Human Wellness, 9*, 103–111.
74. Povey, M. J., & Mason, T. J., (1998). *Ultrasound in Food Processing* (p. 282). Singapore: Springer Science & Business Media.
75. Raso, J., Condon, S., & Trepat, F. S., (1994). Mano-thermosonication: New method of food preservation. In: *Food Preservation by Combined Processes: Final Report for FLAIR Concerted Action 7 (Subgroup B)* (pp. 37–42). Netherlands: European Commission.
76. Reategui, J. L. P., Da Fonseca, M. A. P., Barbero, G. F., Rezende, C. A., & Martinez, J., (2014). Extraction of antioxidant compounds from blackberry bagasse using supercritical CO_2 assisted by ultrasound. *The Journal of Supercritical Fluids, 94*, 223–233.

77. Santos, H. M., & Capelo, J. L., (2007). Trends in ultrasonic-based equipment for analytical sample treatment. *Talanta, 73*, 795–802.
78. Santos, H. M., Lodeiro, C., & Capelo-Martínez, J. L., (2009). The power of ultrasound. In: *Ultrasound in Chemistry: Analytical Applications* (pp. 1–16). KGaA, Weinheim: Wiley-VCH Verlag GmbH & Co.
79. Santos, K. C., Guedes, J. S., Rojas, M. L., Carvalho, G. R., & Augusto, P. E. D., (2021). Enhancing carrot convective drying by combining ethanol and ultrasound as pre-treatments: Effect on product structure, quality, energy consumption, drying and rehydration kinetics. *Ultrasonics Sonochemistry, 70*, E-article 105304.
80. Sara, S., Martina, C., & Giovanna, F., (2014). High pressure carbon dioxide combined with high power ultrasound processing of dry cured ham spiked with *Listeria Monocytogenes*. *Food Research International, 66*, 264–273.
81. Scudino, H., Silva, E. K., Gomes, A., Guimaraes, J. T., Cunha, R. L., Santana, A. S., Meireles, M. A. A., & Cruz, A. G., (2020). Ultrasound stabilization of raw milk: Microbial and enzymatic inactivation, physicochemical properties and kinetic stability. *Ultrasonics Sonochemistry, 67*, E-article 105185.
82. Shamila-Syuhada, A. K., Chuah, L. O., Wan-Nadiah, W. A., Cheng, L. H., Alkarkhi, A. F. M., Effarizah, M. E., & Rusul, G., (2016). Inactivation of microbiota and selected spoilage and pathogenic bacteria in milk by combinations of ultrasound, hydrogen peroxide, and active lactoperoxidase system. *International Dairy Journal, 61*, 120–125.
83. Silva, J. S. D., Voss, M., Menezes, C. R. D., Barin, J. S., Wagner, R., Campagnol, P. C. B., & Cichoski, A. J., (2020). Is it possible to reduce the cooking time of *Mortadellas* using ultrasound without affecting their oxidative and microbiological quality? *Meat Science, 159*, E-article 107947.
84. Soliva-Fortuny, R., Balasa, A., Knorr, D., & Martin-Belloso, O., (2009). Effects of pulsed electric fields on bioactive compounds in foods: A review. *Trends in Food Science & Technology, 20*, 544–556.
85. Soria, A. C., & Villamiel, M., (2010). Effect of ultrasound on the technological properties and bioactivity of food: A review. *Trends in Food Science & Technology, 21*, 323–331.
86. Su, Y., Zhang, M., Bhandari, B., & Zhang, W., (2018). Enhancement of water removal and the quality of fried, purple-fleshed sweet potato in the vacuum frying by combined power ultrasound and microwave technology. *Ultrasonics Sonochemistry, 44*, 368–379.
87. Tan, M. C., Chin, N. L., & Yusof, Y. A., (2011). Power ultrasound-aided batter mixing for sponge cake batter. *Journal of Food Engineering, 104*, 430–437.
88. Teh, S. S., Niven, B. E., Bekhit, A. E. D. A., Carne, A., & Birch, E. J., (2015). Microwave and pulsed electric field assisted extractions of polyphenols from defatted canola seed cake. *International Journal of Food Science & Technology, 50*, 1109–1115.
89. Wang, J., Liu, Q., Xie, B., & Sun, Z., (2020). Effect of ultrasound combined with ultraviolet treatment on microbial inactivation and quality properties of mango juice. *Ultrasonics Sonochemistry, 64*, E-article 105000.
90. Wei, M. C., Xiao, J., & Yang, Y. C., (2016). Extraction of humulene-enriched oil from clove using ultrasound-assisted supercritical carbon dioxide extraction. *Food Chemistry, 210*, 172–181.
91. Zhang, X., Wang, L., Cheng, M., Wang, R., Luo, X., Li, Y., & Chen, Z., (2015). Influence of ultrasonic enzyme treatment on the cooking and eating quality of brown rice. *Journal of Cereal Science, 63*, 140–146.

92. Zhang, Y., & Abatzoglou, N., (2020). Review: Fundamentals, applications and potentials of ultrasound-assisted drying,.*Chemical Engineering Research and Design, 154*, 21–46.
93. Zou, Y., Zhang, W., Kang, D., & Zhou, G., (2018). Improvement of tenderness and water holding capacity of spiced beef by the application of ultrasound during cooking. *International Journal of Food Science and Technology, 53*, 828–836.
94. Zou, Y., Kang, D., Liu, R., Qi, J., Zhou, G., & Zhang, W., (2018). Effects of ultrasonic-assisted cooking on the chemical profiles of taste and flavor of spiced beef. *Ultrasonics Sonochemistry, 46*, 36–45.

CHAPTER 7

POTENTIAL OF ULTRASOUND TECHNOLOGY IN NUTRACEUTICALS AND PHARMACEUTICALS

SHASHIKANT V. BHANDARI, ANIKET P. SARKATE, PRANALI V. KUTHE, and BHAGWAN K. SAKHALE

ABSTRACT

Ultrasound waves are short waves produced by technological sources and are used in a number of applications in the industrial, medical, environmental, and food sectors. Ultrasound waves can be generated and detected by using different types of transducers that are discussed in this chapter. The chapter also deals with the importance and applications of ultrasound in various fields, such as nutraceuticals, food industry, meat preservation in large scale, pharmaceuticals, etc.

7.1 INTRODUCTION

Acoustics is an interdisciplinary field that involves life and earth sciences, engineering, and the arts. It is the study of elastic waves. The human auditory system can be divided into three branches based on the frequency range and hearing qualities to which it responds: infrasound, sound, and ultrasound:

1. **Infrasound:** It is the branch that deals with frequencies below the human hearing range (0–20 Hz).
2. **Sound:** It is in the audible range for humans (20 Hz–20 kHz). A sound wave is a vibration-induced change in air pressure. A tuning fork, a guitar string, an organ pipe's air column, the head (or rim)

Advanced Research Methods in Food Processing Technologies. Junaid Ahmad Malik, Megh R. Goyal, Preeti Birwal, & Ritesh B. Watharkar (Eds.)
© 2024 Apple Academic Press, Inc. Co-published with CRC Press (Taylor & Francis)

of a snare drum, steam escaping from a radiator, a clarinet reed, the diaphragm of a loudspeaker, the vocal cords, or almost anything that vibrates in a frequency range audible to a listener can cause vibration (roughly 20 to 20,000 cycles per second for humans). The generation of a sound wave requires both a vibratory disturbance and an elastic medium, the most common of which is air [24].

3. **Ultrasound:** It is the branch that deals with elastic waves from 20 kHz up to frequencies associated with wavelengths comparable to intermolecular distances (about 1,012 Hz). Ultrasound waves are very short waves (wavelengths ranging from cm to nm) created by technology sources that are used in a variety of industrial, medicinal, and environmental operations.

Aside from frequency, the range of wave intensity has a significant impact on the generation, propagation, and application of acoustic waves. As a result, a subclassification based on the use of low or high intensity waves should be implemented within each of the three acoustic divisions. As a result, ultrasonography can be divided into two types: low-intensity waves (also known as signal ultrasound) and high-intensity waves (also known as medical ultrasound; often known as power ultrasound). The distinction between low and high intensity waves is difficult to quantify; nevertheless, at intensity values ranging from 0.1 W/cm^2 to 1 W/cm^2, depending on the medium, it can be approximated. Ultrasonic waves can be used for exploration, detection, and information as well as action because they have a short wavelength and a high degree of discrimination and energy concentration. They can also be used for communication, particularly in water, where electromagnetic wavelengths are limited. In exploration, detection, and information, ultrasonic sounds can determine the properties and internal structure of propagation media without changing them. High-intensity ultrasonic waves can produce irreversible changes in the medium on which they act. As a form of communication, an ultrasonic signal can be adjusted and conveyed.

Underwater acoustics is a communication application in which sonic and ultrasonic waves are used to detect submerged objects, as well as for echo ranging, depth sounding, and other activities. Non-destructive testing (NDT), process control, and medical diagnosis are all typical uses for low-intensity ultrasound. Cleaning, drying, mixing, emulsification, degassing, defoaming, atomization, particle agglomeration, sonochemical reactions, welding, drilling, and other high-intensity applications are just a few examples. High-intensity ultrasound is also highly beneficial in medical therapy [6].

This chapter focuses on: (i) various ultrasonic applications in pharmaceutical, nutraceutical, and food processing; and (ii) the transducers required to create and detect ultrasound waves.

7.2 HISTORICAL BACKGROUND

Sound waves have been studied for hundreds of years for a variety of purposes, but the discovery of bat echolocation in 1,790 sparked the development of ultrasonography. The bats make sounds to detect things and food by using their hearing to maneuver in the air by listening for the return of the high-frequency sound. Lazzaro Spallanzani, a biologist, was the first to document this ultrasonic range as a mode of flying.

These high-frequency sound waves pave the way for the evolution of ultrasonic technologies in the future. There were limited advancements in the late 18th and early 19th centuries, but Spallanzani's essay was written 100 years before humans could harness and use ultrasonography.

Physicists Pierre and Jacques Curie improved their understanding of pyroelectricity, which is the electric potential created by a substance in response to a temperature shift [16], in 1880. They also studied crystal structural qualities to demonstrate a piezoelectric action, which served as the scientific foundation for the first transducer. Their device made a high-frequency sound that was subsequently reverberated.

The First World War aided many technological advancements in a range of industries, and ultrasonic technology was of particular interest to the military at the time. The first practical application of ultrasonic and piezoelectric technologies was the development of a sonar system to locate submerged submarines during World War I.

The promise of this section of physics and engineering was employed in a variety of sectors after it was discovered and developed in the early 1900s.

7.3 GENERATION AND DETECTION OF ULTRASONIC WAVES: BASIC TRANSDUCER TYPES

Any device that can generate and/or detect ultrasonic waves is known as an ultrasonic transducer. A transducer is a device that changes the form of energy. The most common conversion in transmitters is from electrical to ultrasonic energy, while the most common conversion in receivers is from ultrasonic to electrical energy.

The most common electrical transducers are piezoelectric, magnetostrictive, capacitive or electrostatic, and electromagnetic. Other mechanically actuated transducers, such as whistles and sirens, exist, but they are mostly used as historical markers. Piezoelectric transducers, which are by far the most common transducers in ultrasonics, are based on the piezoelectric effect.

1. **Magnetostrictive Transducers:** These use the magnetostriction phenomenon, which happens when ferromagnetic materials change dimension in reaction to a magnetic field. The magnetic properties of the material will change if it is distorted as a result of an external perturbation. Classic materials that have this effect include iron, nickel, cobalt, and their many alloys, as well as ceramic materials containing cubic ferrites. Since the 1970s, new magneto-strictive materials based on rare earth compounds have been developed with huge magneto-strain and high energy density.
2. **Electrostatic Transducers:** These also known as capacitive transducers, are flat condensers with one stiff electrode and a thin membrane between them. The membrane moves at the same frequency as the alternate voltage when a bias voltage is superimposed on it. These transducers have been used since the 1950s, but recent advances in micromachining techniques for transducer production have accelerated their development for high-frequency imaging applications [25].
3. **Electromagnetic Transducers:** These use the well-known interaction of a permanent magnet's magnetic field and the opposing electric current in a moving coil to convert electric oscillations into ultrasonic vibrations. In noncontact ultrasonic NDT of metallic or ferromagnetic materials employing electromagnetic acoustic transducers, the electromagnetic mechanism is now frequently used (EMAT: electromagnetic acoustic transduces).
4. **Mechanical Transducers:** These are used exclusively as transmitters and produce large-amplitude vibrations. They don't require electricity to function; instead, air or liquid jets force the cavity to vibrate at its natural frequency (whistles) or to be interrupted by the rotation of holes in a rotor, which produces puffs of fluid at a certain frequency, resulting in sound waves (sirens). Such transducers are rarely used and have all but vanished [30].
5. **Narrowband and Broadband Ultrasonic Transducers:** The two main types of ultrasonic transducers are narrowband and broadband ultrasonic transducers. The bandwidth of a transducer is defined as

the relationship between the difference between the frequencies on each side of the center frequency, where the amplitude is 0.707 times the amplitude of the center frequency, and the center frequency. The bandwidth factor is the inverse of the quality factor (Q). Narrowband transducers are often used to provide high-intensity ultrasound in the low frequency range (20–100 kHz).

6. **Broadband Transducers (30–70% Bandwidth):** These are widely used in exploration, detection, and information applications ranging from NDT and medical imaging, which employ very short ultrasonic pulses (typically three or four cycles) and frequencies in the 0.5–50 MHz range, to determining the propagation characteristics of a medium over a wide frequency range.
7. **A Sandwich Transducer:** It is a halfwave resonant length expander that, in its most basic form, consists of a disc, or paired discs, of piezoelectric ceramics sandwiched between two similar metal blocks. The resonance frequency is determined by the dimensions and acoustic qualities of the metal pieces and ceramics.

Mechanical forces induced by particle motion are employed in machining, soldering, welding, cutting, metal shaping, and other high-intensity ultrasonic energy applications in solids. In comparable processing applications, the sandwich transducer is used, but it is followed by a special-shaped metallic transmission line that functions as a mechanical amplifier and ensures a high displacement at the working end. Mechanical amplifiers or horns are resonant vibrating solid bars of various shapes, often stepped, conical, or exponential, which show differing displacements at their two ends inversely proportional to their areas according to momentum conservation [11].

Power transducers with micromachined silicon horns have recently been presented in biomedical applications. Sandwich transducers are commonly employed for high-intensity applications in liquid vessels of various geometries with a large number of sandwich transducers mounted to the bottom and/or walls. Ultrasonic reactors of this type can also be employed as flow systems [36].

Broadband transducers are required for applications involving brief pulses and high resolution. These transducers generate and/or receive an ultrasonic signal, which is subsequently used to detect an obstruction, measure a physical quantity, or sense a system's condition. Medical diagnostics and NDT are the two main application areas. Temperature, fluid velocity, density, viscosity, thickness, location, and other variables can all be measured using this device. Typical frequencies range from hundreds of kilohertz to several megahertz.

7.4 BASIC PRINCIPLE OF ULTRASONIC WAVE PROPAGATION

The creation of vibrations from the source, which provides the elastic energy, as well as particle motion in the medium through which the waves travel, are all part of ultrasound propagation. Ultrasound as a wave motion is governed by the well-known acoustical wave equations, which are derived from the equation of motion, equations of conservation (mass and energy), and the equation of state or constitutive equations. Ultrasonic waves can propagate in a variety of ways, depending on the nature of the medium and the characteristics of vibration generation. The two most frequent types of propagation waves are longitudinal (compressional) and transverse (shear).

In longitudinal or compressional waves, particles oscillate in the propagation direction, but in transverse waves, particles oscillate in the normal propagation direction. Transverse waves can only propagate in media with form elasticity (solids and fluids), but longitudinal waves can propagate in any material with bulk elasticity (solids and some viscoelastic liquids) [18]. In addition to longitudinal and transverse waves, there are waves that involve both longitudinal and transverse motion, such as flexural waves, surface waves, and torsional waves.

For fluids in general, the speed of sound c is given by the Newton–Laplace equation:

$$C = [K/\rho]^{0.5} \quad (1)$$

where; K is the isentropic bulk modulus (or the modulus of bulk elasticity for fluids); and ρ is the density.

7.5 APPLICATIONS OF ULTRASOUND TECHNOLOGY

7.5.1 COMMON PHARMACEUTICAL AND NUTRACEUTICAL APPLICATIONS

7.5.1.1 CLEANING

Material cleaning is one of the oldest and most well-known applications of high-intensity ultrasound. The cleaning activity of ultrasonic energy is mostly due to cavitation and streaming. As the perforations collapse, shock waves and extremely high temperatures are produced, allowing the unclean material to separate from the filthy solid [17]. This type of action can occur even in the tiniest pores of a solid object, resulting in extremely efficient cleaning. Furthermore, ultrasonic energy generates agitation in the liquid, which aids in contaminant dispersion. The market for nutraceutical products such as dietary supplements, minerals, vitamins, herbals, and nutrients has risen as consumers become more health conscious.

As a result of the company's quick expansion into the food and beverage business, demand has skyrocketed. Massive swings in product development and capacity are driving manufacturing, but allergies, viruses, bacteria, and other pollutants are also on the rise. Cleaning is necessary to keep up with the equipment, and it occasionally slips through the cracks [10].

Increased efficacy, decreased surface residue, and easier cleaning validation are just a few of the benefits of aqueous cleaning in pharmaceutical manufacture. Aqueous cleaning can achieve crucial clean standards on practically any pharmaceutical manufacturing equipment, from tablet presses to stainless steel mixing tanks. Tablets, capsules, fluids, and intermediates, among other healthcare goods, must be thoroughly cleaned [2].

7.5.1.2 ATOMIZATION

Ultrasonic energy may cause liquid atomization. Due to radiation pressure and the creation of capillary waves on the liquid's surface, ultrasound can cause tiny droplets to form. The frequency of a droplet is related to its size. Fogs or mists with a fine and homogenous droplet size distribution are created using ultrasound.

Nutrients can be delivered to the target location in the body using atomized particles as delivery methods to keep them fresh during processing and storage. By improving the design and functionality of functional food ingredients, these atomized particles contribute to the creation of new products and delivery strategies [33]. Encapsulation is a critical use of ultrasonic atomization, particularly in the food and pharmaceutical industries, where a variety of fluids must be treated to create the right composites. Ultrasonic nozzles that can handle a variety of fluids require emulsion creation before drying [14].

7.5.1.3 MIXING, HOMOGENIZATION, AND EMULSIFICATION

High-powered ultrasound can produce uniform solid/liquid and liquid/liquid dispersions. The primary components that facilitate this process are cavitation and streaming. Bubble collapse creates enormous stresses, which press down on the particles, causing them to shrink and combine. Without the use of additives, ultrasound has been used to form stable emulsions of immiscible liquids and particle dispersions in liquids.

Ultrasound mixing, homogenization, and emulsification in solid and liquid suspensions, as well as the use of higher static pressure conditions (up to a few bar) and ultralow US frequency (well below 20 kHz), appear to be very promising according to detailed comparisons on sludge breakup and organic dissolution [8].

Other uses include nanosuspension for nutraceutical delivery, SLNs (solid lipid nanoparticles), and NLCs (nanostructured lipid carriers) for nutraceutical delivery, and polymer nanoparticles (NPs) for nutraceutical delivery [5].

7.5.1.4 DEFOAMING

Foams are created in a variety of processes and are frequently an undesirable by-product that makes process control and product creation more difficult. One common example is the fermentation industry, where foam is one of the most serious problems. Chemical antifoaming agents are used in traditional methods for managing foams in the food and beverage sectors, which can lead to product contamination. The use of high-intensity ultrasonic waves to break down foam bubbles is an innovative, clean, and effective method. Although the mechanisms causing acoustic foam disintegration are unknown, experimental evidence suggests that radiation pressure and resonance frequency have a role. Ultrasonic defoaming devices with frequencies of 21 and 26 kHz are already on the market.

Foams are employed in a wide range of industries, including food production and cosmetics. Intensive foaming has negative consequences in the majority of technological processes, such as decreased use of useful volume of technological equipment, violation of manufacturing rules and sterility of biotechnological processes, increased product losses and equipment productivity decline, and environmental pollution. Mechanical breakers, lowering packaging container temperatures below ambient temperatures, or adding chemical anti-foams have all been used to handle foam in the food sector in the past.

Foam formation is a typical issue in the dairy processing industry, and it can result in a decrease in output. The presence of dissolved oxygen in dairy products can have a negative impact on the product's final quality by contributing to oxidation reactions within the product, shortening its shelf life. Protein-stabilized foams, on the other hand, include dairy milk foams. In the presence of cream in the form of an emulsion, airborne ultrasonography can only partially reduce dairy foams [12].

Because they don't require much air movement, don't create chemical contamination, and can be used in a sterile atmosphere, high-intensity ultrasonic waves are an intriguing technique of breaking foam. It's ideal for usage in the food and pharmaceutical industries because of this. An ultrasonic defoaming device was created using a novel type of focused ultrasonic generator. This new approach has been utilized to successfully reduce foam production in high-speed carbonated beverage bottling and canning processes. It has also been possible to scale up such systems to control foam.

7.5.1.5 DRYING AND DEWATERING

Dehydration is a method of food preservation that has traditionally been achieved using either hot air, which can cause goods to deteriorate, or freeze drying, which retains food quality but is expensive.

Ultrasonic dehydration, which uses airborne radiation or direct contact with food to remove moisture, has proven to be a feasible alternative for removing moisture from food. The food item is subjected to high ultrasonic stresses, resulting in a "sponge effect" and rapid moisture migration through natural channels. Furthermore, creating ultrasonic cavitation inside the liquid may help separate the closely bound moisture. Ultrasound has a lot of promise for speeding up low temperature drying. Food goods' quality is preserved by ultrasonic drying [15].

In the pharmaceutical industry, ultrasound dewatering is employed in wastewater treatment plants. Ultrasonic waves provide a continuous field for electro osmotic dewatering and particle consolidation. Ultrasonic waves also assist in the liberation of dissolved gas and the release of trapped liquid in solids [34].

7.5.1.6 ULTRASOUND-ASSISTED SUPERCRITICAL FLUID EXTRACTION (USFE)

SFE (supercritical fluid extraction) is a separation method based on supercritical interaction of an extractable chemical-containing product with a solvent. Although this strategy is thought to be quite useful, its dynamics are slow. The action of ultrasound's radiation pressure, microstreaming, and agitation improves the solvent's penetration into the product, which is a beneficial impact of SFE. Ultrasound has been shown to be useful in the extraction of a number of chemicals using SFE. Examples of applications include herbal and oil extraction, protein extraction, bioactive extraction from plant materials such as polyphenols, anthocyanins, tartaric acid, fragrance compounds, polysaccharides, and functional compounds, and bioactive extraction from animal materials (Table 7.1) [39].

TABLE 7.1 Examples of Ultrasonic-Assisted Extraction [13]

Materials	Bioactive Compounds	Method and Conditions	Applications
Forsythia suspense plant	Phillyrin	UAE MetOH 20% 60°C, 60 min	Used as medicine due to its anti-inflammatory, antioxidant, antiviral, and vasorelaxant activities.
Hawthorn seeds by-product	Flavanoids	UAE EtOH 72%, 65°C, 37 min, 40 W	Health-promoting compounds negatively associated with coronary heart diseases.
Litchi seeds by-product	Polysaccharides	UAE Water, 45 min, 222 W	Food and biomedical applications. Posse's antitumoral, antioxidant, and hypoglycemic properties.
Nannochloropsis oculate biomass Alga	Lipids	UAE 5% biomass, 30 min, 1,000 W	Feedstock for biodiesel production.
Penggan peel by-product	Hesperidin	UAE MetOH, 40°C, 60 kHz	Food and pharmaceutical industry. Possess antioxidant, anti-inflammatory, and anti-allergic activities.
Prunella vulgaris L. plant	Flavonoids	UAE EtOH 41%, 79°C, 30.5 min	Medical applications. Used for alleviating sore throat, reducing fever, and accelerating wound healing.
Spirulina platensis alga	Beta-carotene	UAE n-heptane, 30C, 167 W/cm^2	Pharmaceutical industry. It helps to protect against cancer, diabetes, and other chronic diseases.

Source: Adapted from Ref. [13].

7.5.1.7 BIOREMEDIATION: ULTRASONIC INACTIVATION OF MICROORGANISM

Ultrasonic energy delivered at a high intensity into a biological medium has the potential to destroy cells. The primary mechanisms that result in mechanical and thermal impacts are cavitation and streaming. As a result, ultrasound can be utilized to speed up and optimize bioprocesses such food sterilization, heat treatment (thermo-sonication), and modifying enzyme characteristics [27].

The most energy-efficient and effective strategies to inactivate bacteria are thermosonic (heat with sonication), manosonic (pressure with sonication), and manothermosonic (heat with pressure and sonication).

Reduced flavor loss, greater homogeneity, and significant energy savings are among the advantages of ultrasound pasteurization versus heat pasteurization. The success of an ultrasonic therapy is determined by the type of bacteria being examined. Other factors to consider are the amplitude of the ultrasonic waves, the exposure time, the volume of food being processed, the composition of the food, and the treatment temperature. Ultrasonography has bactericidal properties when suspended in culture media. According to Lillard, salmonellae adhering to broiler skin were reduced after 30 minutes of sonication in peptone at 20 kHz.

Dolatowski & Stasiak's research revealed that ultrasound processing has a significant impact on meat microbiological contamination. High-intensity ultrasound has a wide range of possible applications in the food industry. In the food industry, both high- and low-frequency ultrasound have been found to have great potential for changing or describing food quality [9].

7.5.1.8 PARTICLE AGGLOMERATION

It is possible to collect suspended airborne particles, particularly very minute particles, by agglomerating them. When an aerosol is exposed to high-intensity ultrasonic waves, it can generate collisions and agglomerations among suspended particles, resulting in larger particles that are easier to collect or precipitate.

When it comes to adding moisture and heat to flour, flexible flour agglomeration gives you a lot of options. Gums that are smooth Agglomeration can also be used by manufacturers to keep lumps out of gum solutions.

Using a high-intensity ultrasonic technique, starch solutions are created. To ensure lump-free dispersibility and quick viscosity, rapid hydration rate, and smooth look and texture, some food applications require "instant starches, either cold-water-swelling or classic drum-dried pregelatinized" [38].

7.5.1.9 SONOCHEMICAL PROCESSES

Streaming causes cavitation and agitation, which results in extremely high temperatures and pressures, which are employed to drive sonochemical processes. Sonochemistry is a vast field that studies the nature and applications of sonochemical processes in chemical synthesis, electrochemistry, water treatment, and nanomaterial fabrication.

7.6 APPLICATIONS OF ULTRASOUND IN PROCESSING OF MEAT

Meat is a perishable food product with precise quality specifications. As a result, using the right processing techniques is critical. Although intensive treatments can extend shelf life, quality suffers as a result. Ultrasound can help with a variety of meat processing operations, extending shelf life while preserving quality. Some of the ways that ultrasonography can be used in meat and fish products are discussed in subsections.

7.6.1 FREEZING

One of the most common methods for preserving meat and fish items is freezing. Ice crystallization is the most important factor in frozen product quality. Meat structure is destroyed, and sensory characteristics are lost due to a lack of homogeneous crystallization.

Sono-crystallization is a phenomenon in which acoustic cavitation free promotes nucleation. Furthermore, heat and mass transport can be enhanced due to the presence of microstreaming and general turbulence. Under the effect of ultrasound, traditional cooling produces a higher-quality product with a substantially shorter dwell time. Furthermore, because there is more nucleation, the final size of the ice crystals is smaller, resulting in less cell injury [4].

7.6.2 THAWING

The change of frozen food into an unfrozen state at melting temperature is referred to as thawing. It is tough to implement at the industrial level because the technique generally takes longer. Due to its sluggishness, it can cause physicochemical changes in food as well as microbial deterioration. The use of ultrasound for thawing food was attempted around 50 years ago, but industrial acceptability was impeded by issues, such as low penetration in frozen food, localized heating, and high-power requirements. As an acoustic wave at 1,500 Hz at 60 W was used in water immersion thawing on cod blocks, the energy demand was reduced by up to 71% when compared to simple immersion thawing [32].

For frozen beef, pig, and fish that had been thawed to a depth of 7.6 cm within, the frequency should be adjusted to frequency (500 kHz) and intensity (0.5 Wcm^2).

7.6.3 TENDERIZATION

Tenderness is the most important of various meat quality characteristics. After an animal is slaughtered, the body's homeostasis is disrupted, which causes endogenous proteases enzymes to break down the animal's muscles, resulting in tenderization.

Several technologies are used to tenderize meat, including electric simulation, blade tenderization, high-pressure grinding (1,000 to 8,000 pressures), and the hydro-dyne process [7].

Chemical treatments, such as injecting a brine containing sodium chloride, calcium chloride, polyphosphates, and acids, as well as biochemical ones, such as monitoring the activity of exogenous (e.g., bromine, papain) or endogenous enzymes over time, are also used to tenderize meat [3].

Meat tenderization is influenced by sarcomere length, intramuscular connective tissue, and the muscle's proteolytic impact. Traditional weak acid methods involving vinegar, lemon juice, salt, and enzymatic preparation are used to improve meat softness by ensuring that the pH is optimal for enzyme action. Ultrasound was first used in the meat business in the 1950s for beef tenderization, but there was no effect on pH. It was later discovered that the tenderization effect of ultrasound on meat is due to lysosomal rupture and disruption of myofibrillar protein (MP). Stable cavitation and collapse cavitation, which is the principle of the ultrasonic process, are the causes of lysosomal rupture.

It can be concluded because the tenderization of meat increases due to collagen fragmentation, which is confirmed in the microstructure examination of meat, and there is no pH change when the treatment is done at a lower intensity, but there is an increase in pH when the treatment is done at a higher intensity. According to Dolatowski et al. [9], the treated meat's water retention capacity rises following treatment. Even though many studies have been done on ultrasonic beef tenderization, more study is needed, as evidenced by an ongoing project in Spain called "Non-invasive high-power ultrasound (HPU) processing method for meat tenderization"[9].

7.6.4 MARINATION

Marination is the process of soaking meat in liquids before cooking it. The solution's constituents, which are largely salt and phosphate, have an impact on the marination process. According to Smith & Young [41], roughly half of all poultry meat is marinated, which increases color, texture, flavor, juiciness, and saleable yield, resulting in a higher profit margin for the seller. Immersion, injection, and vacuum tumbling have all been used in the past, but all three processes have the problem of requiring longer and resulting in inconsistent distribution due to damage to the meat structure [21].

7.6.5 COOKING

Several physicochemical changes occur during the cooking of beef, making it more pleasant and digestible. Lower water absorption, on the other hand, can result in a harder texture. Cooking for prolonged periods of time results in increased energy consumption as well as nutrient and sensory losses. Ultrasound can be employed as an aiding technology to help solve this situation [28]. Because cavitation can result in hydrogen bodings with water, ultrasound treatment will offer the product a smoother texture while also enhancing the water absorption of the MP. The nutrient is maintained in meat due to greater water binding, which minimizes leaching losses during cooking [31].

When foods are exposed to high temperatures in a traditional cooking process, the outside may become overcooked while the interior remains undercooked, resulting in a loss in product quality. Ultrasound has the capacity to increase heat transfer characteristics, which is a critical necessity to avoid such issues, and it has been used in the kitchen. A patent outlines a cooking vessel that uses ultrasonic to apply to heated oil to achieve better and more even overall frying while also claiming to save energy.

7.7 APPLICATIONS OF ULTRASOUND TECHNOLOGY IN NUTRACEUTICALS

While modern processed foods and traditional foods are both popular in supermarkets today, their production, processing, and packaging methods have evolved and rationalized to an unprecedented degree. These technologies are primarily designed to minimize processing time, save energy, and increase the shelf life and quality of food items. Thermal technologies (radiofrequency (RF) and microwave (MW) heating), vacuum cooling technology, high pressure processing, and pulsed electric field (PEF) technology are some of the novel technologies that have the potential to produce high-quality and safe food products. However, current limitations such as high investment costs, complete control of process variables, regulatory approval, and, most importantly, consumer acceptance have been delaying a wider implementation.

Ultrasound (US) has been the topic of research and development in the food business in recent years. Ultrasound has a lot of attention since it can provide enterprises with practical and dependable ultrasound equipment. Its growth as a green innovative technology has drawn attention to its importance in environmental sustainability in recent years. There are three different types of ultrasound applications: apply directly to the product, connecting to the device, and immersion in a high-intensity ultrasonic bath (Table 7.2).

The development of ultrasound technology is only recent for exploitation of power ultrasound. In this sense, special attention has been paid to its use in the recovery of NBCs from different natural sources. Ultrasound-assisted extraction (UAE) has been proposed as a timid alternative to conventional SE, providing higher recovery of targeted compounds with lower solvent consumption and/or faster analysis and bioactivity properties. Its better extraction efficiency is related to the phenomenon called acoustic cavitation. When the ultrasound intensity is sufficient, the expansion cycle can create cavities or microbubbles in the liquid. Once formed, bubbles will absorb the energy from the sound waves and grow during the expansion cycles and recompress during the compression cycle. Further, bubbles may start another rarefaction cycle or collapse leading shock waves of extreme conditions of pressure and temperature (several 100 atmospheres and around 5,000 K of temperature) [40].

7.7.1 FILTRATION

Separating solids from liquids is a common operation in the food industry, whether it's to make a solid-free liquid or a solid separated from its mother liquor. However, one of the most serious issues is the deposition of solid materials on the filtration membrane's surface. By shattering the concentration polarization and cake layer at the membrane surface using ultrasonic radiation, the flow can be increased without altering the membrane's intrinsic permeability.

TABLE 7.2 Applications of High-Intensity Ultrasound in Food Processing

Ultrasound Principle	Application	Conventional Treatment	Food Product	Advantages
Cavitation phenomenon	Emulsification	Mechanical	Emulsions (ketchup, mayonnaise, etc.)	Less time Emulsion stability
Cavitation phenomenon	Cutting	Knives	Fragile products (cake, cheese, etc.)	Less time Reducing products losses Accurate and repetitive cutting
Cavitation phenomenon	Defoaming	Thermal Chemical Electrical Mechanical	Carbonated drinks Fermented products (beer)	Less time Improving hygiene
Compression-rarefaction	Degassing	Mechanical	Chocolate Fermented products (beer)	Less time Improving hygiene
Increasing mass transfer	Pickling/marinating	Brine	Vegetables Meat Fish Cheese	Less time Improving organoleptic quality Product stability
Uniform heat transfer	Drying	Atomization Freezing Hot gas stream Pulverization	Dehydrated products like fruits and vegetables	Less time Improving organoleptic quality Improving heat transfer

TABLE 7.2 *(Continued)*

Ultrasound Principle	Application	Conventional Treatment	Food Product	Advantages
Uniform heat transfer	Cooking	Stove	Meat	Less time
		Flyer	Vegetables	Improving heat transfer and organoleptic quality
		Water bath		
Uniform heat transfer	Freezing/ crystallization	Freezer	Meat	Less time
		Freezing by immersion, by contact	Vegetables	Small crystals
			Fruits	Improving
			Milk products	Rapid temperature decreasing
Vibrations	Filtration	Filters (membranes semi-permeable)	Liquids (juices)	Less time
				Improving filtration
Vibrations	Demolding	Greasing molds	Cooked products (cake)	Less time
		Teflon molds		Reducing product losses
		Silicon molds		

Source: Adapted from Ref. [4]. Copyright © 2010 Elsevier B.V.

The cleaning is based on the liquid jet, and other cavitational mechanisms result in particle liberation from the obstructed membrane. Furthermore, the optimal ultrasonic intensity is critical for preventing filter degradation. Fruit extracts and beverages can also be produced using ultrasound. In the case of apple pulp juice extraction, traditional belt vacuum filtration reduces moisture content from 85% to 50%, whereas electroacoustic technology reduces moisture content by 38%. The use of ultrasound in conjunction with membrane filtration has also been studied and shown to be beneficial. The filtering of whey solutions was aided by ultrasonic irradiation at low power levels.

Ultrasound assisted in preventing filter blockage and flow through it by lowering the compressibility of both the initial protein deposit and the expanding filter cake, resulting in a considerable increase in flow rate. Furthermore, combining a filter with ultrasounds extends the filter's life by preventing clogging and caking at the filter's surface, which prevents clogging and caking. All of the aforementioned factors are important in commercializing filtration technologies, and several businesses are already producing ultrasonic filtration systems as a supplement to conventional (vibratory) screens. While there is still work to be done in the laboratory and on the commercial level, there is still a demand for it. There is also a set amount of time that must be spent on treatment [1].

7.7.2 DEGASSING/DEAERATION

A mixture of gasses, such as dissolved oxygen, carbon dioxide, and nitrogen gas, exists in a liquid. Boiling and lowering pressure are two popular methods for degassing, however ultrasonography has the advantage of a minimal temperature change. When ultrasound, such as an ultrasonic cleaning bath, is utilized with regular tap water inside, degassing is a very obvious phenomenon. It happens when acoustic waves combine the rapid vibrations of gas bubbles, causing the bubbles to develop to a size large enough to allow them to rise up through the liquid against gravity until they reach the surface.

This technique can be used in the food sector to degas carbonated liquids like beer (defibbing) before bottling. The goal of carbonated drink processing is to remove air from the liquid surface in order to prevent germs and oxygen from causing organoleptic damage to the product. Degassing is accomplished by attaching a transducer to the outside of the bottle. When compared to mechanical agitation, ultrasonic agitation reduces the number of broken bottles and liquid overflow. The use of low-intensity ultrasound during the fermentation of sake, beer, and wine resulted in a 36–50% reduction in the time required. Degassing with ultrasonically aided degassing is very quick in watery systems, but it's much more difficult in viscous liquids like melted chocolate [4].

7.7.3 DEPOLYMERIZATION

The degradation of polymers is one of the oldest applications of ultrasound. The depolymerization process is triggered by cavitation and involves two mechanisms: mechanical degradation of the polymer due to a collapsed cavitation bubble, and chemical degradation due to a chemical reaction between the polymer and high-energy molecule, such as hydroxyl radicals produced by cavitation.

The use of ultrasound to depolymerize starch is a hot topic in the food business, with studies dating back to 1933. Sonication is a common method for depolymerization because it produces superior rates, yields, and chemo-, regio-, and stereoselectivity, and it may be done with simple immersion probes working at 20 kHz. As a result, sonication can improve a variety of whey protein properties, including solubility and foaming ability.

Low-power ultrasound reduces the viscosity of polymeric liquids temporarily; however, high powers cause depolymerization and a permanent alteration in their rheology. In the food processing industry, such permanent alterations are useful. Indeed, present ultrasonic frequency, ultrasound power, temperature, processing time, and other food processing factors should be studied to determine not only their positive but also negative effects on food quality and structure.

Ultrasonic irradiation has a lot of promise for converting biomass raw materials like polymeric carbohydrates into usable lower-weight molecules. A recent study comparing ultrasound to other cutting-edge technologies like c-radiation and MW heating revealed that ultrasonication is the most efficient way to depolymerize Xyloglucan and Hyaluronan [29].

7.7.4 DEMOLDING AND EXTRUSION

Generally, the industrial cooking of foods leads to adhesion of the products to the cooking vessel or in other operations it must detach from its mold. Ease of removal makes the cleaning and recycling of the container far easier. To remove the cooked product from the mold is difficult because of product/mold adhesion caused by the cooking. To counteract this difficulty in industrial processing of molded food products, the molds are fabricated with a surface coating made of a thin layer of silicone or PTFE (polytetrafluoroethylene) or by a coating of white grease. However, it is necessary to replace the mold covering periodically because the shelf-life of the thin lubricating layer is relatively short. Such operations are expensive and, if not 100% successful, i.e., incomplete coverage, the containers will arrive at the cleaning post with product still adhering to them and so disrupt the production line. At present, to solve this problem, mechanical methods such as knocking vibration are used to remove the products.

Such activities are costly, and if they are not 100% successful, i.e., the containers come at the cleaning station with product still adhered to them, disrupting the manufacturing line. Currently, mechanical methods such as knocking vibration are utilized to remove the items to alleviate this problem. Food products can be released by coupling the mold to an ultrasonic source as an alternative to these traditional procedures. The device for demolding industrial food items couples the mold and the ultrasonic source in order to improve the removal of the product contained in the latter due to the high-frequency relative movement between the mold's and product's contact surfaces.

This approach eliminates surface coatings and ensures that any residual material in the mold is promptly removed. Extrusion requires a similar quality of ultrasound, namely the capacity to release material from a surface and so reduce drag.

7.7.5 CUTTING

The use of ultrasound in food cutting has enhanced the overall efficiency of food processing. Ultrasonic food cutting technology offers a novel technique to chop or slice a wide range of foods, reducing production time, product loss, and maintenance expenses. Ultrasonic cutting employs a knife-like blade connected to an ultrasonic source through a shaft. The cutting tool itself can take on a variety of configurations, each of which can be considered an acoustic horn and hence a component of the overall ultrasonic resonating device. Cutting with ultrasonic vibration superimposition is a direct rival to technologies like high-velocity water jet cutting and traditional procedures like sawing or knife cutting. Ultrasonic cutting has been shown to have low energy consumption. The properties of ultrasonic cutting vary depending on the type of food and its state, such as frozen or thawed.

The most common use of ultrasound is in the cutting of delicate products. It is useful for delicate and heterogeneous products (cakes, pastry, and bakery products), as well as fatty (cheeses) or sticky products. Another advantage of this approach is that it improves hygiene by preventing product adhesion to the blade and therefore reducing the growth of microorganisms on the surface, i.e., ultrasonic vibrations provide 'auto-cleaning' of the blade. Because of the precision and consistency of the cut, there are fewer losses during the cutting process (due to cracks, crumbs, etc.), and a greater standardization of the weight and dimensions of pieces.

7.7.6 MISCELLANEOUS APPLICATIONS

It is possible to speed up fermentation processes using low-power ultrasound, which has applications in a variety of foods. Ultrasound can also be used on meals to accelerate the aging process. Because of the temperatures and pressures created by ultrasonic waves, free radicals are formed in water. One of the most common uses for ultrasonically aided oxidation is the aging of fermented products like wine and whiskey in alcoholic beverages. Indeed, a device with a frequency range of 20 to 80 kHz has been patented for this purpose.

Another potential application of ultrasound in winemaking is to promote the breakdown of colors and tannins from the fruit's first processing. The use of ultrasound to minimize dough rest time is described in a patent. Indeed, the yeast is subjected to mechanical stress during the mechanical treatment of dough, which lowers its activity afterward. The use of ultrasound should allow for a reduction in the stress and pressures applied to yeast, as well as a reduction in dough rest time.

7.8 APPLICATIONS OF ULTRASOUND IN PHARMACEUTICALS

7.8.1 ULTRASOUND-ASSISTED COMPRESSION OF METAL POWDERS

The static load, temperature, and compaction time have all decreased as a result of this. It was observed that using sonic vibrations (10–20 kHz) in the manufacture of metal compacts improved both density and mechanical strength [37]. Bulk particle movement occurs when metal powder is compressed using ultrasonic vibration, resulting in greater densification and uniformity in the compacted material's structure.

7.8.2 ULTRASOUND-ASSISTED COMPACTION OF CERAMIC POWDERS

Ceramics are brittle materials that can fracture when pre-existing flows or microcracks propagate. The quality of the product is influenced by the electric material utilized in the manufacturing process. As a result, ultrasound was used to assist with pressure during the powder compression stage to reduce the number of pre-existing effects, decrease porosity, enhance compact density, increase compact hardness, and reduce compaction force.

7.8.3 ULTRASOUND-ASSISTED COMPACTION MOLDING OF POLYMER POWDERS

This provides a number of advantages over more traditional methods. Compaction molding of high molecular weight polymers, which are difficult to mold using traditional methods due to their high melt viscosity, and quick and cost-effective manufacturing procedures without the use of external heat sources are two of these approaches. The ability of thermoplastic materials to be molded by ultrasound is largely governed by their capacity to transmit ultrasonic energy, which is known as damping properties. The density, the moduli of dynamic elasticity and shear, the sound velocity, and the resultant wavelength are all crucial parameters in the plasticization process [23].

7.8.4 APPLICATIONS OF ULTRASOUND IN PHARMACEUTICAL TABLETING

1. **To Aid the Direct Compression Method:** Although granulation of powders is typically done before compression, direct compression is the preferred method of tablet manufacturing for a variety of reasons, including simplicity, speed of compression pressure applied processes, cheaper costs, and increased drug stability. In terms of compact density, there has been an increase. Direct compression of a high-dose drug like paracetamol, which has poor compressibility and flow properties, is not practicable. Soft or capping tablets are generated since the diluting capacity of filler-binders is limited, and tablet size and weight are limited. Ultrasound has the capacity to circumvent these limitations, allowing direct compression to be used even when traditional compaction methods have failed [19].

2. **To Improve Powder Flow, Tablet Density, and Uniformity:** Ultrasound use during tableting can increase powder flow, tablet density, and uniformity by "jarring and vibrating" the powder particles and reducing internal friction [35].
3. **To Improve Tablet Mechanical Strength:** When compared to traditional tableting, ultrasound-assisted compression can produce coherent tablets with greater crushing strength and less friability at lower pressures. The temperature rise within the compact produced by ultrasound has been linked to the tablets' improved characteristics. The overall temperature of the tablet was observed to be elevated by around 19, 30, or 368°C after tableting of ibuprofen utilizing 32 MPa compression force and application of 1-, 2-, or 5-seconds ultrasound (amplitude 7 mm) in tests. Increased temperature increases the level of plasticity and stress relaxation within the tablet during traditional compression, but elasticity decreases, resulting in an increase in tablet tensile strength. This could also apply to ultrasonic-assisted compression, in which ultrasound boosts the temperature inside the compact, allowing for plastic deformation and tension release, as well as enhancing particle-particle interaction, leading to stronger tablets. Furthermore, fusion bonding may contribute to the tablets' greater mechanical strength. The manufacturing of stronger tablets is aided by the local temperatures achieved during ultrasound-assisted compression.

7.8.5 APPLICATIONS OF ULTRASOUND METHOD IN PHARMACEUTICAL POWDER COMPRESSION

Direct compression is the favored method of tablet production for a variety of reasons, including convenience of use, speed of processing, lower costs, and improved medication stability. Direct compression of a highly dosed medicine with weak compressibility and flow characteristics, such as paracetamol, is not practicable. Compression of a powder, on the other hand, is a difficult operation that is frequently hampered by a variety of issues. These issues have been extensively researched, and they primarily concern volume reduction and the establishment of sufficient strength between the powder particles to maintain tablet integrity. The use of ultrasonic energy has been demonstrated to be effective in reducing and even preventing certain issues [20].

One of the main reasons of tablet flaws such as friability, adhering, capping, and lamination is poor compatibility of pharmaceutical powders. When compared to traditional tableting, ultrasounds improve the properties of the compression process, resulting in optimum mechanical strength of the compacts at lower pressures. Ultrasounds raise the temperature inside the tablet, increasing plasticity and interparticle bonding, resulting in stronger, less porous tablets. Levina and Rubinstein used ultrasound-assisted compression to make coherent ibuprofen tablets at pressures as low as 20–30 MPa. These tablets' breaking forces were found to be substantially higher than those achieved using traditional compaction. As the time of the ultrasonic treatment increased, the crushing strength of the pills increased dramatically.

The use of ultrasounds during compaction encourages particle rearrangement in the powder bed, resulting in a higher apparent density, a bigger contact area, and, as a result, a higher degree of bonding. Furthermore, melting at the interparticle contact locations creates solid bridges between particles. As a result, tablets have a lower elastic recovery and are more likely to cap. Furthermore, trapped air in the powder is more easily released during ultrasonic vibration.

7.8.6 APPLICATIONS OF ULTRASOUND IN PREPARATION OF SOLID DISPERSIONS

In recent decades, the creation of poorly soluble medications has grown, necessitating the development of formulations that improve their solubility and, as a result, their absorption. Solid dispersion is one of the most useful solubility improvement technologies, and it can be created via solvents, melting, and/or mechanical means. In the absence of organic solvents or evaporation apparatus, the ultrasound-assisted compression technique has been explored as an alternative way for improving the bioavailability of poorly soluble medicines through direct processing of the powder mixture. The amorphization of ketoprofen, a medication with a low water solubility, by a US compaction procedure. After a few hours of US treatment, the medication appeared as a soft and dense paste (due to melting at 96 o C), and the crystallinity was regained. USAC (ultrasound assisted compaction) compacted binary combinations of ketoprofen and Eudragit® S100, a thermoplastic acrylic polymer soluble at pH > 7.

Based on the excipient/drug ratio and the US energy source, DSC experiments demonstrated a decrease in crystallinity of ketoprofen. As the molten phase adsorbed on the surface of the excipient particles, Eudragit® delayed or stopped the regeneration of the drug crystallinity, keeping the drug in a glassy condition. The scientists suggest that this relationship could be useful in the development of pH-controlled dosage forms that release the drug into the intestine's physiological medium. Another evidence of significant physical changes in solids compacted under US conditions was published. In solid dispersions containing indomethacin (a weakly water-soluble medication) and low melting excipients such as polyethylene glycols (PEGs) of various molecular weights, US-discharge during compaction reduced the crystallinity of both drug and excipient. The drug's dissolving rate was sped up because the indomethacin was partially dissolved in the excipient. The degree of back-crystallization, which reduces the dissolution rate as a function of material aging, is determined by the type of PEG used [26].

The physical state of indomethacin underwent significant changes, with the drug spreading as a thin layer over the excipient particle surface, giving it a glossy aspect. In the case of lactose, ultrasounds had no effect on phase transition. In comparison to a typical compaction, dissolution profiles revealed an enhanced release of the medication from the systems treated with ultrasounds. Feasibility of employing a US-assisted compression process to manufacture solid dispersions using fenofibrate (a poorly water-soluble medication) and several polymers as carriers (Eudragit® EPO, Kollidon® VA 64, Soluplus®, HPMC 2910/5, Povacoat® MP). The ultrasonic energy was insufficient to cause the crystalline drug material to completely amorphized. When compared to the untreated physical mixture of drug and excipient, the dissolving behavior was improved, but the dissolution profiles were identical to those produced by adding the polymeric carrier directly to the dissolution medium.

Furthermore, the scientists have stressed the need of monitoring the US compaction process, citing the variable sonotrode to die distance and ultrasonic application time as critical criteria for the results' repeatability. The distance between the sonotrode and the die was kept constant in this investigation at 16.08 mm, so it's not unexpected that ultrasonic energy had no effect at this distance. According to Ueda et al. US compaction can be used to improve the physical properties of a medicine through the thermal effect, and it also has the benefit of requiring just mg of sample and a few tens of seconds each compaction.

In order to determine the ideal setup conditions for creating miscible amorphous drug-polymer dispersions, the authors looked into the relationship between US compaction parameters and the polymer glass transition. Model polymers included PVP90, PVP30, PVPVA, Kollicoat® IR, Kollidon® SR, and Soluplus®.

7.8.7 APPLICATION OF ULTRASOUND IN THE FORMULATION OF SUSTAINED-RELEASE DOSAGE OF DRUGS

For the creation of sustained-release oral dosage formulations, US compaction looks to be a promising approach. Although sustained-release tablets provide a number of clinical benefits, including improved patient compliance and fewer side effects, they are relatively expensive per unit dose. Furthermore, it is frequently difficult to create tablets with an adequate weight, and patients may have trouble ingesting the dosage form. The use of an ultrasound-assisted powder compression technology could reduce the cost and weight of tablets. Several researchers have looked into the use of ultrasound-assisted compression as a method for quickly preparing controlled-release drug delivery systems via ultrasound-assisted compression of theophylline sustained-release matrices employing Eudragit® RL and Eudragit® RS.

According to the scientists, ultrasonic caused the acrylic polymers that covered the theophylline particles to melt, resulting in slower drug release from the matrices. The release rates appeared to rise as the formulations' theophylline content increased. Eudragit® RL and theophylline mixtures compacted using ultrasound-assisted compression at energies ranging from 15 to 150 J. A glassy matrix of amorphous excipient with embedded islands of theophylline crystals was visible in SEM microphotographs. *In vitro* dissolution tests revealed that compacts made using ultrasounds had a longer drug release time. The *in vitro* release rate of theophylline was found to be a function of ultrasound intensity and could be linked to tablet porosity, according to these researchers.

The Percolation Theory was proposed by our research group as a theoretical paradigm for interpreting the technological and biopharmaceutical distinctions between ultrasonic compressed and regular tablets. This theory, which is based on statistical physics, assumes that the system is based on a regular lattice. The "percolating cluster" and the "percolation threshold" are two of the theory's most intriguing concepts. The medicine and/or excipient in a matrix system produce a "percolating cluster," which is formed by particles of the same component that "contact" each other from one side to the other of the tablet, based on their respective volume ratios.

The "percolation threshold" for a medication and excipient is the concentration (v/v) over which a percolating cluster develops for the first time, and it usually reflects a system critical point. Close to this stage, significant changes in some system features, such as the active agent's release mechanism or the tablet structure, can be detected. The percolation threshold of several pharmacological systems has been estimated by several authors. To examine the impact of the compression process on drug release control, researchers used direct compression and ultrasound-assisted compression to create matrix systems from binary mixtures of KCl (potassium chloride – a model drug) and Eudragit® RS-PM with varied drug contents (10–90% w/w) [22].

7.9 SUMMARY

Ultrasound waves are used in a variety of industrial, medicinal, and environmental operations. The primary transducers are classified into narrow-band and broadband transducers. The ultrasonic waves are produced and detected using piezoelectric, magnetostrictive, capacitive or electrostatic, and electromagnetic electrical transducers. Ultrasonic waves find wide applications in pharmaceuticals for drying of granules, agglomeration process, preparation of solid dosage forms like tablets, powders, preparation of liquid dosage forms like suspension, controlled release oral suspension, emulsion, etc. Ultrasound plays important role in preparation of nutraceutical products, in food industry for the preservation, processing, and other development of food products.

KEYWORDS

- **dosage-form**
- **infrasound**
- **nutraceuticals**
- **pharmaceuticals**
- **transducers**
- **ultrasound**

REFERENCES

1. Alfonso-Muniozguren, P., Serna-Galvis, E. A., Bussemaker, M., Torres-Palma, R. A., & Lee, J., (2021). Review on pharmaceuticals removal from waters by single and combined biological, membrane filtration and ultrasound systems. *Ultrasonics Sonochemistry, 76*, E-article 105656.

2. Alcanox, (2008). *Applications in Pharmaceutical Manufacturing* (pp. 1–10). https://vertassets.blob.core.windows.net/download/b3a46178/b3a46178-db7d-440b-960c-0e979b8a36b2/applications_in_pharma_updated.pdf (accessed on 09 February 2023).
3. Brennan, J. G., & Grandison, A. S., (2011). Freezing; Chapter 4. In: Pardo, J. M., & Niranjan, K., (eds.), *Food Processing Handbook* (2nd edn., pp. 1–559). [online]; https://vdoc.pub/documents/food-processing-handbook-second-edition-6cg3ic822kn0 (accessed on 09 February 2023).
4. Chemat, F., Zill-e-Huma, B., & Khan, M. K., (2011). Applications of ultrasound in food technology: Processing, preservation and extraction. *Ultrasonics Sonochemistry, 18*(4), 813–835.
5. Chen, J., & Hu, L., (2020). Nanoscale delivery system for nutraceuticals: Preparation, application, characterization, safety, and future trends. *Food Engineering Reviews, 12*, 14–31.
6. Curie, P., (2007). History of medicine. *Current Oncology, 14*(2), 74–82.
7. Delgado, A. E., Zheng, L., & Sun, D. W., (2008). Influence of ultrasound on freezing rate of immersion-frozen apples. *Food and Bioprocess Technology, 2*(3), 263–270.
8. Delmas, H., & Barthe, L., (2015). Ultrasonic mixing, homogenization, and emulsification in food processing and other applications. *Power Ultrasonics, 25*, 757–791.
9. Dolatowski, Z. J., Stadnik, J., & Stasiak, D., (2007). Applications of ultrasound in food technology. *Acta Scientiarum Polonorum, 6*(3), 89–99.
10. Dougmac, (2021). *Nutraceuticals Container Washing Equipment*. https://www.dougmac.com/industries-served/nutraceuticals (accessed on 09 February 2023).
11. Falou, O., (2011). Modeling high frequency ultrasound scattering from cell sand. *Ultrasound Contrast Agents*, 80–96.
12. Gallego-Juárez, J. A., Rodríguez, G., Riera, E., & Cardoni, A., (2015). Ultrasonic defoaming and debubbling in food processing and other applications. In: *Power Ultrasonics* (pp. 793–814). Woodhead Publishing.
13. Gil-Chávez, G., Villa, J. A., Fernando, A. Z. J., Basilio, H. J., Sepulveda, D., Yahia, E. M., & González, G. A., (2013). Technologies for extraction and production of bioactive compounds to be used as nutraceuticals and food ingredients: An overview. *Comprehensive Reviews in Food Science and Food Safety, 12*(1), 5–23.
14. Gogate, P. R., (2015). The use of ultrasonic atomization for encapsulation and other processes in food and pharmaceutical manufacturing. In: *Power Ultrasonics* (pp. 911–935). Woodhead Publishing.
15. Golla, P. S., Johnson, H. W., & Senthilnathan, P. R., (1992). Application of electro acoustics for dewatering pharmaceutical sludge. *Environmental Progress, 11*(1), 74–79.
16. Graff, K. F., (1981). History of ultrasonics. In: Mson, W. P., (ed.), *Physical Acoustics* (Vol. 15, pp. 1–97). London: Elsevier B.V.
17. Hayward, L. H., Hunt, M. C., Kastner, C. L., & Kropf, D. H., (1980). Blade tenderization effects on beef longissimus sensory and Instron textural measurements. *Journal of Food Science, 45*(4), 925–935.
18. Hoskins, P., Martin, K., & Thrush, A., (2010). *Diagnostic Ultrasound Physics and Equipments* (2nd edn., p. 254). London – UK: Cambridge University Press.
19. Levina, M., Rubinstein, M. H., & Rajabi-Siahboomi, A. R., (2000). Principles and application of ultrasound in pharmaceutical powder compression. *Pharmaceutical Research, 17*(3), 257–265.

20. Lieberman, H. A., & Lachman, L., (1980). Pharmaceutical dosage, forms: Tablets. *Journal of Pharmaceutical Sciences, 70*(3), 340–349.
21. McDonnell, C. K., Lyng, J. G., & Allen, P., (2014). The use of power ultrasound for accelerating the curing of pork. *Meat Science, 98*(2), 142–149.
22. Millán-Jiménez, M., Galdón, E., Ferrero, C., & Caraballo, I., (2017). Application of ultrasound-assisted compression in pharmaceutical technology. Design and optimization of oral sustained-release dosage forms. *Journal of Drug Delivery Science and Technology, 42*, 119–125.
23. Nayar, S. K., & Benatar, A., (1989). Ultrasonic molding of plastic powders. *ASM International*, 139–145.
24. Newman, P. G., & Rozycki, G. S., (1998). The history of ultrasound. *Surgical Clinics of North America, 78*(2), 179–195.
25. Oralkan, O., Ergun, A. S., Johnson, J. A., & Karaman, M., (2002). Capacitive micromachined ultrasonic transducers: Next generation arrays for acoustic imaging, ultrasonics, ferroelectrics and frequency control. *IEEE, 49*(11), 1596–1610.
26. Pereira, S. V., Colombo, F. B., & De Freitas, L. A. P., (2016). Ultrasound influence on the solubility of solid dispersions prepared for a poorly soluble drug. *Ultrasonics Sonochemistry, 29*, 461–469.
27. Pham, T. D., Shrestha, R. A., Virkutyte, J., & Sillanpää, M., (2009). Recent studies in environmental applications of ultrasound. *Canadian Journal of Civil Engineering, 36*, 1849–1858.
28. Pohlman, F. W., Dikeman, M. E., Zayas, J. F., & Unruh, J. A., (1997). Effects of ultrasound and convection cooking to different end point temperatures on cooking characteristics, shear force and sensory properties, composition, and microscopic morphology of beef longissimus and pectoralis muscles. *Journal of Animal Science, 75*(2), 386–391.
29. Prajapat, A. L., Subhedar, P. B., & Gogate, P. R., (2016). Ultrasound assisted enzymatic depolymerization of aqueous guar gum solution. *Ultrasonics Sonochemistry, 29*, 84–92.
30. Preumont, A., & Mokrani, B., (2014). Electromagnetic and piezoelectric transducers. In: *Active and Passive Vibrator Control of Structures* (pp. 213–248). Springer, Vienna.
31. Pukale, D. D., Maddikeri, G. L., Gogate, P. R., Pandit, A. B., & Pratap, A. P., (2015). Ultrasound assisted transesterification of waste cooking oil using heterogeneous solid catalyst. *Ultrasonics Sonochemistry, 22*, 278–286.
32. Qiu, L., Zhang, M., Chitrakar, B., & Bhandari, B., (2021). Application of power ultrasound in freezing and thawing processes: Effect on process efficiency and product quality. *Ultrasonics Sonochemistry, 70*, E-article 105230.
33. Raval, D., Kabariya, J., Hazra, T., & Ramani, V., (2019). Review on electro spraying technique for encapsulation of nutraceuticals. *International Journal of Communication Systems, 7*(5), 1183–1187.
34. Riera, E., García-Pérez, J. V., Cárcel, J. A., Acosta, V. M., & Gallego-Juárez, J. A., (2011). Computational study of ultrasound-assisted drying of food materials. *Innovative Food Processing Technologies: Advances in Multiphysics Simulation*, 265–301.
35. Rodriguez, L., Cavallari, C., Passerini, N., Saettone, M. F., Fini, A., & Caputo, O., (1998). Evaluation of theophylline tablets compacted by means of a novel ultrasound-assisted apparatus. *International Journal of Pharmaceutics, 170*, 201–208.
36. Rozenberg, L. D., (2014). *Sources of High-Intensity Ultrasound* (p. 284). Cham: Springer.

37. Siwkiewicz, Z., & Stolarz, S., (1986). The effect of ultrasonic vibrations of the die on the density distributions of slender iron powder compacts. *International Journal of Powder Metallurgy, 18*, 407–408.
38. *Using Agglomeration.* [Online]. https://www.naturalproductsinsider.com/archive/using-agglomeration (accessed on 09 February 2023).
39. Vilkhu, K., Mawson, R., Simons, L., & Bates, D., (2008). Applications and opportunities for ultrasound assisted extraction in the food industry: A review. *Innovative Food Science & Emerging Technologies, 9*(2), 161–169.
40. Zhao, L., Zhao, G., Chen, F., Wang, Z., Wu, J., & Hu, X., (2006). Different effects of microwave and ultrasound on the stability of (all-E)-astaxanthin. *Journal of Agricultural and Food Chemistry, 54*(21), 8346–8351.
41. Smith, D. P., & Young, L. L. (2007). Marination pressure and phosphate effects on broiler breast fillet yield, tenderness, and color. Poultry Science, 86(12), 2666–2670.

CHAPTER 8

PRINCIPLES OF ULTRASONICATION PROCESSING IN THE FOOD INDUSTRY

ROSHNA KHAWAS and PRANJAL DUTTA

ABSTRACT

Food spoilage causes undesirable changes in nutritional and sensory characteristics of food due to infestation by insects, growth and activity of microorganisms, moisture loss, chemical reactions, or enzymatic changes. Thermal treatments are used to inactive bacteria and increase the shelf-life of foods, and these are robust methods that can cause loss of food nutrients, volatile compounds, and flavors. In recent years, non-thermal technology has gained more importance due to the increase in demand for fresh-like food products with minimal changes in food preserving all its important nutrients and properties. Ultrasonication is a non-thermal technology having the potential alternative to thermal processing technologies. It is based on the frequency produced by sound waves with a range greater than human hearing capacity. It produces fresh quality foods with high nutritive value due to retention of nutrients and sensory properties.

8.1 INTRODUCTION

Food spoilage occurs due to physical, chemical, or microbial activities. Texture, color, nutritive values, and edibility of foods are vulnerable to spoilage [1]. Food preservation includes various processing steps to maintain the food quality at the desired level. Traditional thermal processing of food includes pasteurization, sterilization, and drying. The food incorporates

Advanced Research Methods in Food Processing Technologies. Junaid Ahmad Malik, Megh R. Goyal, Preeti Birwal, & Ritesh B. Watharkar (Eds.)
© 2024 Apple Academic Press, Inc. Co-published with CRC Press (Taylor & Francis)

many nutrients having functional properties, such as pigments, antioxidants, bioactive compounds, vitamins, and minerals [26]. Thermal processing has some disadvantages, such as loss of flavor, color, freshness, and some nutritional properties. Innovative approaches are required for retention of these nutrients in food products.

Ultrasonication is an emerging non-thermal process to increase the shelf-life of foods by inactivating microorganisms or enzymes with high retention of nutrients and sensory attributes of processed foods. Ultrasounds are sound waves having a range greater than human hearing capacity (20 kHz or more). Ultrasonication is a versatile and evolving technology to decrease the operating time and cost. It guarantees food safety and quality [46]. When ultrasound is transmitted across a biological structure, it causes compressions and rarefaction of the medium particles, which result in high impartment of energy. A number of biochemical, chemical, and physical effects can be observed that are independent of the sound wave amplitude applied and frequency used [14].

Over the past few years, ultrasound technology has expanded in nearly every field including mineral processing, medical scanning ultrasonic therapy, non-destructive testing (NDT), surface cleaning, industrial welding, environmental decontamination applications, nanotechnology (NT), and food industries. Ultrasound method is also used as a processing agent in foam formation, mixing of materials, or agglomeration, destruction, and precipitation of airborne powders, enhanced extraction of the valuable compound from vegetables and food products, increase efficiency of filtration, improvement of drying and extraction techniques in solid materials [47].

Fruit and vegetable qualities have been enhanced with ultrasound treatment in pre- and post-harvest periods. Various versatile and profitable benefits obtained from the ultrasonic technology in the food industry further include non-destructive analytical method assessment of opaque fluid properties and foreign materials identification in food via food container walls and for investigating physicochemical and structural qualities of fluid foods by combining nuclear magnetic resonance (NMR) and low power ultrasound with spectroscopy [5]. In possible future, applications of ultrasound technology in the food industry have huge scope due to the higher purity of the final products, eliminating undesirable sensory quality in less time and with less energy compared to conventional processes [47].

This chapter focuses on the basic principles of ultrasonication, mechanisms of microbial inactivation, and how the technology works on microorganisms, spores, and enzymes. It also includes applications of ultrasonication in the food industry, and changes in food quality. It presents methods and equipment, along with the advantages and limitations of ultrasonication.

8.2 AIMS OF ULTRASONICATION

The applications of ultrasonication in food industries have two main targets: (i) assist traditional food processes; and (ii) replace thermal technologies. In the latter case, this technology enhances the efficiency of the process and improves the drawbacks of conventional food processing technologies [50]. The principle aim of this technology is to extend the shelf-life of food, improve its quality, save energy, and reduce the processing time [16].

8.3 HISTORY OF ULTRASONICATION

In 1790, the first detailed experiment was performed by Lazzaro Spallanzani on bats, which indicated that non-audible sound might exist with the discovery of echolocation used by the bats. Lazzaro performed numerous tests on bats to reveal that they use sound waves to navigate instead of sight [31]. In 1877, Jacques and Pierre discovered the piezoelectric effect, which was utilized in the design of the transducer for ultrasonic waves in the air and water. The successful use of piezoelectricity in the generation and detection of ultrasound waves in deep seawater was followed by further development [27]. Low-intensity ultrasound methods were used 60 years ago, to characterize foods; however, the potential of this method has been evaluated lately. Since the first real-world application of piezoelectricity in 1917 was trying to detect submarines using sound navigation and ranging technology also known as Sonar [27, 47].

In 1894, Thornycroft & Barnaby observed implosion of bubbles and cavities in water due to vibrations generated in the propulsion of missiles propelled by a destroyer, a phenomenon is known as cavitation [47]. Ultrasound is being rapidly developed in the field of research and successfully providing many uses in food industries, such as modification and analysis of food samples. Numerous areas have been recognized with great potential for future development, such as: drying, degassing, bleaching, crystallization, filtration, extraction, meat tenderization, freezing, homogenization, etc. [14]. The ongoing demand for low and high-frequency applications will bring many new and beneficial opportunities in the coming days.

8.4 WORKING PRINCIPLE OF ULTRASOUND TECHNOLOGY

8.4.1 GENERATION OF ULTRASOUND

Piezoelectric transducers are generally used for generating cavitation. The excitation frequencies associated with cavitation phenomena usually range from 20 kHz to 60 kHz. Liquid is unable to follow the oscillation of sound due to inertia; and therefore, low-pressure regions frequently occur in anti-nodes around the solid surfaces and also in the liquid [20]. The occurrence of cavitation and the formation of gas bubbles in the medium are due to pressure change in these regions. During the expansion cycle, these bubbles will have a large surface area that increases the diffusion of gas allowing bubbles to expand. At a certain point, ultrasonic energy delivered is not enough to hold the vapor phase within bubbles and thus condensation occurs rapidly. These condensed molecules collide and implode to create shock waves to generate high pressure and temperature about 50 MPa and 5,500°C, respectively. These implosions resulting in pressure change are the main bactericidal effects in ultra-sound method [26].

8.4.2 MECHANISM OF MICROBIAL INACTIVATION

Proper information on the physiological behavior of microorganisms against inactivation agents is important for the improvement of safe foods. It is essential in identifying critical factor, predicting microbial inactivation under wider range of conditions and also helps in understanding the kinetics of inactivation and effect of environmental factor on resistance [37]. Treatment with ultrasound can generate various bioeffects on cells that are *in vitro*, such as: alteration in cell structures, alteration in cell functions and cell lysis. Specifically, ultrasound may lead to modification of membrane and can cause damage to the reversible and irreversible cell.

During the US process, temporary opening and resealing of plasma membrane occurs, termed as sonoporation [56]. Production of free radicals, localized heating, and thinning of cell membrane are main mechanisms behind the destruction of microbial [26]. Cavitation is generated when ultrasound is applied to a liquid medium, and this phenomenon is responsible for the lethal effect required for microbial inactivation. The implosion of bubbles under high frequency sound wave generates high temperature and pressure, and free radicals and strong mechanical forces are formed [37]. Free radicals (which are produced from thermolysis of water, such as hydrogen peroxide (H_2O_2) and hydroxyl radical (OH^-)) are the main factors that cause cell membrane rupture and ultimately cell death [56].

8.5 ULTRASOUND METHODS IN THE FOOD INDUSTRY

Ultrasound is widely evolving technology in the field of the food industry and its application is constantly increasing in various food processing operations. However, ultrasound alone might not be sufficient for maximum destruction of microbial; but when combined ultrasound with heat or pressure or both may give effective results [46].

8.5.1 ULTRASONICATION (US)

In this method, ultrasound is applied at a low temperature. It is mainly used for products, which are heat sensible, where there is a concern about loss of nutrients, denaturation of protein, non-enzymatic browning, etc. To inactivate microorganisms and stable enzymes, it requires longer treatment time and thus requires high energy. Depending on an applied ultrasonic power and operating time, there might be an increase in temperature and accordingly needs to improve the process [47].

8.5.2 THERMOSONICATION (TS)

This method is a combination of ultrasonic power and thermal energy in which the food product is exposed to the medium heat. The effect of combining heat with sonication for microbial inactivation was found to be more effective than sonication without heat [4]. As an outcome of additional heat, the ultrasound generates an excessive amount of cavitation, which gives effective results on inactivation of microbial than the heat alone. Therefore, the combination of low-frequency ultrasound with mild heat will help in reducing the processing time and temperature by 55% and 16%, respectively [47].

8.5.3 MANOSONICATION (MS)

This method is a combination of ultrasound and pressure. High inactivation efficiency can be obtained by combining medium pressure and low temperature in comparison with ultrasound alone [46]. This is due to free radical production and higher bubble explosion. A study showed a reduction in D-value of *Listeria monocytogenes* in a citrate phosphate buffer from 5.70 to 2.50 minutes, when there was an increase in pressure from 0 to 200 kPa with ultrasound at 20 kHz and 90 μm. Another study reported that the D-value of *Yersinia enterocolitica* in a citrate phosphate buffer decreased from 1.52 to 0.28 minutes, when the pressure was increased from 0 to 300 kPa in a combination with ultrasound at 20 kHz and 150 μm. Hydrostatic pressure is responsible for this phenomenon as it enhances some effects, such as: the production of free radicals and sonoluminescence (the puzzling light emitted when small gas bubbles imploded in a liquid) [28].

8.5.4 MANOTHERMOSONICATION (MTS)

It combines and synergizes the ultrasound with moderate temperature and pressure to inactivate enzymes and microorganisms. This technique has been successful for the preservation of food owing to its potential to inactivate microorganisms and endogenous enzymes, while retaining nutrients and flavor [43]. In this method, the applied temperature and pressure will maximize the cavitation and give greater efficiency. MTS method inactivates various enzymes at low temperature and less time as compared to thermal treatments at the same temperature. Microorganisms that have high thermoresistant enzymes (such as: polyphenol oxidase, lipoxygenase (LOX), and peroxidase (PO)) are reported to be inactivated by MTS [47].

8.6 COMPONENTS OF ULTRASONICATION METHOD

8.6.1 ULTRASONIC GENERATORS

An ultrasonic generator stimulates the transducers. The generator receives the electrical energy from the power source and converts it into a suitable frequency and sends it to the transducers to stimulate it at the desired frequency [32]. The power requirements to drive transducers may vary, but usually, the ultrasonic generator needs to supply frequencies to the transducer at a voltage, which is higher than that delivered from the power source to accomplish the intended operation [18].

8.6.2 TRANSDUCER

The transducer is a piezoceramic material; when stimulated by an electric signal, it immediately changes its shape. When excited by a high-frequency electronic generator, the transducer vibrates at the resonant frequency and induces amplified vibrations of the diaphragm. This causes a series of compressions and rarefactions in the liquid in the tank [32]. There are three types of ultrasonic transducers which are discussed in subsections.

8.6.2.1 LIQUID-DRIVEN TRANSDUCERS

It is a liquid whistle, which generates vibration at ultrasonic frequencies when liquid is forced through a thin metal blade, and a high degree of mixing is obtained due to the cavitation effects in the liquid and rapidly alternating pressure [7]. At this point, cavitation mixing occurs in two ways:

- Firstly, via venture effect, on exiting the orifice the liquid rapidly grows into a larger volume; and
- Secondly, when the process material flows over the blade causing vibration.

The connection between blade and orifice is critically controlled to enhance the activity of the blade. Different shapes and sizes of the orifice determine the required operating pressure and output. The velocity can be changed to achieve the degree of dispersion and the required particle size [41].

8.6.2.2 MAGNETOSTRICTIVE TRANSDUCERS

Magnetostriction effect is a phenomenon in which internal stresses are created when a ferromagnetic material (such as cobalt, nickel or iron) is exposed to an external magnetic field, causing the change in shapes and dimensions of material [12]. A magnetostrictive transducer is a transducer, which utilizes the magnetostrictive phenomena and represents interconnection between mechanical and magnetic fields in ferromagnetic materials [33]. A large amount of driving force is delivered as the construction of the system is extremely robust. However, external cooling is required due to loss of electrical energy in the form of heat, which is about 40%. Its maximum operating frequency is limited to 100 kHz [46].

8.6.2.3 PIEZOELECTRIC TRANSDUCERS

It is an electrostrictive device that utilizes ceramic substances, such as barium titanate (BT), lead metaniobate (PN), and lead zirconate titanate (PZT). It is extremely effective (the transfer of acoustic energy is about 80–95%) but less robust as compared to magnetostrictive devices. It is unable to tolerate high temperature for a long period (not >85°C) [7]. Conversion of electro-mechanical energy occurs in a piezoelectric plate. To maximize the reflections at the interface in between the transducer and the nearby medium, single or multiple matching layers can be placed in front of it. To damp the transducer so that a wider range of frequency can be achieved, a backing is attached at the back of the transducer [29].

8.7 APPLICATIONS OF ULTRASONIC METHOD IN THE FOOD INDUSTRY

8.7.1 DEFOAMING

Foam is a multiphase media normally produced as an undesirable by-product by various industrial methods (Figure 8.1). Conventional defoaming methods use chemical, mechanical, electrical or thermal effects. The most effective method is a chemical method with antifoam agents, but they contaminate the process as well [21]. However, high power ultrasonic defoaming systems have proven to be very effective, it creates an ultrasonic static wave in the air to generate nodes and antinodes. The nodes attract matter and so do the foam bubbles, which implode as a result of the compression force generated [39].

8.7.2 DEGASSING/DEAERATION

A liquid incorporates gases in a blended state, such as dissolved nitrogen, oxygen, and carbon dioxide. Boiling and reducing pressure are two common ways for degassing, and slight changes in temperature are beneficial for ultrasound process [11]. Degassing is more noticeable, when a cleaning bath of ultrasonic field is frequently used inside with tap-water. It takes place due to the vibration of gas bubbles produced by acoustic waves. These bubbles expand till they reach the surface across the liquid against gravity. The foremost target of this process is to eliminate the air from the liquid before they create the problem like bubble formation during the processing of carbonated beverages. This method is mainly used for degassing of carbonated drinks, such as beer, Coca-Cola, Pepsi, etc. [46].

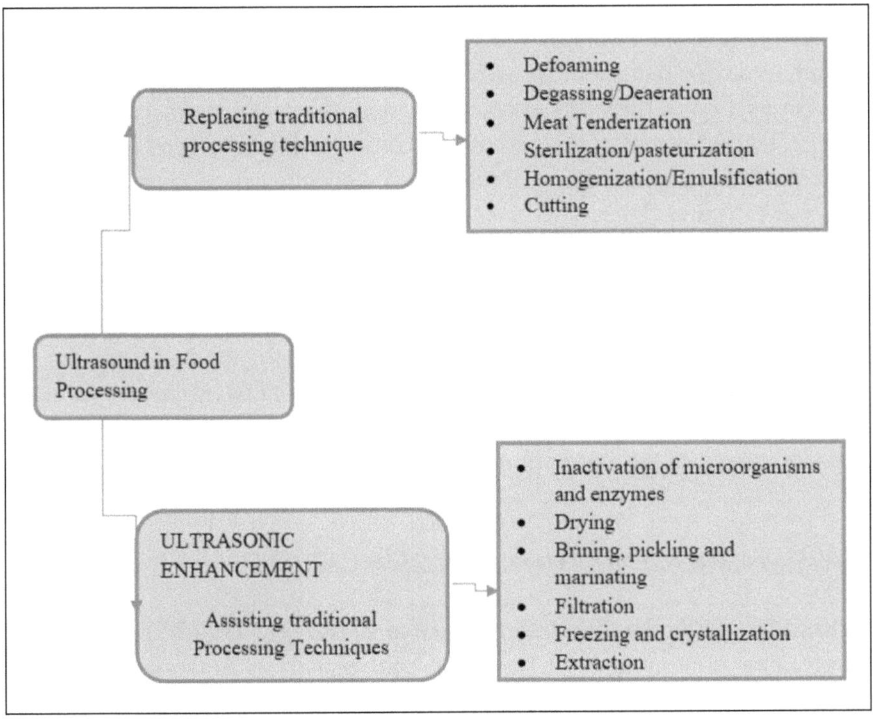

FIGURE 8.1 Classification of applications of ultrasound in food processing.

Source: Adapted with permission from Ref. [58]. Rights managed by Taylor & Francis

8.7.3 MEAT TENDERIZATION

Mechanical pounding is the traditional method for tenderization of meat that enhances the quality of meat. Application of ultrasound for meat tenderization has been beneficial and effective. It is done in two ways: it enhances enzymatic reactions through the biochemical effect or breaks the uprightness of muscular cells [11]. Tenderization of meat with ultrasound improves the properties of meat, such as: cohesiveness, tenderness, and water binding capacity [40]. Tenderness of meat leads to consumer's acceptance since inconsistent tenderness is the major problem in the meat sector. Moreover, ultrasound serves as an anti-microbial tool in the meat technology and has delivered great potential to improve the meat quality [5].

8.7.4 STERILIZATION/PASTEURIZATION

Bacterial spores are found to be more resistant as compared to vegetative cells to the specific approach to obtain sterilization. Sterilization means the inactivation of all forms of life in food. A process may induce sterility of vegetative cells but not of spores. Since bacterial spores are highly resistant life forms, any sterility process cannot ignore the possible presence of spores [51]. Thermal pasteurization and sterilization are mostly used methods till-date for inactivation of microorganisms and enzymes in food products; but it has some disadvantages, such as loss of nutrients, development of undesirable flavor and quality deterioration of food. Ultrasound can effectively inactivate enzymes (such as polyphenol oxides, pectin methyl esterase, peroxides, etc.), which are responsible for spoilage of fruit and vegetable juices. There is a growing interest in the dairy industry for application of ultrasound in pasteurization. It has been effective for destroying of *Listeria monocytogenes*, *Pseudomonas fluorescent*, and *E. coli* without any undesirable effect on the casein content of total protein of pasteurized milk [11].

8.7.5 HOMOGENIZATION/EMULSIFICATION

Ultrasonic cavitation for treatment of a two-phase liquid system can be performed continuously in a tank. The cavitation process results in a mechanism, which is effective in dispersion and homogenization in high instantaneous localized pressure pulse [48]. The emulsion obtained from this process includes distribution of sub-micron particles. This method consumes less energy in comparison with older traditional methods. This method is widely used in food industries for various products, such as mayonnaise, tomato ketchup, fruit juices, etc. [46].

8.7.6 ULTRASONIC CUTTING

Ultrasonic cutting is a size-reduction process, which utilizes the vibrational energy of power ultrasound that is created with the movement of the conventional blade [34]. Ultrasonic cutting is achieved by locating the nanoscopic feed motion of the product or the cutting device with a tiny vibration of the cutting tool. The cutting tool and the product interact with each other creating a number of effects. The concentration of primary energy in the separation zone and the moderation of contact friction through the tool flank rise up from the cyclic loading, which results in advantages, such as: smooth cut surface, reduced product deformation and reduced cutting force (CF) [19].

8.7.7 INACTIVATION OF MICROORGANISMS AND ENZYMES

Lowering the rate of food spoilage, primarily from the growth of spoilage microorganisms and enzyme activity and reducing pathogenic microbial numbers to an acceptably low level is the key element in a successful food preservation strategy [55]. Heat treatments are the most common methods for inactivation of microbial in food products. Thermal technology can inactivate some spores and vegetative microorganism but cannot inactivate heat resistant organisms [47]. Ultrasonication process inactivates microorganism including heat resistant microbial without causing any change in sensory and nutritional properties of food products. The ultrasonication required to achieve a 5-log reduction in microbial counts is based on guidelines by FDA (Food and Drug Administration).

Earlier, the electronic industry used ultrasound as disinfection treatment; but now in food industries, it is used as an alternative sanitization process [38]. The efficiency of enzymes inactivation by ultrasound is based on several factors, such as temperature, ultrasonic power, treatment time, frequency, antioxidant, pH, and free radicals. The ultrasonic inactivation of enzymes is usually due to protein denaturation, either by the formation of free radicals or by the formation and implosion of cavitating bubbles due to shear forces [25]. The decrease in the rate of enzyme inactivation is related to an increase in protein concentration and can be denatured by the presence of 2-mercaptoethanol. Based on the results obtained, it is concluded that the mechanism of enzymes inactivation is chemical rather than mechanical process [10].

8.7.8 ACOUSTIC DRYING

When drying is assisted by ultrasound, it is known as acoustic drying and has high commercial potential. Ultrasound generates nanoscopic channels in the material that permit easy vapor transportation from center to surface. In the drying process, sonication treatment allows low-temperature application than the traditional drying technology. To prevent alteration of color, flavor, and nutritional values, acoustic drying technology can be used to dry heat-sensitive foods [54]. Ultrasonic treatment, when used with hot-air drying process, can lower the input power to about 20% by increasing heat and mass transfer in comparison with non-treated ultrasonic drying methods. It can also result in a decrease in drying time and decrease in the level of final moisture by 16% and 6%, respectively [44].

8.7.9 PICKLING, BRINING, AND MARINATING

Pickling, brining, and marinating methods are applied to different types of meat and vegetables [46]. Ultrasound enhances both internal and external mass transfer. Brining is a mass transfer process that occurs in two ways: the transfer of water from meat to brine and transfer of solutes from brine to meat [11]. It also improves the water tissue and transfer of salt, which reduces the pickling time and helps in uniform salting [54].

8.7.10 FILTRATION

The procedure of separating solids from liquids is an essential process in the food industry, either to develop a solid-free liquid or to develop separated solid out of its mother liquor; but the major obstacle in this process is the settling of solid substances on the membrane filtration surface [50]. The vibrational energy created by ultrasound permits the particles to be kept in suspension and move continuously, facilitating the passage of solvent into pores and freeing the surfaces of the filters. It also produces a frictionless surface, which permits tiny particles and the liquid to pass through more smoothly, thus reducing process time and improving the flow [22].

8.7.11 FREEZING AND CRYSTALLIZATION

In the food industry, freezing is an essential method of food preservation, which helps in maintaining the quality of food materials and extends its shelf-life. It transforms water into ice crystals that preserve the food structure. Crystallization occurs in two stages: the initial stage is nucleation and the second stage is crystal growth, whose number and size will be determined by the freezing rate. Small ice crystals are formed with rapid freezing rates, whereas large ice crystals are formed with slow freezing rates, which deteriorate the quality of food and damage its physical structure (i.e., texture) [2]. Ultrasound aids in the crystallization process by controlling the nucleation and growth of crystals in frozen foods. According to laboratory investigation, power ultrasound can stimulate the freezing process of food products, mainly by increasing the heat and mass transfer process [50].

8.7.12 EXTRACTION

Extraction with ultrasound is termed as green extraction technology. The extraction process, when assisted by ultrasound, can reduce total extraction time, the energy consumed and the amount of solvent used and CO_2 emissions. Ultrasound aids extraction by reducing total processing time with high reproducibility [54]. The penetration of the solvent into the cells is enhanced by mechanical effects, it improves the cavitation and the transfer effect, which results in breakage of cell-wall causing it to release its contents into the medium. Higher yields are achieved in a shorter period and low temperature with the help of this technology. The technology can also modify and gives products of high quality, which are not possible using traditional thermal technology [22].

8.8 CLASSIFICATION OF APPLICATIONS

8.8.1 BASED ON THEIR MODE OF APPLICATION

8.8.1.1 DIRECT APPLICATION TO THE PRODUCT

In dehydration by air-borne ultrasonic radiation, the major obstacle is ill-match between acoustic impedance that causes low penetration of the acoustic energy in the food product. A new procedure is introduced to increase the impact of ultrasound on food dehydration process in which the application of ultrasonic vibration is in direct contact with the food product and it is applied together with a static pressure [23]. A study was carried out to observe main physical parameters and their relative impact in the process applying a power capacity of about 100 W and a stepped-plate transducer operating at 20 kHz. The results obtained using direct contact showed that new technology provides a very effective method for food dehydration [3].

8.8.1.2 COUPLING WITH THE DEVICES

Various couplants (such as water, glycerin, grease) are employed to transfer the ultrasonic energy along the conjoint between the test material and transducer. It excites the pulse-echo signal amplitude, which is referred to as signal height or echo height. This ultrasonic pulse-echo method is widely used for characterizing microstructure, detecting flaws in materials, assessing mechanical properties and measuring ultrasonic velocity. Appropriate selection of couplants is done by testing the condition or the environment, such as water; because corrosion water is not appropriate for the inspection of carbon steel; and to inspect the inclined place it may require viscous couplants [35].

8.8.1.3 SUBMERGENCE IN AN ULTRASONIC BATH

In this method before applying the sound waves in an ultrasonic cleaning tank of water, the complete submergence of the object to be cleaned is a must. Normally, the majority of baths work at approximately 40 kHz, and the transducer elements are fixed to the underneath of the tank. For such baths, the highest intensities are achieved throughout the bath depth at fixed levels [45]. Ultrasonic cavitation is the main mechanism responsible for energy transfer, in accordance with which bubbles are formed and implode creating shock waves. It can make this bath suitable for applications, such as cleaning and sterilization by forcing it to release surface impurity. It can also be used for cell separation, cell lysis (of fragile cells) and liquid degassing [15].

8.8.2 BASED ON INTENSITY OF SOUND WAVES

8.8.2.1 LOW POWER (HIGH FREQUENCY)

Low power ultrasound uses high frequency (2–10 MHz). It is largely used in medicine and chemical analysis as a diagnostic tool because of nondestructive quality of low-power ultrasound [30]. It mainly observes the composition and physicochemical properties of food products during their storage and various types of processing that improve their quality and control their properties [46].

8.8.2.2 HIGH POWER (LOW FREQUENCY)

High power ultrasound applications involve low frequency (20–60 kHz) and high power (10–300 W) [42]. Application of high-power ultrasound (HPU) generally depends on the effect induced by complex vibration in the propagating media to produce ultrasonic cavitation in liquids or biological tissue [30]. It involves biochemical, mechanical, and physical changes during various processes, such as: emulsification, freezing, drying, extraction, and inactivation of pathogens from the food products [46].

8.9 EFFECTS OF ULTRASONICATION

8.9.1 MICROORGANISMS INACTIVATION

Ultrasonic cavitation is caused by the changes in pressure that is responsible for the mechanism of ultrasound for microorganism inactivation. Localized changes during the cavitation process cause thinning and disruption of cell membranes, damages to DNA by producing free radicals and cell-wall breakdown. Structure of the membrane varies with the types of microorganisms, such as, Gram-negative bacteria consists of an outer membrane with thinner cell-wall, whereas Gram-positive bacteria lack membrane and have thicker cell-wall [16]. Gram-negative bacteria are less resistant to ultrasound in comparison with Gram-positive bacteria, as thicker cell walls of Gram-positive bacteria provide them with a better protection against ultrasound impacts [57].

8.9.2 SPORES INACTIVATION

Spores are generally found to be more resistant to ultrasonic waves. In food preservation method, inactivation of heat resistant fungal spores and bacteria is a great challenge due to their ability to survive against traditional thermal treatment, causing food-borne spoilage and diseases. In such cases, combination of hurdle technologies involving HPU, such as, thermosonication (TS), manosonication (MS), manothermosonication (MTS) are used [17]. The MS treatment at higher temperature (MTS) and combination of heat with ultrasound (TS) has been found to be effective on spore inactivation than traditional thermal treatment alone [17, 49].

8.9.3 *ENZYMES INACTIVATION*

Various enzymes can easily be inactivated by using ultrasonic energy. An acoustic streaming and shock waves are generated in the medium by the cavitation effect, which produces sufficient chemical, mechanical, and thermal effects for enzymes inactivation [25]. The application of ultrasound modifies the secondary and tertiary structure of protein due to Vander Waals interaction occurring in polypeptide chains and breakdown of bonding between hydrogens. During application of ultrasound, free radical production is necessary for enzyme inactivation. Barteri et al. [8] concluded that enzymes inactivation is caused by the formation of disulfide-linked aggregates due to the generation of free radicals during ultrasonic treatment [16].

8.9.4 *CHANGES IN THE QUALITY OF FOODS*

Ultrasound causes changes in the functional, chemical, and physical properties of food products and therefore can improve its quality [13]. These days, the main target of food technology is to reduce the loss of nutrients and other important sensory properties during storage and processing. Ascorbic acid (AA) is generally regarded as an indicator of nutrients quality, as it is not stable up against thermal treatment during storage and processing of foods.

Tiwari et al. [53] found that ultrasonic treatment results in higher retention of vitamin C of orange juice as compared to heat treatment. It is reported that fruit juices processed with the application of ultrasound have a minimal impact on its quality, such as, blackberry juice, orange juice, strawberry juice and guava juice [16]. In a meat, water retention capacity is an essential parameter with economic importance; and some authors have reported that ultrasonicated meat are similar to a meat in initial phase of postmortem with increased water-holding capacity; they also reported increase in the aging rate of meat because of changes in structure of myofibrillar proteins (MPs). Food viscosity can either decrease or increase depending on the ultrasound intensity while the effect being permanent or temporary.

In ultrasound-treated cornstarch granules, Huang et al. [24] found a decrease in viscosity due to the degradation caused by a partial cleavage of the glycosidic linkages resulting in a decrease in the molecular weight of the starch molecules. However, Bates et al. [9] reported that the reverse case is also possible in some vegetable purees. Due to an excessive penetration of moisture into the fiber network enabled by ultrasound treatment, there is an increase in the viscosity of tomato puree [52].

Ultrasonication technology is also effective in enhancing the hydrolysis of protein enzymatically and producing protein conjugates [38]. Various studies indicated that the physical and chemical effects of ultrasonication depend on exposure time, the food volume and food composition, the temperature reached by the tissues and the amplitude of the ultrasonic waves [13].

8.10 ADVANTAGES OF ULTRASONICATION METHOD

- Efficiency of the process and the drawbacks of thermal treatments are improved [50];
- It guarantees food safety and its quality [26];
- It reduces the operating time and expenses [26];
- Ultrasonic sound waves are safe and non-toxic [38];
- Ultrasonication process inactivates enzymes and microorganism including heat resistant microbial without causing any change in sensory and nutritional properties of food products [38];
- Ultrasound treated food have characteristics like untreated fresh product [22];
- Ultrasound when combined with heat or pressure or both may give effective results [46];
- Ultrasound-assisted washing of fresh produce has shown enhanced sanitation efficacy mainly due to the physical effects of acoustic cavitation [6].

8.11 LIMITATIONS OF ULTRASONICATION PROCESS

- Budding technology [47];
- High capital investment [36];
- Require skilled labor to operate;
- Results in modifications in physicochemical parameters and degradation of some compounds [50];
- Still under research and development [22];
- The free radicals formed during cavitation may cause harmful effect on the consumer [47].

8.12 SUMMARY

Considering all the characteristics, it can be concluded that ultrasonication technology has the ability to enhance the quality of food products, extend its shelf-life with minimal changes in nutritional and sensory properties. It may not be able to completely replace the traditional method of food preservation, but it enhances the efficiency when combined with other food preservation methods. In ultrasound technique, food treatment process can be done in a short time with high reproducibility, giving higher purity of the final product. More research studies are still needed for further development of this technology. However, continuous efforts are being made to improve fully automated ultrasound systems and to combine it with the food production lines to save energy, reduce cost and ensure the safety and the production of high value food products.

KEYWORDS

- **contamination**
- **degradation**
- **frequency**
- **infestation**
- **preservation**
- **ultrasound**

REFERENCES

1. Amit, S. K., Uddin, Md. M., Rahman, R., Rezwanul, S. M., & Khan, M. S., (2017). Review on mechanics and commercial aspects of food preservation and processing. *Agriculture and Food Security, 51*(6), 1–22.
2. Awad, T. S., Moharram, H. A., Shaltout, O. E., Asker, D., & Youssef, M. M., (2012). Application of ultrasound in analysis, processing and quality control of food. *Food Research International, 48*(2), 410–427.
3. Ahmed, J., & Rahman, M. S., (2012). *Handbook of Food Process Design* (Vol. 2, p. 1600). New Jersey-US: Blackwell Publishing Ltd.
4. Aadil, R. M., Zeng, X. A., Zhang, Z. H., Wang, M. S., Han, Z., Jing, H., & Jabbar, S., (2015). Thermosonication: A potential technique that influences the quality of grapefruit juice. *International Journal of Food Science, 50*(5), 1065–1282.

5. Boateng, E. F., & Nasiru, M. M., (2019). Application of ultrasound in meat processing technology. *Food Science and Technology, 7*(2), 11–15.
6. Bermudez-Aguirre, D., (2017). *Ultrasound Advances in Food Processing and Preservation* (p. 556). London-UK: Academic Press.
7. Brennan, J. G., (2006). *Food Processing Handbook* (p. 602). Weinheim-Germany: Wiley-VCH.
8. Barteri, M., Diociaiuli, M., Pala, A., & Rotella, S., (2004). Low frequency ultrasound induces aggregation of porcine fumarase by free radicals' production. *Biophysical Chemistry, 111*(1), 35–42.
9. Bates, D. M., Bagnall, W. A., & Bridges, M. W., (2006). Method of treatment of vegetable matter with ultrasonic energy. *US Patent Application Publication, 0110503*(A1), 1–5.
10. Coakley, W. T., Brown, R. C., James, C. J., & Gould, R. K., (1973). The inactivation of enzymes by ultrasonic cavitation at 20 kHz. *Archives of Biochemistry and Biophysics, 159*(2), 722–729.
11. Chemat, F., Huma, Z. E., & Khan, Md. K., (2011). Application of ultrasound in food technology: Processing, preservation and extraction. *Ultrasonic Sonochemistry, 18*(4), 813–835.
12. Collins, D., (2018). *How do Magnetostrictive Sensors Work?* (p. 3). Linear Motion Tips. https://www.linearmotiontips.com/how-do-magnetostrictive-sensors-work/ (accessed on 09 February 2023).
13. Carillo-Lopez, L. M., Alarcon-Rojo, A. D., Luna-Rodriguez, L., & Reyes-Villagrana, R., (2017). Modification of food systems by ultrasound. *Journal of Food Quality, 2017*(5794931), 1–12.
14. Dolatowski, Z. J., Stadnik, J., & Stasiak, D., (2007). Application of ultrasound in food technology. *Agricultural University of Lublin, 6*(3), 89–99.
15. Driscoll, A. O., (2018). *Differences Between Ultrasonic Baths and Ultrasonic Homogenizers* (p. 3). Homogenizers. http://homogenizers.net/blogs/blog/differences-between-ultrasonic-baths-and-ultrasonic-homogenizers (accessed on 09 February 2023).
16. Ercan, S. S., & Soysal, C., (2013). Use of ultrasound in food preservation. *Natural Science, 5*(8A2), 5–13.
17. Evelyn, & Silva, F. V. M., (2018). Perspectives of high-power ultrasound in food preservation. *Material Science and Engineering, 345*(1), 1–6.
18. Fuchs, J., (2012). *Ultrasonic-Ultrasonic Generator-Introduction* (p. 2). Cleaning Technologies Group. https://techblog.ctgclean.com/2012/01/ultrasonics-ultrasonic-generators-introduction/ (accessed on 09 February 2023).
19. Feng, H., Barbosa-Canovas, G. V., & Weiss, J., (2011). Ultrasound technologies for food and bioprocessing; Chapter 9. In: Schneider, Y., Zahn, S., & Rohm, H., (eds.), *Ultrasonic Cutting of Foods* (pp. 211–237). New York–US: Springer.
20. Gugulothu, S. K., Vijaya, K. P. R., & Deekshith, P., (2012). Exploring cavitation phenomenon with and without ultrasonic transducer. *Procedia Engineering, 38*, 154–164.
21. Gallego-Juarez, J. A., Rodriguez, C. J. C., Riera, E., Cardoni, A., & Graff, K. F., (2015). Ultrasonic defoaming and debubbling in food processing and other applications; Chapter 26. In: Gallego-Juarez, J. A., & Graff, K. F., (eds.), *Power Ultrasonic Application of High-Intensity Ultrasound* (pp. 793–814). Kidlington-UK; Woodhead Publishing.
22. Gallego-Juarez, J. A., Rodriguez, C. J. C., Moraleda, G., & Yang, T. S., (1999). New high-intensity ultrasonic technology for food dehydration. *Drying Technology; An International Journal, 17*(3), 597–608.

23. Gallo, M., Ferrara, L., & Naviglio, D., (2018). Application of ultrasound in food science and technology. *Multidisciplinary Digital Publishing Institute (MDPI) Journals, 7*(164), 1–18.
24. Huang, Q., Li, L., & Fu, X., (2007). Ultrasound effects on the structure and chemical reactivity of cornstarch granules. *Starch, 59*(8), 371–378.
25. Islam, Md. N., Zhang, M., & Adhikari, B., (2014). The inactivation of enzymes by ultrasound–a review of potential mechanics. *Food Reviews International, 30*(1), 1–21.
26. Jan, A., Sood, M., Sofi, S. A., & Norzom, T., (2017). Non-thermal processing in food applications. *International Journal of Food Science and Nutrition, 2*(6), 171–180.
27. Jabar, F. H. A., Mohammad, J. I., Zain, F. M., & Hasan, A. B., (2018). Data exfiltration of ultrasonic signal in computer security system. *Indonesia Journal of Electrical Engineering and Computer Science, 10*(2), 480–497.
28. Jia, J., Liu, D., & Ma, H., (2019). Advances in food processing technology; Chapter 7. In: Liu, D., Ma, X., Wang, W., Zou, M., Wang, D., & Ling, J., (eds.), *Research Progress on Power Ultrasound Technology* (pp. 149–187). Hangzhou-China: Springer Nature Singapore Pvt. Ltd. and Zhejiang University Press.
29. Jensen, H., (1986). *Calculations for Piezoelectric Ultrasonic Transducers* (Vol. 536, pp. 83–97). Rise National Laboratory, Denmark.
30. Jayasooriya, S. D., Bhandari, B. R., Torley, P., & D'Arcy, B. R., (2007). Effect of high-power ultrasound waves on properties of meat. *International Journal of Food Properties, 7*(2), 301–319.
31. Kane, D., Grassi, W., Sturrock, R., & Balint, P. V., (2004). A brief history of musculoskeletal ultrasound: 'from bats and ships to babies and hips.' *British Society for Rheumatology, 43*(7), 931–933.
32. Kohli, R., & Mittal, K. L., (2009). Development in surface contamination and cleaning; Chapter 6. In: Awad, S. B., & Nagarajan, R., (eds.), *Ultrasonic Cleaning* (Vol. 2, pp. 226–278). New York–USA: William Andrew.
33. Kim, Y. Y., & Kwon, Y. E., (2015). Review of magneto strictive patch transducers and applications in ultrasonic non-destructive testing of waveguides. *Elsevier – Ultrasonics, 62*, 3–19.
34. Kentish, S., & Feng, H., (2014). Applications of power ultrasound in food processing. *Annual Review of Food Science and Technology, 5*, 263–284.
35. Kim, Y. H., & Song, S. J., (2002). A study on the couplant effects in contact ultrasonic testing. *Journal of the Korean Society for Nondestructive Technology, 22*(6), 621–626.
36. Luque De, C. M. D., & Capote, F. P., (2006). *Analytical Application of Ultrasound* (1st edn., p. 350). Kidlington-UK: Elsevier Science.
37. Manas, P., & Pagan, R., (2005). Microbial inactivation by new technologies of food preservation. *Journal of Applied Microbiology, 98*(6), 1387–1397.
38. Majid, I., Nayik, G. A., & Nanda, V. (2015). Ultrasonication and food technology: A review. *Cogent Food & Agriculture, 1*(1), 1071022.
39. Massalfassar, Valencia, (2019). *Ultrasonic Defoaming* (p. 3). Ultratecno Power Ultrasonic Cleaning. https://www.ultratecno.eu/technology/technology-high-power-ultrasound-hpu/ultrasonic-defoaming/ (accessed on 09 February 2023).
40. Madhu, B., Srinivas, M. S., Srinivas, G., & Jain, S. K., (2019). Ultrasonic technology and its application in quality control, processing and preservation of food. *Current Journal of Applied Science and Technology, 32*(5), 1–11.

41. Mason, T. J., & Lorimer, J. P., (1989). An introduction to sonochemistry. *Elsevier – Endeavour, 13*(3), 123–128.
42. O'Daly, B., Morris, E., Gavin, G. P., O'Byrne, J., & McGuinness, G., (2008). High-power low-frequency ultrasound: A review of tissue dissection and ablation in medicine and surgery. *Journal of Material Processing Technology, 200*(1–3), 38–58.
43. Puri, R. K., Gandhi, G., Shashank, C. G., & Kaur, T., (2018). Mano-thermo-sonication in food preservation. *International Journal of Current Microbiology and Applied Sciences, 7*(07), 2743–2753.
44. Pakbin, B., Rezael, K., & Haghighi, M., (2015). An introduction review of application of ultrasound in food drying process. *Journal of Food Processing & Technology, 6*(1), 1–3.
45. Povey, M. J. W., & Mason, T. J., (1997). *Ultrasound in Food Processing* (p. 282). London – UK: Blackie Academic & Professional.
46. Rana, A., Meena, & Shweta, (2017). Ultrasonic processing and its use in food industry. *International Journal of Chemical Studies, 5*(6), 1961–1968.
47. Ravikumar, M., Suthar, H., Desai, C., & Gowda, S. A. J., (2017). Ultrasonication: An advanced technology for food preservation. *International Journal of Pure & Applied Bioscience, 5*(6), 363–371.
48. Rani, B., Singh, U., Prasad, M., Chauhan, A. K., & Maheshwari, R., (2012). Utilization of ultrasound technological advances in the food industry. *International Research Journal of Pharmacy, 3*(3), 125–127.
49. Raso, J., Palop, A., Pagan, R., & Condon, S., (1998). Inactivation of *Bacillus subtilis* spores by combining ultrasonic waves under pressure and mild heat treatment. *Journal of Applied Microbiology, 85*(5), 849–854.
50. Sani, A., & Sardarodiyan, M., (2016). Ultrasound applications for the preservation, extraction, processing and quality control of food. *Biotechnology an Indian Journal (BTAIJ), 12*(4), 162–174.
51. Sierra, G., & Boucher, R. M. G., (1971). Ultrasonic synergistic effects in liquid-phase chemical sterilization. *American Society for Microbiology, 22*(2), 160–164.
52. Soria, A. C., & Villamid, M., (2010). Effect of ultrasound on the technological properties and bioactivity of food. *Trends in Food Science & Technology, 21*(7), 323–331.
53. Tiwari, B. K., O'Donnell, C. P., Muthukumarappan, K., & Cullen, P. J., (2009). Ascorbic acid degradation kinetics of sonicated orange juice during storage and comparison with thermally pasteurized juice. *LWT–Food Science and Technology, 42*(3), 700–704.
54. Unver, A., (2016). Applications of ultrasound in food processing green chemistry & technology letters. *GIAP Journals, 2*(3), 121–126.
55. Villamiel, M., Montilla, A., Garcia-Perez, J. V., Carcel, J. A., & Benedito, J., (2017). Ultrasound in food processing; Chapter 9. In: Arroyo, C., & Lyng, J. G., (eds.), *The Use of Ultrasound for the Inactivation of Microorganisms and Enzymes* (pp. 258–278). Chichester-UK: Willey Blackwell.
56. Wang, P., Li, Y., Wang, X., Guo, L., Su, X., & Liu, Q., (2017). Membrane damage effect of continuous wave ultrasound on K562 human leukemia cells. *American Institute of Ultrasound in Medicine, 31*, 1977–1986.
57. Zaran, H., Jambrak, A. R., Vukusic, T., & Visnja, S., (2016). Effects of the high-power ultrasound on microorganisms in fruit juices. *MOJ Food Processing & Technology, 2*(5), 1, 2.
58. Tao, Y., & Sun, D. W. (2015). Enhancement of food processes by ultrasound: a review. Critical reviews in food science and nutrition, 55(4), 570–594.

CHAPTER 9

POTENTIAL OF THERMOSONICATION PROCESSING IN THE FOOD INDUSTRY

SANDHYA SINGH, BALDEV SINGH KALSI, and
MOHAMMED SHAFIQ ALAM

ABSTRACT

Thermosonication is a potential substitution to conventional thermal treatments. Thermosonication is the process of utilization of a combination of ultrasound (sonication) and heat. It can amplify the rate of inactivation of microbes and enzymes. This chapter summarizes the principle and mechanism of thermosonication. Further, the documentation of the effect and application of thermosonication on different food products has also been presented. Thermosonication requires further research studies for the purpose of industrial scale implementation along with the in-depth and clear interpretation of the effect of thermosonication on properties of food.

9.1 INTRODUCTION

The traditional thermal techniques like pasteurization and sterilization are being utilized to inactivate the microbes and enzymes that establish safety and longer shelf-life of food products. However, these techniques have drawback of exposure to high temperature for many processed foods which leads to unfavorable change of nutritional properties (vitamins and proteins) and sensory properties (texture, color, and smell). This urges the need of new targeted preservation methods that will have similar lethal effect on microbes and enzymes as the thermal treatments. Ultrasound is one of the promising

Advanced Research Methods in Food Processing Technologies. Junaid Ahmad Malik, Megh R. Goyal, Preeti Birwal, & Ritesh B. Watharkar (Eds.)
© 2024 Apple Academic Press, Inc. Co-published with CRC Press (Taylor & Francis)

non-thermal technologies that have the significance in increasing the shelf of the products by inactivating the microbes and enzymes. Ultrasound is apparently the easy and flexible technology for producing the extracts and destroying the cells [23, 50]. Because ultrasound alone has limited effect on the inactivation of microorganisms, other preservation techniques can also be used in combination with this [12].

Thermosonication (TS) is ultrasound treatment combined with moderate heat. This combination method hastens the process of sterilization of food and consequently reduces the time and intensity of heat treatment along with deterioration of food quality. Thermosonication is an exceptional alternative to thermal treatment; and it minimizes quality loss of the tomato juice [69] and watercress [17].

The purpose of this chapter is to explore the effect of the combination of ultrasound with moderate heat (also known as thermosonication) on the shelf-life of different food products while maintaining their quality parameters. This chapter also includes the principle, mechanism, and application of thermosonication. This also comprises a detailed review of the effects of thermosonication on the different quality parameters of food products in addition to the destruction of spoilage causing microorganisms and enzymes.

9.2 PRINCIPLE AND MECHANISM OF THERMOSONICATION

Ultrasound is described as a pressure wave with a frequency of 20 kHz or more [16]. Commonly, ultrasound generating apparatus uses the frequencies in range of 20 kHz–10 MHz, which is above the range of human hearing (>16 kHz) [11]. When the ultrasound is applied to the food, the compression and rarefaction of the medium particles produce large amount of energy. This leads to a complex phenomenon known as cavitation, which mainly includes creation, expansion, and bursting of bubbles (Figure 9.1) [14, 56]. The collapse of bubbles produces intense temperature (5,000 k) and pressure (1,000 atm) in the cavitation zone [63]. This breaks the compounds in the liquid leading to dispersion of particles and cell disruption with a localized effect of sterilization or pasteurization [2]. The moderate thermal treatment combined with cavitation shows additional effect on heat-shocked cell-wall, which increases the rate of inactivation of microorganisms and enzymes [18].

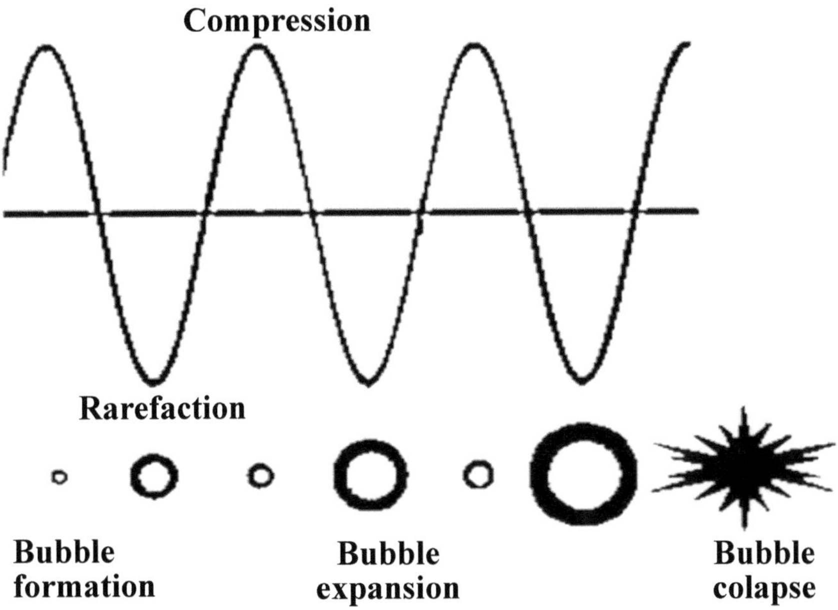

FIGURE 9.1 Ultrasound leading to cavitation.

Source: Adapted **from Ref. [63]**

9.3 APPLICATIONS OF THERMOSONICATION IN FOOD INDUSTRY

9.3.1 INACTIVATION OF MICROORGANISMS

The thermosonication treatments are quite effective on microbial inactivation in different foods. The ultrasound's amplitude, sonication time and thermal temperature affects the destruction of microbial [2]. The most acceptable theory about the inactivation of microorganisms is the phenomenon of cavitation, which can be categorized as stable cavitation and transient cavitation. The transient type involves mechanical effects like high pressure and temperature in localized area, while stable type shear stress is caused by steady implosion and blast bubbles surrounding cell [10]. The combination of heat with ultrasound has been known to increase the effectiveness of inactivation of microorganism in contrast to ultrasound alone. The carrot juice containing pulp showed complete inactivation of microorganisms (total bacteria count and yeast and molds count) when it was thermosonicated at 60°C for 5 minutes [3].

Escherichia coli (*E. coli*) and *Saccharomyces cerevisiae* (*S. cerevisiae*) were reduced by 5 log cycles in blood fruit (*Haematocarpus validus*) juice, when it was treated with thermosonication 50°C for 21 minutes. Though similar reduction was obtained for pasteurization, yet the fruit quality was retained maximum for thermosonicated juice samples than pasteurized ones when stored for 10 days [60]. *S. cerevisiae* inactivation was higher for beer when treated with thermosonication than the thermal processing. In addition, the processing time was also reduced. At temperature of at 55°C, thermosonication took only 26s in comparison to 38 minutes for thermal processing [43].

The treatment of ultrasound alone showed a lower impact on the inactivation of *Alicyclobacillus acidoterrestris.* The synergetic effect was seen, when apple juice was treated with ultrasound at 95°C, because the inactivation of spore was higher than the thermally treated juice alone at 95°C [66]. While investigating the effect of dual frequency thermosonication at 20/40 kHz on orange juice, spores and vegetative cells of *Alicyclobacillus acidoterrestris* were completely destroyed while retaining the juice quality [68].

The heat resistant molds (such as: *Byssochlamys nivea* (*B. nivea*) and *Neosartorya fischeri* (*N. fischeri*)) are generally linked with the deterioration of pasteurized fruit products and have the capacity to produce mycotoxins. The research study was conducted for the effect of thermosonication (24 kHz) at 75°C on *B. nivea* in strawberry puree and *N. fischeri* in apple juice and the study showed that thermosonication was more resistant to spores of *N. fischeri* than the *B. nivea* [62].

The effect of Thermosonication on camu-camu (*Myrciaria dubia*) nectar was not possible to confirm, because of the small value of pH in addition to low contamination of the pulp [24]. Nisin-assisted thermosonication at 52°C can significantly decrease the microbial load in apple juice [38].

9.3.2 ENZYME INACTIVATION

The enzymes are inactivated primarily due to the phenomenon of cavitation in ultrasound method. These intense conditions break the hydrogen bonds and van der Waal's forces in the polypeptide chain, which leads to the loss of the biological activity of enzymes. Moreover, the extreme pressure and temperature near cavitation generates hydroxyl and hydrogen-free radicals from water molecules. There is a possibility of reaction of these independent radicals with amino acids, which help in the stability of enzymes, binding of substrate and action of catalyst and thus changing their biological activity [42]. The alone application of ultrasound below the temperature of 50°C is not efficient in terms of inactivation of several of food enzymes due to their heat-resistance [10].

Table 9.1 illustrates the inactivation of enzymes with thermosonication in different food products. The inactivation of enzymes gets masking effect from lower temperature, while the reduction in relative enzyme activity is due to exposure to ultrasound with increasing time [19]. PME activity in mango and jackfruit smoothie was reduced and was significantly influenced by amplitude and temperature conditions [8]. Combining thermosonication with another non-thermal technology (like high pressurized carbon) can be used to inactivate the PME and PPO enzymes in apple juice [31].

The POD causes quality loss in fruits and vegetable because it deteriorates the phenolic substances and cause discoloration [15]. Moreover, POD is linked with the production of off-flavors in raw and unbalanced fruits [45]. Thermosonication showed increment rate in inactivation of peroxidase (POD) in contrast to conventional thermal treatment [6, 22]. Moreover, both PPO and POD enzymes should be taken into consideration for the thermosonication, because these show incredibly alike resistance [21]. POD in carrot juice was inactivated to a greater degree by thermosonication and the major change was observed in dual-frequency 22/40 kHz [29]. The enzymes (like peroxidases (POD), poly phenol oxidase (PPO) in addition to pectin methyl esterase (PME)) have high activity in *Haematocarpus Validus* (blood fruit).

During processing, the releasing of these enzymes lowers the quality in low time duration. The inactivation of these enzymes in blood juice can be achieved by thermosonication treatment at a temperature of 55°C for about 5 min with 1.4 W/ml of ultrasound power load. The inactivation of PME, POD, and PPO was about 92.88%, 92.81% and 91.6%, respectively [55].

9.3.3 QUALITY PRESERVATION

The star fruit juice thermosonicated at 44 kHz had considerable rise in the quality parameters (such as: antioxidant activity, total phenolic content, ascorbic acid (AA), total flavonoid content, and cloud index as well as browning index) as compared to samples pasteurized at 90°C [45].

TABLE 9.1 The Inactivation of Enzymes Using Thermosonication

Type	Enzyme	Ultrasonic Equipment	Experimental Conditions	Results	References
Apple juice	PME PPO	13 mm probe, 750 W	20 kHz, 25 to 100% amplitude; Temperature: 44 to 67°C for enzyme inactivation kinetics	Almost inactivation of PPO (residual activity: 3 ± 1%) at 67°C; PME was more resistant.	[32]
Blood fruit juice	PME POD PPO	2.2, 1.4 W/ml	20 kHz at 18% amplitude at temperature: 44°C and 55°C; Time: 5 and 10 minutes	Reduction in residual activity of PME, POD, and PPO enzymes to 6.50%, 6.18%, and 5.38%, respectively.	[55]
Custard apple juice	POD PME	67.84 W/cm²	20 kHz, at temperature of 30±1°C; Time: 0 to 40 minutes	0% residual activity; complete inactivation of POD and PME was achieved within 40 min and 50s.	[19]
Mango juice	PPO POD PME	250 W	Temperature: 25, 45, 65, and 95°C; Time: 10 min	Residual activity of PME, PPO, POD; and was 2.24, 3.47 and 1.61%, respectively at 95°C	[20]
Spinach juice	POD PPO	600 W	30 kHz, 50% duty cycle; Temperature 60±1°C; Time: 20 minutes	Notable inactivation in POD and PPO	[41]
Strawberry nectar	PPO	150 W	20 kHz, Temperature: 25–75°C; Time: 0.1–15 min	About 90% inactivation was achieved	[25]

9.3.3.1 EFFECT ON PHYSICOCHEMICAL PROPERTIES

The effect on titratable acidity (TA), pH, and total soluble solids (TSS) of food products is summarized in Table 9.2. Non-significant change in TSS, pH and TA was reported by most authors when food product was treated with thermosonication. However, the thermosonication treatment on camu-camu nectars showed an increase in TA, which could be associated with the rupturing of cell membrane and releasing of organic acids [24]. The TSS of the tangerine juice was increased during the thermosonication at 60°C, which could be associated with intracellular compounds being released due to the cavitation [7].

TABLE 9.2 Effects of Thermosonication on the Physical Properties of Food Products

Product	Ultrasonic Conditions	Results
Camu-Camu nectar	40 kHz; Temperature: 40, 50, 60°C; Time: 30 or 60 min	No significant alteration in pH, titratable acidity TA) was increased [24].
Haematocarpus Validus (blood juice)	2.2 or 1.4 W/ml, 20 kHz at 18% amplitude; Temperature: 44 and 55°C; Time: 5 and 10 minutes	No significant change in TSS, pH, TA [55].
Hog plum juice	400 W, 40 kHz; Temperature: 40 to 60°C; Time: 5 to 30 minutes	No Significant change in pH, TSS, and TA [47]
Star fruit juice	44 kHz, full power of 600 W; Temperature: 25–45°C; Time: 15, 30, 45, and 60 minutes	No significant change in pH, TSS, and TA [45].
Tangerine juice	35 kHz frequency; Temperature: 50 or 60°C; Time: 5 and 10 min	No significant alteration in pH and TA; increment in TSS [7].

9.3.3.2 COLOR

Color is considered as an essential quality indicator for the consumer's first choice and serves as quality checking for microbial growth during different stages of processing and storage of fruits and vegetables (FAV) [2, 10]. The change in color of different food products is shown in Table 9.3. The increase in color values could be due to the reason that the cavitation might have precipitated unstable suspended particles to some extent, which instigated additional colored compounds [47]. While the decrease in color values at high temperatures and extended time of heating could be correlated with Maillard reaction [65]. The decrease in the values of a^* and b^* of camu-camu nectar could also be related to the degradation of anthocyanins [24].

TABLE 9.3 Effects of Thermosonication on the Color of Food Products

Product	Ultrasonic Conditions	Result
Camu-Camu nectars	40 kHz; Temperature: 40, 50, 60°C; Time: 30 or 60 min	Increase in *L**; decrease in *a** and *b** [24]
Haematocarpus Validus juice	2.2, 1.4 W/ml, 20 kHz at 18% amplitude; Temperature: 40 and 55°C; Time: 5 and 10 minutes	Increase in *L**, *a** and *b** [55].
Hog plum juice	400 W, 40 kHz; Temperature: 40, 50, 60°C; Time: 5, 10, 20, 30 minutes	Increase in *L** and *b**; but decrease in values at 60°C [47]
Tangerine juice	35 kHz frequency; Temperature: 50 or 60°C; Time: 5 min or 10 min	Increases in the *L** for 50/5 minutes [7].
Tomato juice	250 W at ultrasonic frequencies: 0, 35, or 130 kHz; Temperature: 20, 40, or 60°C; Time: 0, 5, or 10 minutes	Increase in *L** and *a**; and decrease in *b** [37].

9.3.3.3 ASCORBIC ACID (AA)

AA (Table 9.4) is a form of unstable compound, which gets oxidized without difficulty on extended heating and high heat treatment [2].

Retention of AA is utilized as a quality indicator. Table 9.4 recapitulates the effect of thermosonication on AA of food products. The thermosonication of hog plum (*Spondias mombin* L.) juice at 40°C and 50°C showed significant increase in AA; and a decrement at 60°C [24]. The increase in AA might be associated with the effective elimination of dissolved oxygen that leads to stability of AA during the phenomenon of cavitation [1]. Similar results were observed in camu-camu nectars [24]. Another theory for the increase of AA after thermosonication is the rupturing of cell membrane during cavitation phenomenon, which could lead to larger amount of bioactive compounds and low temperature leading to less degradation of these compounds [5].

9.3.3.4 BIOACTIVE COMPOUNDS

The effect of thermosonication on bioactive compounds can be arranged in descending order based on their importance: ultrasound frequency, temperature, and exposure time [10, 44]. Table 9.5 depicts the effect of different thermosonication treatments on different biocompounds of food products. The bioavailability of compounds has been improved by thermosonication.

TABLE 9.4 Effects of Thermosonication Treatment on the Ascorbic Acid of Food Products

Product	Ultrasonic Conditions	Result
Camu-Camu nectars	40 kHz; Temperature: 40, 50, 60°C for 30 or 60 min	Significantly increase in contrast to control only at 40°C [24].
Haematocarpus Validus juice	2.2, 1.4 W/ml, 20 kHz at 18% amplitude; temperature: 44 and 55°C for 5 and 10 minutes.	Decrease at 55°C and increase at 44°C [55].
Hog plum juice	40 kHz, 400 W; Temperature: 40 to 60°C; Time: 5 to 30 minutes	Significantly increase as compared to control at 40 and 50C; and decrease at 60°C [47].
Quince (*Cydonia Oblonga*) juice	26 kHz, amplitudes: 40%, 45%, 50%, 55% and 60%; Temperatures: 30, 35, 40, 45 and 50°C; Time: 2, 4, 6, 8 and 10 minutes).	Decrease in ascorbic acid [71].
Star fruit juice	44 kHz, full power of 600 W; Temperature: 25, 45°C; Time: 15, 30, 45, and 60 min.	Increase in ascorbic acid [45].
Strawberry nectar	150 W, 20 kHz; Temperature: 25–75°C; Time: 1, 15 min.	At low temperature: ascorbic acid was increased, and with an increase of ultrasound energy density at elevated temperature, ascorbic acid was decreased [25].
Tomato paste	87.52 W/cm^2, 500 W, 20–25 kHz, Temperature: 22 and 65°C; Time: 10 min.	Thermosonicated samples showed higher ascorbic acid than the thermally treated sample [28].

The increase in total phenolic compounds and carotenoids may be associated with the destruction of cell membrane and eventually induce these compounds [10]. This increase in total phenolic content may be due to the hydroxyl group's inclusion to the aromatic compounds [45]. The flavonoid increases in Camu-Camu nectars; but with a high temperature of 60°C, the value of flavonoid was decreased. It was reported that the decrease in flavonoid due to elevated temperature may be associated with excessive cavitation, which has a negative effect on the amount of biocompounds [52]. An increase of DPPH (2,2-diphenyl-1-picrylhydrazyl) or antioxidant activity was reported in star fruit under thermosonication, which may be associated with generation of hydroxyl ions along with hydroxylation of flavonoids [13, 45].

9.3.4 EXTRACTION OF BIOCOMPONENTS

Thermosonication is a potential technology for the extraction of important biocomponents from the food and vegetables and its waste materials. Table 9.6 shows the effect of thermosonication on the extraction of essential components from different foods. The combination of ultrasound with temperature gave a remarkable yield of anthocyanin from the pomace of black carrot [4]. Ultrasound-assisted extraction (UAE) or thermosonication extraction has shown better results as compared to conventional methods, such as Soxhlet extraction. The yield and quality of castor oil extracted using thermosonication was superior compared to Soxhlet extraction. Moreover, the extraction time was reduced from 8 hours to 35 minutes and yield was increased from 57% to 62% [39]. The extraction of blood fruit juice using thermosonication at optimum conditions of temperature, ultrasound power, and time of 59.9C, 118 Watt/cm^2, and 7 minutes, respectively, gave more than 70% juice yield [60].

The extraction of pectin from different types of varieties (White and pink or red) of grapes showed higher content of galacturonic acid in conventionally heated samples than in the thermosonicated samples [36]. The thermosonication used to extract the bioactive compounds from grape peels showed superior results compared to conventional aqueous extraction. The thermosonicated samples showed higher values of antioxidant capacity (AC), anthocyanin, and total phenolic for about 1.4, 1.3, and 1.2 times, respectively in comparison to traditional aqueous extraction method [58].

TABLE 9.5 Effects of Thermosonication Treatments on the Different Compounds of Food Products

Product	Ultrasonic Conditions	Result
Almond-based beverage	13-mm probe, 19 kHz, acoustic power: 4.6, 8.5 and 14.5 W, temperature: 50°C Time: 5, 10, and 15 min.	• Reduction of 10% in total phenolic content in contrast to untreated sample • Increase in total flavonoid content • DPPH was similar to control sample [64]
Blueberry juice	Power: 200, 400, and 600 W, 20 kHz at 50% duty; Temperature: 60C Time: 15 min.	• Increase in total phenolic content and total flavonoid [70].
Camu-Camu nectars	40 kHz; Temperature: 40, 50, 60°C for 30 or 60 min.	• Increase in flavonoids except for temperature of 60°C. • Increase in DPPH and phenolic compounds at 60°C; while at other temperature and increasing time reduced these contents [24].
Haematocarpus validus juice	2.2, 1.4 W/ml, 20 kHz at 18% amplitude, Temperature: 44, 55°C for 5 and 10 minutes	• Significant retention of antioxidant ascorbic acid, polyphenols, flavonoids, flavonols, and anthocyanins. • Increase in these compounds at 44°C and decrease at 55°C [55].
Hog plum juice	40 kHz, 400 W, Temperature: 40–60°C; Time: 5–30 minutes.	• Increase in antioxidant activity, total phenolic content, carotenoids, and flavonoids in comparison to the control [47].
Star fruit juice	44 kHz, full power of 600 W; Temperature: 25, 45°C; Time: 15, 30, 45, and 60 min.	• Increase in antioxidant activity, total phenolic content, flavonoids, and carotenoids [45].

TABLE 9.6 Effect on the Extraction of Different Material from Food Products

Biocomponent (Food Product)	Ultrasonic Equipment	Experimental Conditions	Results
Anthocyanin (black carrot's pomace)	102.4 W	24 kHz, ultrasound density used in 20.5–409.6 J/g; Temperature: 5–50°C.	The yield can be increased at least 20% by thermosonication [4]
Bioactive compounds (violet grape pomace)	750 Watts, 13 mm diameter titanium probe.	20 kHz; Temperature: 25, 55 and 40°C.	Superior extraction of antioxidant capacity, anthocyanins, and total phenolic [58].
Oil (castor)	130 W; a standard probe length of 113 mm and diameter of 6 mm.	20 kHz; Temperature: 70°C.	The yield was increased, and extraction time was reduced [39].

9.3.5 APPLICATIONS OF THERMOSONICATION IN DAIRY INDUSTRIES

9.3.5.1 MILK QUALITY AND SAFETY

Persons with lactose intolerance problems or allergies of milk protein have propensity to drink vegetable-based milks, such as hazelnut milks. The traditional pasteurization at 85°C deteriorates the nutritional and structural properties of milk. However, the thermosonication for 25 minutes at 60% amplitude or thermosonication for 15 min at 80% amplitude was able to completely inactivate the total aerobic mesophilic bacteria and yeast mold [34]. Moreover, the goat milk thermosonicated (20 kHz) at 400 W showed a considerable inactivation of microbes (near to the value of 2.3 log CFU/ml) and better results in contrast to control milk samples and pasteurized milk samples [54].

Thermosonication of human milk showed higher inactivation as compared to traditional heat treatment for of *Staphylococcus aureus, Escherichia coli, and Salmonella* spp. at 60°C, 48°C, and 65°C, respectively [49]. In contrast to traditional pasteurization, thermosonication showed a significant decrease in the residual activity of POD and LOX enzymes. The inactivation of LOX and POD of almond milk was reduced by increasing the processing time and temperature. The residual activity of LOX and POD was reduced to 5.12% and 6.37% from 42.2 and 88.1%, respectively when thermosonicated at 60°C for 40 minutes at constant frequency of 40 kHz [40].

Milk contains Plasmin, which is secreted from mammary cells [27]. The system of plasmin activity comprises of plasmin (PL), plasminogen (PG), plasminogen activators (PA), plasminogen activator inhibitors, and plasmin inhibitors. Among these, PL and PG are heat stable and thus are not affected by pasteurization, though resistant to treatments of HTST (high-temperature, short-time) and UHT (ultra-high temperature) to some extent [33]. The total plasmin activity can be reduced to about 94% in fresh skim milk and cream when thermosonicated (20 kHz) at an average power of 115 W for 3 min [67].

9.3.5.2 HOMOGENIZATION

Homogenization is a process of reducing fat globules of milk for better product development, like stabilizing the milk against creaming and partial coalescence and changing viscosity [51]. Thermosonication can be quick technology for the homogenization of camel milk cream as it has the capacity for reducing the fat globule size and increase of other factors like stability of emulsion, hardness, and consistency. The optimization study on camel milk cream showed that at 81 W, 4°C for 5.9 minutes showed minimum values of fat globule's size and adhesiveness along with highest values of high emulsion stability, hardness, consistency, and dynamic viscosity [34]. Similarly, thermosonication (20 kHz) of goat milk at 300 W was able to reduce the size of fat from 3.5 to 0.4 micrometer [53]. It concludes that thermosonication has a potential for utilization in the milk industry.

9.3.5.3 OTHER APPLICATIONS IN DAIRY PRODUCTS

The milk cream is used to make butter, which is a water-in-oil emulsion. Thermosonicated cream's butter has improved the spreadability and stability of oxidation as compared to the pasteurized cream's butter. Moreover, the butter's moisture content along with water activity and pH was reduced after thermosonication [61].

Yoghurt prepared from thermosonicated milk showed a microstructure of honeycomb under electron microscopy demonstrating a superior porous character as compared to yogurt prepared from conventional heat treatment at 90°C. In addition, these cultures of yogurt also showed superior values of gelation pH, viscosities, and water holding capacities (WHC) [57]. While analyzing texture profile and microstructure of goat milk [54], thermosonicated samples had improved firmness along with adhesiveness and deluxe network structure than those samples from milk homogenization. The thermosonication (24 kHz) treatment in milk helps the yogurt-drinks to have increased viscosity and lessened separation of serum in contrast to yogurt of milk thermally processed at 90°C for 10 minutes. The samples thermosonicated at 150 W showed completely no serum separation [30]. Similar results were observed for ayran, which is a yogurt-based drink. The viscosity and coefficient of consistency was improved, and serum separation was decreased when ayran was treated with thermosonication (35 kHz) [26].

The pre-treatment with thermosonication (20 kHz) at 250 W for 25 minutes of exposure time to whey led to maximum recovery of lactose of >90%. In this case, thermosonication inactivated the microorganisms; and viscosity of whey was decreased leading to superior recovery [35]. The whey-based orange juice drink treated with thermosonication showed superior results in inactivating the microorganism along with increment in antioxidant and higher retention of AA as compared to HTST [48].

The thermosonication can be used to minimize the fat content in regularly-eaten milk products (e.g., Fresh cheese). Thermosonication has been used to make low-fat cheese, which showed similar texture profiles with respect to high-fat cheese; in addition, no changes were observed in values of pH, TA and yield [9].

9.4 SUMMARY

The combined application of ultrasound cavitation and temperature has wide potential and abundant advantages over the thermal pasteurization alone. Thermosonication reduces the processing time. Various microorganisms and enzymes can easily be inactivated by thermosonication while maintaining the bioactive compounds and structural properties. This technology is fast, efficient, and potential substitution for a conventional technology (such as: thermal processing) to enhance the quality and shelf-life of food products. However, research efforts are required to improve processing equipment to achieve wider applications of this technology at the industry level.

KEYWORDS

- **enzyme**
- **enzyme inactivation**
- ***Escherichia coli***
- **non-thermal technologies**
- **pectin methyl esterase**
- **thermosonication**
- **ultrasound**

REFERENCES

1. Aadil, R. M., Zeng, X. A., Zhang, Z. H., Wang, M. S., Han, Z., Jing, H., & Jabbar, S., (2015). Thermosonication: Potential technique that influences the quality of grapefruit juice. *International Journal of Food Science & Technology, 50*(5), 1275–1282.
2. Abdullah, N., & Chin, N. L., (2014). Application of thermosonication treatment in processing and production of high quality and safe-to-drink fruit juices. *Agriculture and Agricultural Science Procedia, 2*, 320–327.
3. Adiamo, O. Q., Ghafoor, K., Al-Juhaimi, F., Babiker, E. E., & Ahmed, I. A. M., (2018). Thermosonication process for optimal functional properties in carrot juice containing orange peel and pulp extracts. *Food Chemistry, 245*, 79–88.
4. Agcam, E., Akyıldız, A., & Balasubramaniam, V. M., (2017). Optimization of anthocyanins extraction from black carrot pomace with thermosonication. *Food Chemistry, 237*, 461–470.
5. Aguilar, K., Garvín, A., Ibarz, A., & Augusto, P. E., (2017). Ascorbic acid stability in fruit juices during thermosonication. *Ultrasonics Sonochemistry, 37*, 375–381.
6. Ali, G., Russly, A. R., Jamilah, B., Azizah, O., & Mandana, B., (2011). Effect of heat and thermosonication on kinetics of peroxidase inactivation and vitamin-C degradation in seedless guava (*Psidium Guajava* L.). *International Food Research Journal, 18*(4), 1289–1294.
7. Alves, L. D. L., Dos, S. R. L., Bayer, B. L., Devens, A. L. M., Cichoski, A. J., & Mendonça, C. R. B., (2020). Thermosonication of tangerine juice: Effects on quality characteristics, bioactive compounds, and antioxidant activity. *Journal of Food Processing and Preservation, 44*(12), E-article 14914.
8. Amador-Espejo, G. G., Chávez-Ocegueda, J., Cruz-Cansino, N., Suárez-Jacobo, A., Gutiérrez-Martínez, P., Valencia-Flores, D., & Estrada, R. V., (2020). Thermosonication parameter effects on physicochemical changes, microbial and enzymatic inactivation of fruit smoothie. *Journal of Food Science and Technology, 57*(5), 1680–1688.

9. Amador-Espejo, G. G., Ruiz-Lopez, I. I., Gibbens-Bandala, P. J., Delgado-Macuil, R. J., & Ruiz-Espinosa, H., (2021). Thermosonicated whey protein concentrate blends on quality attributes of reduced fat panela cheese. *Ultrasonics Sonochemistry,* 105621.
10. Anaya-Esparza, L. M., Velázquez-Estrada, R. M., Roig, A. X., García-Galindo, H. S., Sayago-Ayerdi, S. G., & Montalvo-González, E., (2017). Thermosonication: An alternative processing for fruit and vegetable juices. *Trends in Food Science & Technology, 61*, 26–37.
11. Bhaskaracharya, R. K., Kentish, S., & Ashokkumar, M., (2009). Selected applications of ultrasonics in food processing. *Food Engineering Reviews, 1*(1), 31–49.
12. Butz, P., & Tauscher, B., (2002). Emerging technologies: Chemical aspects. *Food Research International, 35*(2, 3), 279–284.
13. Cervantes-Elizarrarás, A., Piloni-Martini, J., Ramírez-Moreno, E., Alanís-García, E., Güemes-Vera, N., Gómez-Aldapa, C. A., Zafra-Rojas, Q. Y., & Del Socorro Cruz-Cansino, N., (2017). Enzymatic inactivation and antioxidant properties of blackberry juice after thermo-ultrasound: Optimization using response surface methodology. *Ultrasonics Sonochemistry, 34*, 371–379.
14. Chandrapala, J., Oliver, C., Kentish, S., & Ashokkumar, M., (2012). Ultrasonics in food processing: Food quality assurance and food safety. *Trends in Food Science & Technology, 26*(2), 88–98.
15. Chen, T., Li, B., Shu, C., Tian, J., Zhang, Y., Gao, N., Cheng, Z., Xie, X., & Wang, J., (2021). Combined effect of thermosonication and high hydrostatic pressure on bioactive compounds, microbial load, and enzyme activities of blueberry juice. *Food Science and Technology International*, Article ID: 10820132211004316.
16. Chen, Z., (2017). Microbial inactivation in foods by ultrasound. *Journal of Food: Microbiology, Safety & Hygiene, 2*(1), E-article 1000e102.
17. Cruz, R. M., Vieira, M. C., Fonseca, S. C., & Silva, C. L., (2011). Impact of thermal blanching and thermosonication treatments on watercress (*Nasturtium Officinale*) quality: Thermosonication process optimization and microstructure evaluation. *Food and Bioprocess Technology, 4*(7), 1197–1204.
18. Cruz, R. M., Vieira, M. C., & Silva, C. L., (2008). Effect of heat and thermosonication treatments on watercress (*Nasturtium officinale*) vitamin C degradation kinetics. *Innovative Food Science & Emerging Technologies, 9*(4), 483–488.
19. Dabir, M. P., & Ananthanarayan, L., (2017). Effect of thermosonication on peroxidase, pectin methyl-esterase activities and on bioactive compounds in custard apple juice. *Journal of Food Measurement and Characterization, 11*(4), 1623–1629.
20. Dars, A. G., Hu, K., Liu, Q., Abbas, A., Xie, B., & Sun, Z., (2019). Effect of thermosonication and ultra-high pressure on the quality and phenolic profile of mango juice. *Foods, 8*(8), 298–302.
21. De Matos, R. M., Valdramidis, V. P., Nunes, C. A., & De Souza, V. R., (2017). Synergistic effect of thermosonication to reduce enzymatic activity in coconut water. *Innovative Food Science & Emerging Technologies, 41*, 404–410.
22. De Medeiros, J. K., Sarkis, J. R., Jaeschke, D. P., & Mercali, G. D., (2021). Thermosonication for peroxidase inactivation in sugarcane juice. *LWT-Food Science and Technology, 140*, 110730.
23. Demirdöven, A., & Baysal, T., (2008). The use of ultrasound and combined technologies in food preservation. *Food Reviews International, 25*(1), 1–11.
24. Do Amaral, S. F. D. C., Moura, L. G. S., De Oliveira, B. K., Aguiar, J. P. L., Mar, J. M., Sanches, E. A., Dos, S. F. F., et al., (2019). Thermosonication applied on camu–camu

nectars processing: Effect on bioactive compounds and quality parameters. *Food and Bioproducts Processing, 116*, 212–218.
25. Dündar, B., Ağçam, E., & Akyıldız, A., (2019). Optimization of thermosonication conditions for cloudy strawberry nectar with using of critical quality parameters. *Food Chemistry, 276*, 494–502.
26. Erkaya, T., Başlar, M., Şengül, M., & Ertugay, M. F., (2015). Effect of thermosonication on physicochemical, microbiological and sensorial characteristics of Ayran during storage. *Ultrasonics Sonochemistry, 23*, 406–412.
27. Fox, P. F., & Kelly, A. L., (2006). Indigenous enzymes in milk: Overview and historical aspects. *International Dairy Journal, 16*(6), 500–516.
28. Gao, R., Wu, Z., Ma, Q., Lu, Z., Ye, F., & Zhao, G., (2021). Effects of breaking methods on the viscosity, rheological properties and nutritional value of tomato paste. *Foods, 10*(10), 2395.
29. Guo, Y., Wu, B., Guo, X., Liu, D., Wu, P., Ma, H., & Pan, Z., (2021). Ultrasonication and thermosonication blanching treatments of carrot at varying frequencies: Effects on peroxidase inactivation mechanisms and quality characterization evaluation. *Food Chemistry, 343*, E-article 128524.
30. Gursoy, O., Yılmaz, Y., Gokce, O., & Ertan, K., (2016). Effect of ultrasound power on physicochemical and rheological properties of yoghurt drink produced with thermosonicated milk. *Emirates Journal of Food and Agriculture, 28*(4), 235–241.
31. Illera, A. E., Beltrán, S., & Sanz, M. T., (2020). Enzyme inactivation and changes in the properties of cloudy apple juice after high-pressure carbon dioxide and thermosonication treatments and during refrigerated storage. *Journal of Food Processing and Preservation, 44*(7), E-article 14521.
32. Illera, A. E., Sanz, M. T., Benito-Román, O., Varona, S., Beltrán, S., Melgosa, R., & Solaesa, A. G., (2018). Effect of thermosonication batch treatment on enzyme inactivation kinetics and other quality parameters of cloudy apple juice. *Innovative Food Science & Emerging Technologies, 47*, 71–80.
33. Ismail, B., & Nielsen, S. S., (2010). Invited review: Plasmin protease in milk: Current knowledge and relevance to dairy industry. *Journal of Dairy Science, 93*(11), 4999–5009.
34. Kashaninejad, M., & Razavi, S. M. A., (2020). Influence of thermosonication treatment on the average size of fat globules, emulsion stability, rheological properties and color of camel milk cream. *LWT-Food Science and Technology, 132*, E-article 109852.
35. Khaire, R. A., & Gogate, P. R., (2018). Intensified recovery of lactose from whey using thermal, ultrasonic and thermosonication pretreatments. *Journal of Food Engineering, 237*, 240–248.
36. La Cava, E. L., Gerbino, E., Sgroppo, S. C., & Gómez-Zavaglia, A., (2018). Characterization of pectins extracted from different varieties of pink/red and white grapefruits (*Citrus Paradisi*) by thermal treatment and thermosonication. *Journal of Food Science, 83*(6), 1613–1621.
37. Lafarga, T., Ruiz-Aguirre, I., Abadias, M., Viñas, I., Bobo, G., & Aguiló-Aguayo, I., (2019). Effect of thermosonication on the bioaccessibility of antioxidant compounds and the microbiological, physicochemical, and nutritional quality of an anthocyanin-enriched tomato juice. *Food and Bioprocess Technology, 12*(1), 147–157.
38. Liao, H., Jiang, L., Cheng, Y., Liao, X., & Zhang, R., (2018). Application of nisin-assisted thermosonication processing for preservation and quality retention of fresh apple juice. *Ultrasonics Sonochemistry, 42*, 244–249.

39. López-Ordaz, P., Chanona-Pérez, J. J., Perea-Flores, M. J., Sánchez-Fuentes, C. E., Mendoza-Pérez, J. A., Arzate-Vázquez, I., Yáñez-Fernández, J., & Torres-Ventura, H. H., (2019). Effect of the extraction by thermosonication on castor oil quality and the microstructure of its residual cake. *Industrial Crops and Products, 141*, E-article 111760.
40. Manzoor, M. F., Siddique, R., Hussain, A., Ahmad, N., Rehman, A., Siddeeg, A., Alfarga, A., et al., (2021). Thermosonication effect on bioactive compounds, enzymes activity, particle size, microbial load, and sensory properties of almonds (*Prunus Dulcis*) milk. *Ultrasonics Sonochemistry, 78*, 105705.
41. Manzoor, M. F., Xu, B., Khan, S., Shukat, R., Ahmad, N., Imran, M., Rehman, A., et al., (2021). Impact of high intensity thermosonication treatment on spinach juice: Bioactive compounds, rheological, microbial, and enzymatic activities. *Ultrasonics Sonochemistry, 78*, 105740.
42. Mawson, R., Gamage, M., Terefe, N. S., & Knoerzer, K., (2011). Ultrasound in enzyme activation and inactivation; Chapter 14. In: Feng, H., Barbosa-Canovas, G., & Weiss, J., (eds.), *Ultrasound Technologies for Food and Bioprocessing* (pp. 369–404). New York, NY: Springer.
43. Milani, E. A., & Silva, F. V., (2017). Ultrasound assisted thermal pasteurization of beers with different alcohol levels: Inactivation of *Saccharomyces cerevisiae* ascospores. *Journal of Food Engineering, 198*, 45–53.
44. Nafar, M., Emam-Djomeh, Z., Yousefi, S., & Hashemi, R. M., (2013). Optimization study on the ultrasonic treatments for *saccharomyces cerevisiae* inactivation in red grape juice with maintaining critical quality attributes. *Journal of Food Quality, 36*(4), 269–281.
45. Nayak, P. K., Chandrasekar, C. M., & Kesavan, R. K., (2018). Effect of thermosonication on the quality attributes of star fruit juice. *Journal of Food Process Engineering, 41*(7), E-article 12857.
46. O'donnell, C. P., Tiwari, B. K., Bourke, P., & Cullen, P. J., (2010). Effect of ultrasonic processing on food enzymes of industrial importance. *Trends in Food Science & Technology, 21*(7), 358–367.
47. Oladunjoye, A. O., Adeboyejo, F. O., Okekunbi, T. A., & Aderibigbe, O. R., (2021). Effect of thermosonication on quality attributes of hog plum (*Spondias Mombin L.*) juice. *Ultrasonics Sonochemistry, 70*, 105316.
48. Oliveira, G. A., Guimarães, J. T., Ramos, G. L. P., Esmerino, E. A., Pimentel, T. C., Neto, R. P., Tavares, M. I. B., et al., (2022). Benefits of thermosonication on orange juice whey drink processing. *Innovative Food Science & Emerging Technologies, 75*, E-article 102876.
49. Parreiras, P. M., Nogueira, J. A. V., Da Cunha, L. R., Passos, M. C., Gomes, N. R., Breguez, G. S., Falco, T. S., et al., (2020). Effect of thermosonication on microorganisms, the antioxidant activity and the retinol level of human milk. *Food Control, 113*, 107172.
50. Piyasena, P., Mohareb, E., & McKellar, R. C., (2003). Inactivation of microbes using ultrasound: A review. *International Journal of Food Microbiology, 87*(3), 207–216.
51. Postelmans, A., Aernouts, B., Jordens, J., Van, G. T., & Saeys, W., (2020). Milk homogenization monitoring: Fat globule size estimation from scattering spectra of milk. *Innovative Food Science & Emerging Technologies, 60*, 102311.
52. Qiao, L., Sun, Y., Chen, R., Fu, Y., Zhang, W., Li, X., Chen, J., Shen, Y., & Ye, X., (2014). Sonochemical effects on 14 flavonoids in citrus: Relation to stability. *PloS One, 9*(2), E-article 87766.

53. Ragab, E. S., Lu, J., Pang, X. Y., Nassar, K. S., Yang, B. Y., Zhang, S. W., & Lv, J. P., (2019). Effect of thermosonication process on physicochemical properties and microbial load of goat's milk. *Journal of Food Science and Technology, 56*(12), 5309–5316.
54. Ragab, E., Yacoub, S., Nassar, K., Zhang, S., & Lv, J., (2021). Textural and microstructural properties of set yoghurt produced from goat milk treated by homogenization and thermosonication. *Alexandria Science Exchange Journal, 42*(4), 985–995.
55. Raju, S., & Deka, S. C., (2018). Influence of thermosonication treatments on bioactive compounds and sensory quality of blood fruit (*Haematocarpus validus*) juice. *Journal of Food Processing and Preservation, 42*(8), E-article 13701.
56. Ravikumar, M., Suthar, H., Desai, C., & Gowda, S. A., (2017). Ultrasonication: An advanced technology for food preservation. *International Journal of Pure & Applied Bioscience, 5*(6), 363–371.
57. Riener, J., Noci, F., Cronin, D. A., Morgan, D. J., & Lyng, J. G., (2009). The effect of thermosonication of milk on selected physicochemical and microstructural properties of yoghurt gels during fermentation. *Food Chemistry, 114*(3), 905–911.
58. Romanini, E. B., Rodrigues, L. M., Finger, A., Chierrito, T. P. C., Da Silva, S. M. R., & Madrona, G. S., (2021). Ultrasound assisted extraction of bioactive compounds from BRS violet grape pomace followed by alginate-Ca^{2+} encapsulation. *Food Chemistry, 338*, E-article 128101.
59. Sasikumar, R., Chutia, H., & Deka, S. C., (2019). Thermosonication assisted extraction of blood fruit (*Haematocarpus validus*) juice and process optimization through response surface methodology. *The Journal of Microbiology, Biotechnology and Food Sciences, 9*(2), 228–235.
60. Sasikumar, R., Pradhan, D., & Deka, S. C., (2019). Effects of thermosonication process on inactivation of *Escherichia coli* and *Saccharomyces cerevisiae* and its survival kinetics modeling in blood fruit (*Haematocarpus validus*) juice to extend its shelf-life. *Journal of Food Processing and Preservation, 43*(11), E-article 14220.
61. Sert, D., & Mercan, E., (2020). Characterization of physicochemical, microbiological, thermal, oxidation properties and fatty acid composition of butter produced from thermosonicated cream. *International Dairy Journal, 109*, E-article 104777.
62. Silva, F. V., (2017). Resistance of *Byssochlamys Nivea* and *Neosartorya Fischeri* mold spores of different ages to high pressure thermal processing and thermosonication. *Journal of Food Engineering, 201*, 9–16.
63. Soria, A. C., & Villamiel, M., (2010). Effect of ultrasound on the technological properties and bioactivity of food: A review. *Trends in Food Science &Technology, 21*(7), 323–331.
64. Strieder, M. M., Neves, M. I. L., Belinato, J. R., Silva, E. K., & Meireles, M. A. A., (2022). Impact of thermosonication processing on the phytochemicals, fatty acid composition and volatile organic compounds of almond-based beverage. *LWT – Food Science and Technology, 154*, E-article 112579.
65. Tiwari, B. K., Patras, A., Brunton, N., Cullen, P. J., & O'donnell, C. P., (2010). Effect of ultrasound processing on anthocyanins and color of red grape juice. *Ultrasonics Sonochemistry, 17*(3), 598–604.
66. Tremarin, A., Brandão, T. R., & Silva, C. L., (2017). Application of ultraviolet radiation and ultrasound treatments for *Alicyclobacillus Acidoterrestris* spores' inactivation in apple juice. *LWT-Food Science and Technology, 78*, 138–142.

67. Vijayakumar, S., Grewell, D., Annandarajah, C., Benner, L., & Clark, S., (2015). Quality characteristics and plasmin activity of thermosonicated skim milk and cream. *Journal of Dairy Science, 98*(10), 6678–6691.
68. Wahia, H., Zhou, C., Fakayode, O. A., Amanor-Atiemoh, R., Zhang, L., Mustapha, A. T., Zhang, J., et al., (2021). Quality attributes optimization of orange juice subjected to multi-frequency thermosonication: *Alicyclobacillus acidoterrestris* spore inactivation and applied spectroscopy ROS characterization. *Food Chemistry, 361*, E-article 130108.
69. Wu, J., Gamage, T. V., Vilkhu, K. S., Simons, L. K., & Mawson, R., (2008). Effect of thermosonication on quality improvement of tomato juice. *Innovative Food Science & Emerging Technologies, 9*(2), 186–195.
70. Wu, Y., Xu, L., Liu, X., Hasan, K. F., Li, H., Zhou, S., Zhang, Q., & Zhou, Y., (2021). Effect of thermosonication treatment on blueberry juice quality: Total phenolics, flavonoids, anthocyanin, and antioxidant activity. *LWT-Food Science and Technology, 150*, E-article 112021.
71. Yıkmış, S., Aksu, H., Çöl, B. G., & Alpaslan, M., (2019). Thermosonication processing of quince (*Cydonia Oblonga*) juice: Effects on total phenolics, ascorbic acid, antioxidant capacity, color and sensory properties. *Ciência e Agrotecnologia* (Science & Agrotechnology), *43,* E-article 19919.

CHAPTER 10

ROLE OF ULTRASOUND-ASSISTED MECHANICAL SHEAR FORCE IN THE PRESERVATION OF FOODS

AMIT KUMAR MUKHERJEE

ABSTRACT

The non-thermal techniques of food processing and preservation are gaining popularity because they can produce foods without loss of nutrients and sensory quality. Ultrasound technology, a non-thermal processing technique, is a promising alternative to thermal processing of foods. Both low intensity and high intensity ultrasonic waves have applicability in enhancing the performance of food unit operations and inactivation of enzymes, microbial cells, and spores. It is successfully being applied in drying, extraction, filtration, freezing, depolymerization, defoaming, and degassing, crystallization, and emulsification operations. When applied in combination with other non-thermal techniques and at higher temperatures, ultrasound performance is highly satisfactory. While other non-thermal techniques of food preservation suffer from drawbacks including cost of installation and inability to perform independently, ultrasound has been found to be free from those shortcomings if applied with care. With the improvement of transducers and better process design, ultrasound can emerge out as an alternative to thermal processing of foods.

10.1 INTRODUCTION

The history of food preservation is much older than the famous discovery of the beginning of the 19th century by Nicolas Appert–the technique of food

Advanced Research Methods in Food Processing Technologies. Junaid Ahmad Malik, Megh R. Goyal, Preeti Birwal, & Ritesh B. Watharkar (Eds.)
© 2024 Apple Academic Press, Inc. Co-published with CRC Press (Taylor & Francis)

preservation by canning. There is evidence of postharvest operations and drying of grains by the ancient men. Fermentation is another such technique, which has been practiced in many parts of the world. The food preservation in modern time faces challenges due to expectations by human in terms of security and protection from diseases, newer evolving lifestyle, growing population, depleting non-renewable energy sources and environmental concern.

The thermal processing of foods has been highly effective in destroying pathogenic microorganisms of food, and the technique was unquestionable as long as there was little concern over restoration of nutritional quality of food materials and ample supply of thermal energy at a cheaper cost. The industries today use fossil fuels as the major sources of energy, while many of these produce huge volumes of effluents with higher BOD (biochemical oxygen demand) and COD values in due course of cleaning, pre-treatment, and processing of different foods.

Thermal processing is still common in the industries for medium- and low-acid foods and beverages. The popularity of pasteurization and sterilization lies in the fact that they can ensure destruction of undesired and harmful microorganisms without much complicated operations and higher expenditure. Another advantage of thermal decontamination of food is its applicability in all types of foods and beverages. On the other hand, many non-thermal treatments, which have been found to be useful (Table 10.1) in food processing, are either limited in application to foods with certain specific textural and structural properties or costlier to adopt for the small- and medium-scale industries. The use of ultrasound for in-place cleaning (CIP) of processing equipment in food processing plants can minimize the production of huge volumes of wastewater throughout the year.

Recent development in alternative processing techniques for foods have overcome a number of issues related to the use of thermal energy, wastewater generation leading to environmental pollution, nutritional losses, textural changes, phase separation and use of chemical preservatives. However, none of these have proved to be fully efficient for application in different types of food materials; while some of these are highly expensive and even unsafe.

High-pressure processing (HPP) technology holds a promising place in food preservation as long as it is affordable for the industry and unquestionable in terms of performance in the reduction of microbial load to the desired level. Generation of pressures above 100 MPa is often a constraint for the large-scale operation and installation of plant machinery and equipment. The pressure requirement is not equivalent for decontaminating all types of microbial cells in food materials subjected to processing. In most of the cases, pressure in combination with heat has been found to be effective. Though, the quality of processed foods is found to be better than one produced by thermal treatment [54]. The high pressure may cause physical injury to the vegetative cells [33].

TABLE 10.1 Applications of Thermal and Non-Thermal Treatments in Food Matrices

Processing Types	Applications
Thermal Processing:	
• Pasteurization	• Disinfection of milk, juices, fluid materials. Killing of pathogens and food spoilers.
• Sterilization	• Applicable to all foods and equipment as required. Kills vegetative cells and spores.
• Microwave heating	• Pasteurization of milk and juices. Drying, tempering, blanching, cooking, pasteurization, sterilization, and baking of foods [42].
• Ohmic heating	• Milk, fruits, vegetables, meat, and seafood [34].
Non-Thermal Processing:	
• Irradiation	• Sprouting inhibition, destruction of all types of microbes in foods of plant and animal origin and spices.
• High pressure	• Fruits and vegetables, juices, milk, beverages, meat, etc.
• Pulsed electric field (PEF)	• Drying and extraction of different foods [16]; processing of fish, meat, beverages, juices, and other foods [6].
• Pulsed UV light	• Used in surface disinfection and decontamination of liquid foods, water, and air [24].

The doses of treatment used in some non-thermal processing are limited and many are only effective in reducing microbial population by 3 to 4 log cycles. The lethality of some common thermal and non-thermal treatments toward microorganisms and their limitations are listed in Table 10.2. Apart from the simplicity in application and advantage of being the most convincing means of food decontamination to the consumers, all existing thermal treatments in Table 10.1 suffer from a common drawback, such as: loss of sensory quality and nutrients of the food. In the beginning of 20th century, the application of ionizing radiation in food preservation has been an effective tool for food decontamination. Ionizing radiation attacks the DNA of the microbial cell.

Radurization (pasteurization by ionizing radiation) of fish, meat, and poultry products is common with doses between 1 and 5 kGy. For sterilization of foods (radappertization), and destruction of spores (like that of *Clostridium botulinum*), a higher dose of radiation (about 50 kGy) is needed [48]. Radappertization is almost limited to meat and fish products [14]. There are

TABLE 10.2 Lethality and Limitations of Different Thermal and Non-Thermal Treatments in Food Systems

Processing Types	Lethality	Limitations
Thermal Processing:		
• Pasteurization	• Killing of almost all pathogens, yeasts, and molds [4].	• High thermal energy requirement, loss of sensory quality and nutrients.
• Sterilization	• Completely destroys all viable microorganisms [4]; destruction of microbial spores.	• Requires thermal energy, loss of sensory quality and some nutrients.
• Microwave heating	• 2–4 log cycle reduction at 720–900 W for 60–90 s [1].	• Difficulties in measuring temperature profile and distribution; does not ensure restoration of sensory quality and nutrients [1].
• Ohmic heating	• Suitable for blanching, thawing, and pasteurization, sterilization, evaporation, dehydration, extraction, etc. [2]	• Initial cost is higher than conventional heating, granular foods are not uniformly heated [2]; toxic effect on meat and meat products by contact of electrodes [2], some pathogens are resistant [32].
Non-Thermal Processing:		
• Irradiation	• Effective against pathogens, only 2–3 log cycle reduction of spores below 10 kGy (kilo gray) [40].	• Limited doses can be applied (up to 10 kGy) [14]. Requires additional measures for safety from hazard.
• High pressure	• Inactivation of vegetative cells starts above 200 MPa at ambient temperature and spores of yeasts and molds at 300–400 MPa, bacterial spores are inactivated above 1,000 MPa and at > 75°C [13].	• Expensive, sterilization is questionable [18]; much effective in combination with heat [54].
• Pulsed electric field (PEF)	• Varying sensitivity for different microbes, effective in killing vegetative cells under optimum pH, ionic strength, electrical conductivity, etc. Bacterial spores are resistant [3].	• Electrical field strength cannot be increased infinitely for spark generation, not suitable for low-moisture food decontamination and not applicable to foods containing air bubbles [6]; high initial cost and no effects on enzymes and spores [35].
• Pulsed UV light	• UV lights of 220–300 nm (UV-C) wavelengths are germicidal, higher doses can kill bacterial and fungal spores [24].	• *Aspergillus* species are resistant and requires very high doses [24], photo-reactivation, superficial decontamination, and shadowing (the microbes) in solid and opaque foods, production of ozone [23].

several regulations for the use of ionizing radiations in foods, and the practice of radiation processing of food items is often considered hazardous. Despite several drawbacks including those related to loss of sensory and nutritional qualities, irradiation technique has dominated alternative food processing due to its ability to destroy food pathogens.

The late emergence of other cold pasteurization and sterilization techniques could not become unquestionably popular, because these require aid of higher temperature than the ambient temperature for an effective outcome. When high intensity electric pulses of short duration are created in the food with the electrodes, then the charge induced within the molecules of the microbial cells leads them to move toward the cell membrane. This creates stresses and causes lysis of the cells by electroporation of the membrane.

The inactivation of microbial cells by PEF of desired intensity is irreversible, due to the loss of semipermeable character of the cell membrane under the electric field [29]. Sale & Hamilton [57] first observed this phenomenon in a number of bacterial species [25]. They also found that the spores of *Bacillus cereus* and *Bacillus polymyxa* species were resistant to electric field strength up to 30 kV/cm. Application of 5–55 kV/cm PEF for a few milliseconds is common for microbial inactivation of foods [28]. Varying sensitivities of microbes toward PEF treatment pose difficulty in optimization and generalization of the process parameters. The food constituents, textural, and other physical properties (such as: water activity, pH, ionic strength, electrical conductivity, etc.) can affect the lethality of PEF to the microbes by acting as stress factors. The commercial scale-up of the process requires high initial investment, while high intensity pulse generation causes sparking within the food. However, minimum sensory and nutrient damage may take place in the PEF treatment.

Ultraviolet (UV) light is another promising non-thermal processing tool for foods. The UV-C part of electromagnetic spectrum corresponds to wavelengths between 200 and 280 nm [23]. Short duration pulses of light of this wavelength range are effective in killing vegetative cells of microbes by destroying nucleic acid and DNA. The formation of lethal thymine dimers on DNA and rupture of bacterial cell due to thermal stress developed by overheating causes the inactivation activity [52]. Several researchers have found that *Staphylococcus aureus* cells suffer 7–8 log cycle reduction by 5.6 J/cm^2 per pulse for 5s without a substantial increase in temperature; and 4.8 log cycle reduction of *Aspergillus niger* spores is possible with pulsed UV light [41]. However, pulsed UV treatment does not ensure deactivation of microbial cells in the bulk food materials and in opaque liquids. This is more a surface decontamination tool than anything else. The photons generated by the pulse interact with hydroxyl groups present in the food, which in due course may produce peroxides from lipids containing unsaturated double bonds [19]. Generation of localized heat, formation of ozone and photo-reactivation of microbial cells under visible light are other constraints for pulsed UV treatment.

Several ultrasound-assisted unit operations (such as: drying, osmotic dehydration (OD), extraction, filtration, freezing, crystallization, thawing, degassing, brining (or pickling), foaming, emulsifying, rehydration, and detection of foreign bodies in foods) have shown encouraging results [10, 43]. Besides being a common subsidiary tool for enhancing processing efficiency, ultrasound can also successfully perform destruction of pathogenic and spoiler microbes that are present in foods. High intensity and frequency of waves are more effective in inactivating microbes and their spores. A frequency range of 200–600 kHz can enhance the effect of inactivation [17].

Acoustic wave propagation takes place in liquid and gas media. When a food material is in direct contact with the ultrasound wave, there is no chance of acoustic propagation of wave due to lack of a suitable medium. The effect of ultrasound on food through direct contact is due to acoustic vibration instead of acoustic cavitation [7]. The solid matrix within the food undergoes compression and expansion, like a sponge during the treatment (called "sponge effect") leading to enhanced mass and energy transfer – one criterion for the drying process. Experimental evidence of 99.99% destruction of dried *Bacillus thuringiensis* spores at 93 kHz has been reported by using a newly designed transducer, which can produce waves within a frequency range of nearly 50 kHz to 10 MHz [27].

The reduced efficacy of ultrasound treatment is attributed to the inadequacy of the design of common transducers used in processing and attenuation of the propagating wave through the medium. As the success of a new technique in food preservation depends on its flexibility of application on a large number of different types of foods of solid as well as liquid phases, therefore ultrasound can be useful in large scale food processing with suitable modification of equipment and process parameters. The mechanical shear produced by ultrasound is predominant in solid food substances as an active decontaminating agent, while liquid or fluid materials largely depend on cavitation effect imparted by the wave.

This chapter presents a thorough outline of the mechanism of ultrasound treatment with special emphasis on microbial decontamination of foods through mechanical shear and cavitation. The numerous unit operations of food processing encouraged by ultrasound treatment have also been briefly discussed due to their relevance in this chapter.

10.2 MECHANISM OF ACTION OF ULTRASOUND

The sound wave propagates through gas, liquid, or solid medium. The ultrasound is sound waves of frequencies higher than 20 kHz [26]. Since a long time in the past, use of ultrasound of different frequencies was limited to medical diagnosis, surgery, and treatment and in non-destructive testing (NDT) of materials. The advantage of ultrasound is that its shorter wavelength can detect tiny objects without ionizing the medium, through which it propagates.

The linear wave propagates through a medium by compression and rarefaction (Figure 10.1), which in turn induces vibration to molecules and gas bubbles coming in the path of the wave. The longitudinal induction of particles by ultrasonic waves occurs in fluids (gases and liquids). The amplitude of the longitudinal vibration is altered by a number of processes, such as reflection, refraction, and dispersion. When the medium particles are firmly bound to each other, ultrasound wave induces shear, which is transverse (perpendicular to the direction of propagation of wave) in nature. Another development within the medium is cavitation by the ultrasound.

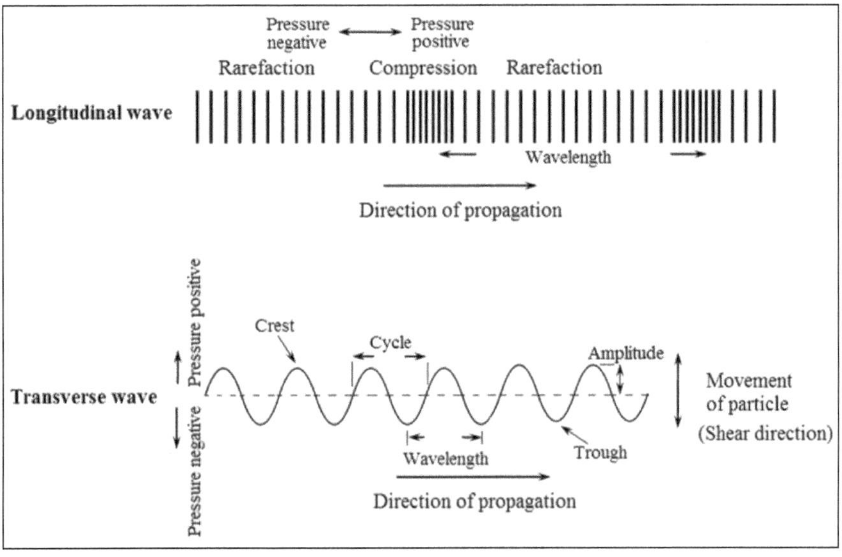

FIGURE 10.1 Longitudinal and transverse sound waves.

Cavitation and sonoporation may occur within and outside the biological cells by the application of ultrasound waves of different frequencies (Figure 10.2). The tiny bubbles of trapped gases of the medium violently collapse as a result of repeated contraction and expansion. Extensive cavitation within a cellular system may lead to sonoporation and release of smaller particles from the cells. Cavitation generates localized high temperature (> 5,000 K) and pressure (70–100 MPa) and turbulence within the medium [30]. The expansion and collapse cycle of the bubbles within the medium may be sinusoidal and may exactly mimic the acoustic wave as they meet. For instance, a 0.5 μm size bubble expands to 30 times and then collapses back to 0.5 μm size again [53].

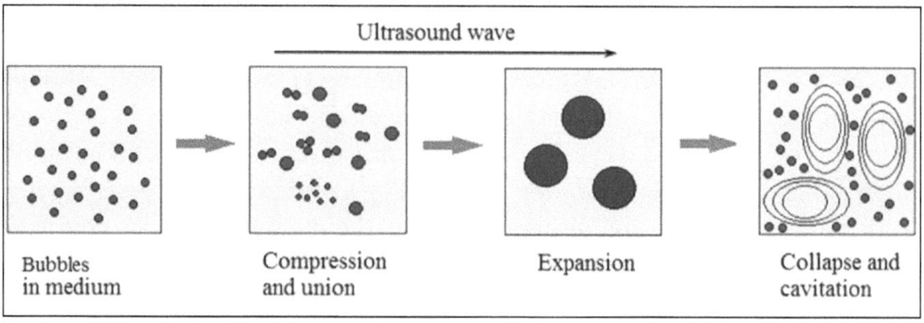

FIGURE 10.2 Compression and expansion of bubbles leading to cavitation.

The combined effect of a large number of cavitation is the capacity to damage microbial cells. Such a change in the physical state is also associated with some chemical changes. Though sound waves are non-ionizing, unlike X-rays, chemical changes within the medium are triggered by this development of higher localized temperatures. Production of primary radicals like H^+ and OH; and formation of oxides of nitrogen [51] have been noticed during ultrasound treatment. Sono-chemical depolymerizations and enzymatic reactions [22], organic syntheses [11] and catalyst-free reactions [8] are also facilitated by the similar treatment.

The development of localized heat and/or high pressure is the cause of most of the transformations of chemical origin at the molecular level. Studies on enzymatic and chemical reactions had ascertained the positive actions of ultrasound. Of these positive roles of sound waves, the majority of incidents were attributed to changes in activation energy and other thermodynamic parameters. Research studies have reported: enhancement of reaction rate of papain [55]; modification of activity, kinetics, thermodynamics, and molecular structure of alkalase [38]; accelerated enzymatic hydrolysis of starch [50], ease of catalytic wet peroxidation of phenol [44], etc. The frequency of sound wave may change on passage through a denser medium and is dependent on the travel time of waves through the medium. Continuous attenuation of waves by striking on the particles lowers the frequencies.

10.3 PRODUCTION OF ULTRASONIC WAVES

In 1917, Langevin produced ultrasonic wave using piezoelectric crystals [47, 58]. Linear electromechanical transformation of electrical energy into mechanical stress and *vice versa* is possible by some crystals lacking in symmetry. On the basis of symmetry elements, crystals can be grouped into 32 classes; and out of these, quartz (SiO_2), tourmaline, Rochelle salt, sucrose, barium titanate ($BaTiO_3$), potassium niobate ($KNbO_3$), zinc oxide (ZnO), lead zirconium titanate, etc., have been found to exhibit piezoelectric behavior.

The piezoelectric ultrasound producer device consists of a piezoelectric material placed between parallel metal plates inducing piezoelectric effect (Figure 10.3); and by the application of electrical energy, deformation of the piezoelectric crystal produces ultrasonic waves. The oscillator frequency can be controlled by the variable capacitor C connected to L_1 in the tank circuit. L_1 plate coil is coupled inductively to grid coil L_2. The piezoelectric material acting as dielectric is connected to the primary L of the transformer. A higher oscillatory electromotive force (e.m.f.) is generated in the coil L, which ultimately guides the movement of the crystal. The battery voltage is amplified through the oscillator and it produces alternating pulses to the piezoelectric material.

Another means of ultrasound generation is magnetostriction which is the phenomenon of changes in length of ferromagnetic materials (iron, nickel, cobalt, etc., and their alloys) in a magnetic field. Among the several advantages, the magnetostrictive patch transducers have been found to be highly sensitive, cost-effective, and durable [31]. However, limited frequency, eddy current and hysteresis losses are the main constraints for the application of magnetostriction.

In the magnetostrictive transducer, a coil is L_1 is wound at the right side of a magnetic metallic rod and is attached to the variable capacitor C (Figure 10.4). The base circuit comprises of a coil L_2 wound on the left side of the rod. The battery generates e.m.f. and a high frequency is produced at L_2 by the oscillator. The output of the current amplifier makes the rod to contract and elongate. This effect in turn produces ultrasonic wave.

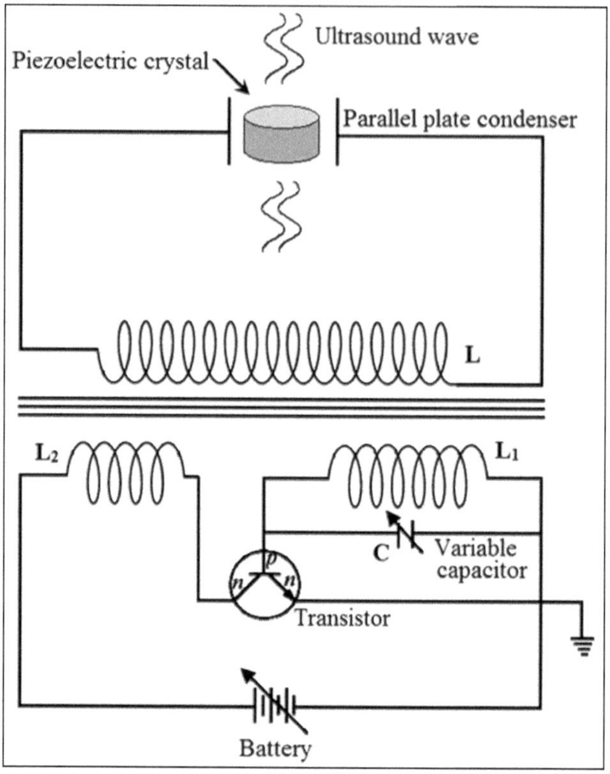

FIGURE 10.3 Circuit diagram of piezoelectric ultrasonic transducer.

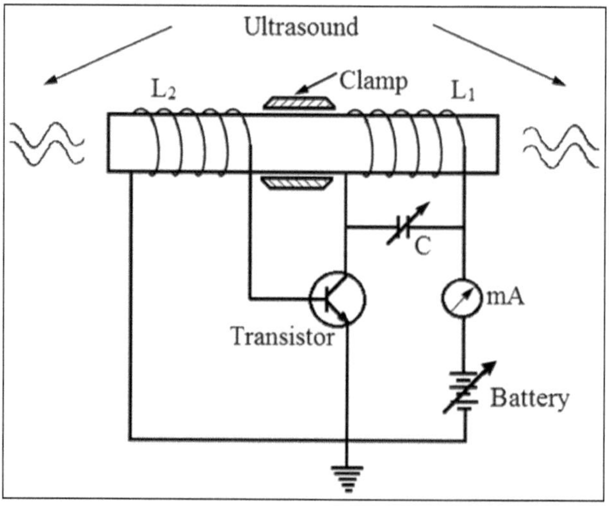

FIGURE 10.4 Magnetostrictive ultrasound generator circuit.

10.4 PIEZOELECTRIC VERSUS MAGNETOSTRICTIVE TRANSDUCERS

Piezoelectric devices are simpler in construction and can generate a wide range of frequencies with high electrical outputs. The extent and size of shear (in solid) and cavitation (in fluid) can be controlled by altering the frequency. Higher frequencies lead to smaller diameter cavitation and also can create greater eddies and turbulence in the fluid media. However, piezoelectric transducers suffer from aging problem, and the strength is lowered after the repeated use.

Two major disadvantages of ultrasound generation magnetostrictive transducers are: (a) the upper limit of frequency available here is only 100 kHz; and (b) the device has an electrical efficiency of only about 60% due to the heat loss [39]. The functioning of these devices is highly temperature-dependent. The extended length of coils is difficult to provide for greater output with magnetostrictive transducers. However, no damage occurs due to long term use of magnetostrictive materials except for the coating substances applied over the coils.

10.5 APPLICATIONS OF ULTRASOUND TECHNOLOGY IN FOOD PROCESSING

Low frequency ultrasonic waves are useful in testing and analyzing food samples. Many unit operations are possible with greater ease and efficiency. For the destruction of microbial population, however, high frequency waves are necessary. Creating an ultrasound environment within various foods requires specialized design of transducers and consideration of compositions and characteristics of non-homogeneous mixtures encountered every time. The recent developments in this field have shown that many food processing operations can be accomplished using this technique [15].

Research studies on ultrasound technology highlighted its increased efficacy when combined with heat and other non-thermal means of processing. Ultrasound is less effective in gaseous media, and high amplitude of vibration is needed to achieve the desired result. The fluctuation in power supply and unstable cavitation associated with attenuation of wave results in inadequate decontamination. The success of ultrasound application at ambient temperature condition is questionable for several foods.

Two types of ultrasonic equipment are common, namely (Figure 10.5): (a) horn type (or probe type); and (b) bath type. Horn type devices are common in the industry and many of these can process up to 1,000 liters of materials [9]. These two types differ in terms of sound emitters [12].

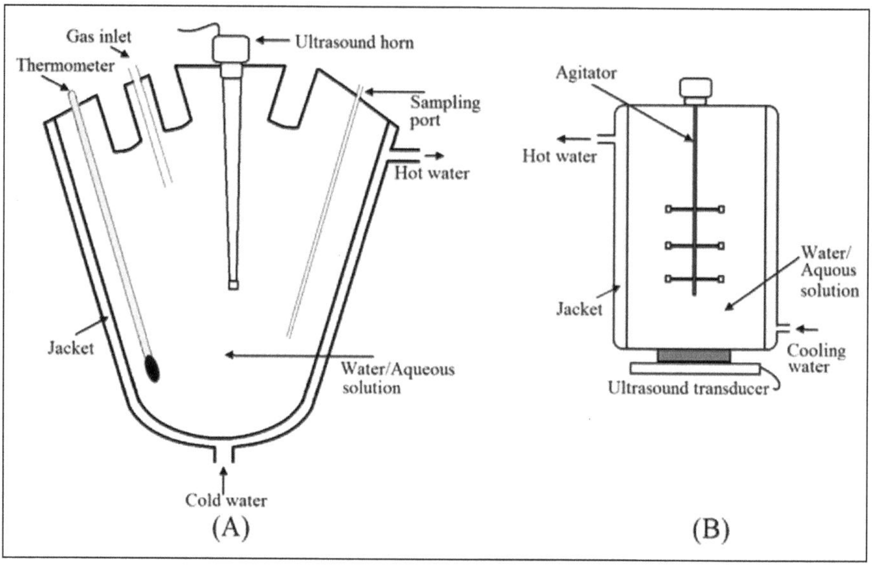

FIGURE 10.5 Ultrasound equipment: (A) horn or probe type; and (B) bath type.

The bath type devices are used in cleaning and degassing operations. When tested on gram-positive bacteria *Lactobacillus rhamnosus* (D-value = 18.1 min) and gram-negative *Escherichia coli* (D-value = 8.3 min), a dose of 20 kHz, 17.6 W of ultrasound for 20 min was effectively able to reduce the population of both species [5]. The authors also reported that almost complete inactivation of these microbial species was possible with higher time exposure (above 22 min) to ultrasound. Villamiel & Jong [49] reported the inactivation of *Pseudomonas fluorescens* and *Streptococcus thermophilus* in Trypticase Soy Broth; and also total bacteria in milk by continuous-flow ultrasonic treatment. Low temperature (50°C) ultrasound pasteurization is possible by restoring the physicochemical properties, color, and flavor of foods [21].

Ultrasound-assisted breakdown of starch and other polymers in solution have been studied [45]. The resulting depolymerization is accompanied with the reduction in viscosity of the solution and permanent change in rheological properties. Depolymerized starch is being used in many food processing applications.

There are requirements of such changes in the juice clarification and preparation of fermentation media, where enzymes and chemical or physical (for example, clarification by centrifugation, etc.) processes were applied. Due to changes in particle size, volume, and other physical properties of proteins by sonication, stable foaming can be produced in the solution. Significant improvement of foaming capacity of egg white protein was observed by ultrasound treatment at 360 W with minor reduction in foam stability [46]. Contrary to this, defoaming of many food materials can be done by controlled ultrasonic treatment.

In many industrial processes, formation of foam causes serious problems in reactions, mixing operation, adsorption, cleaning, extraction, sterilization, etc. Ultrasonic defoaming systems of varying frequency ranges have been manufactured by a Spanish company, PUSONICS [20].

The efficiency of membrane filtration is increased by ultrasound wave treatment. Ultrasound increases the flux by breaking concentration polarization and reducing the cake thickness [36]. The conventional freezing of food is not a steady process with respect to time. Freezing rate curve against time indicates non-linearity due to removal of latent heat by the water and also by different soluble solids at the time of freezing. The actual freezing process is thus delayed. Ultrasound-assisted freezing induces enhanced heat and mass transfer at the ice-liquid interface and increases the rate of freezing [37, 56]. The collapse of bubbles in the medium also helps in the nucleation of the crystals during freezing.

Better extraction of plant-based bioactive components at low temperatures is possible by ultrasound at a cheaper operating cost. Extraction rates of essential oils from spices and solvent extraction of edible oils have been markedly improved by ultrasound. Thus, it is a highly reliable technology, which can enrich and make food processing operations more efficient.

10.6 SUMMARY

Shear forces generated by sound waves have a wide range of applications in the processing of different food materials of varying textural and structural properties. Mechanical shear and cavitation have been successfully applied in the surface biofilm removal from the processing equipment (CIP) and for decontaminating some food materials. These physical tools for the destruction of pathogenic microbes and food spoilers have yielded positive results individually or in combination with other techniques for a number of food materials.

KEYWORDS

- **cavitation**
- **decontamination of food**
- **food preservation**
- **food rheology**
- **mechanical shear**
- **minimal processing**
- **sonication**
- **sterilization**
- **thermal processing**
- **ultrasound**

REFERENCES

1. Ahmed, J., & Ramaswamy, H. S., (2007). Microwave pasteurization and sterilization of foods; Chapter 28. In: Rahman, M. S., (ed.), *Handbook of Food Preservation* (pp. 691–712). Boca Raton – FL, USA: CRC Press.
2. Alkanan, Z. T., Altemimi, A. B., Al-Hilphy, A. R., Watson, D. G., & Pratap-Singh, A., (2021). Ohmic heating in the food industry: Developments in concepts and applications. *Applied Sciences, 11*(6), 1–19.
3. Amiali, M., & Ngadi, M. O., (2012). Microbial decontamination of food by pulsed electric fields (PEFs), Chapter 14. In: Demirci, A., & Ngadi, M. O., (eds.), *Microbial Decontamination in the Food Industry: Novel Methods and Applications* (pp. 407–449). Cambridge – UK: Woodhead Publishing Ltd.
4. Amit, S. K., Uddin, M. M., Rahman, R., Islam, S. R., & Khan, M. S., (2017). A review on mechanisms and commercial aspects of food preservation and processing. *Agriculture & Food Security, 6*(1), 1–22.
5. Ananta, E., Voigt, D., Zenker, M., Heinz, V., & Knorr, D., (2005). Cellular injuries upon exposure of *Escherichia coli* and *Lactobacillus rhamnosus* to high-intensity ultrasound. *Journal of Applied Microbiology, 99*(2), 271–278.
6. Aoude, C., Lammerskitten, A., Parniakov, O., Zhang, R., Grimi, N., El Zakhem, H., & Vorobiev, E., (2022). Equipment and recent advances in pulsed electric fields; Chapter 7. In: Garcia-Vaquero, M., & Rajauria, G., (eds.), *Innovative and Emerging Technologies in the Bio-marine Food Sector: Applications, Regulations, and Prospects* (pp. 149–172). London–UK: Elsevier Inc.
7. Astráin-Redín, L., Alejandre, M., Raso, J., Cebrián, G., & Álvarez, I., (2021). Direct contact ultrasound in food processing: Impact on food quality. *Frontiers in Nutrition, 8*, 1–11.

8. Banerjee, B., (2017). Recent developments on ultrasound assisted catalyst-free organic synthesis. *Ultrasonics Sonochemistry, 35*, 1–14.
9. Bevilacqua, A., Campaniello, D., Speranza, B., Altieri, C., Sinigaglia, M., & Rosaria, C. M., (2019). Two non-thermal technologies for food safety and quality – ultrasound and high-pressure homogenization: Effects on microorganisms, advances, and possibilities: A review. *Journal of Food Protection, 82*(12), 2049–2064.
10. Bhargava, N., Mor, R. S., Kumar, K., & Sharanagat, V. S., (2021). Advances in application of ultrasound in food processing: A review. *Ultrasonics Sonochemistry, 70*, E-article 105293.
11. Bremner, D. H., (1994). Recent advances in organic synthesis utilizing ultrasound. *Ultrasonics Sonochemistry, 1*(2), S119-S124.
12. Chaturvedi, K., Chaturvedi, S., Singha, S., & Das, K., (2021). Ultrasonic in food microbiology: Application and future trends. *Network Biology*, 2021 *11*(4), 315–333.
13. Cheftel, J. C., (1995). High-pressure, microbial inactivation and food preservation. *Food Science and Technology International, 1*(2, 3), 75–90.
14. Chmielewski, A. G., (2007). Practical applications of radiation chemistry. *Russian Journal of Physical Chemistry A, 81*(9), 1488–1492.
15. Condón, S., Mañas, P., & Cebrián, G., (2011). Manothermosonication for microbial inactivation; Chapter 11. In: Feng, H., & Barbosa-Cánovas, G. V., (eds.), *Ultrasound Technologies for Food and Bioprocessing* (pp. 239–285). New York – USA: Springer.
16. Donsì, F., Ferrari, G., & Pataro, G., (2010). Applications of pulsed electric field treatments for the enhancement of mass transfer from vegetable tissue. *Food Engineering Reviews, 2*(2), 109–130.
17. Ercan, S. S., & Soysal, C., (2013). Use of ultrasound in food preservation. *Natural Science*, *5*(8A2), 1–9.
18. Fam, S. N., Khosravi-Darani, K., Massoud, R., & Massoud, A., (2020). High-pressure processing in food. *Biointerface Research in Applied Chemistry, 11*(4), 11553–11561.
19. Fine, F., & Gervais, P., (2004). Efficiency of pulsed UV light for microbial decontamination of food powders. *Journal of Food Protection, 67*(4), 787–792.
20. Gallego-Juarez, J. A., (2010). High-power ultrasonic processing: Recent developments and prospective advances. *Physics Procedia, 3*(1), 35–47.
21. Gallo, M., Ferrara, L., & Naviglio, D., (2018). Application of ultrasound in food science and technology: A perspective. *Foods, 7*(10), 164–170.
22. Gogate, P. R., & Prajapat, A. L., (2015). Depolymerization using sonochemical reactors: A critical review. *Ultrasonics Sonochemistry, 27*, 480–494.
23. Gomez-Lopez, V. M., Ragaert, P., Debevere, J., & Devlieghere, F., (2007). Pulsed light for food decontamination: A review. *Trends in Food Science & Technology, 18*(9), 464–473.
24. Guerrero-Beltran, J. A., & Barbosa-Conovas, G. V., (2004). Advantages and limitations on processing foods by UV light. *Food Science and Technology International, 10*(3), 137–147.
25. Hamilton, W. A., & Sale, A. J. H., (1967). Effects of high electric fields on microorganisms: II. Mechanism of action of the lethal effect. *Biochimica et Biophysica Acta (BBA)-General Subjects, 148*(3), 789–800.
26. Hertzberg, B. S., & Middleton, W. D., (2016). *Ultrasound: The Requisites* (3rd edn., pp. 15, 16). Philadelphia, PA – USA: Elsevier Inc.
27. Hoover, K., Bhardwaj, M., Ostiguy, N., & Thompson, O., (2002). Destruction of bacterial spores by phenomenally high efficiency non-contact ultrasonic transducers. *Materials Research Innovations, 6*(5), 291–295.

28. Jeantet, R., Baron, F., Nau, F., Roignant, M., & Brule, G., (1999). High intensity pulsed electric fields applied to egg white: Effect on *Salmonella enteritidis* inactivation and protein denaturation. *J. Food Prot., 62*, 1381–1386.
29. Jeyamkondan, S., Jayas, D. S., & Holley, R. A., (1999). Pulsed electric field processing of foods: A review. *Journal of Food Protection, 62*(9), 1088–1096.
30. Kentish, S., & Ashokkumar, M., (2011). The physical and chemical effects of ultrasound; Chapter 1. In: Feng, H., & Barbosa-Cánovas, G. V., (eds.), *Ultrasound Technologies for Food and Bioprocessing* (pp. 1–12). New York–USA: Springer.
31. Kim, Y. Y., & Kwon, Y. E., (2015). Review of magnetostrictive patch transducers and applications in ultrasonic nondestructive testing of waveguides. *Ultrasonics, 62*, 3–19.
32. Knirsch, M. C., Dos, S. C. A., De Oliveira, S. A. A. M., & Penna, T. C. V., (2010). Ohmic heating–a review. *Trends in Food Science & Technology, 2 1*(9), 436–441.
33. Knorr, D., (2000). Process aspects of high-pressure treatment; Chapter 7. In: Barbosa-Canovas, G. V., & Gould, G. W., (eds.), *Innovations in Food Processing* (pp. 13–30). Boca Raton, Florida–USA: CRC Press LLC.
34. Kumar, T., (2018). Review on ohmic heating technology: Principle, applications and scope. *International Journal of Agriculture, Environment and Biotechnology, 11*(4), 679–687.
35. Kumar, Y., Patel, K. K., & Kumar, V., (2015). Pulsed electric field processing in food technology. *International Journal of Engineering Studies and Technical Approach, 1*(2), 6–17.
36. Kyllonen, H. M., Pirkonen, P., & Nystrom, M., (2005). Membrane filtration enhanced by ultrasound: A review. *Desalination, 181*, 319–335.
37. Li, B., & Sun, D. W., (2002). Effect of power ultrasound on freezing rate during immersion freezing. *J. Food Eng., 55*, 277–282.
38. Ma, H., Huang, L., Jia, J., He, R., Luo, L., & Zhu, W., (2011). Effect of energy-gathered ultrasound on alcalase. *Ultrasonics Sonochemistry, 18*(1), 419–424.
39. Mason, T. J., (1999). Laboratory equipment and usage considerations. In: Crum, L. A., Mason, J., Reisse, J. L., & Suslick, K. S., (eds.), *Sonochemistry and Sonoluminescence* (pp. 245–258). Dordrecht: Springer Science + Business Media.
40. Mostafavi, H. A., Mirmajlessi, S. M., & Fathollahi, H., (2012). The potential of food irradiation: benefits and limitations; Chapter 2. In: Eissa, A. H. A., (ed.), *Trends in Vital Food and Control Engineering* (pp. 43–68). Rijeka – Croatia: In-Tech.
41. Oms-Oliu, G., Martín-Belloso, O., & Soliva-Fortuny, R., (2010). Pulsed light treatments for food preservation: A review. *Food and Bioprocess Technology, 3*(1), 13–23.
42. Puligundla, P., Abdullah, S. A., Choi, W., Jun, S., Oh, S. E., & Ko, S., (2013). Potentials of microwave heating technology for select food processing applications–a brief overview and update. *Journal of Food Processing & Technology, 4*(11), 1–9.
43. Rastogi, N. K., (2011). Opportunities and challenges in application of ultrasound in food processing. *Critical Reviews in Food Science and Nutrition, 51*(8), 705–722.
44. Rokhina, E. V., Repo, E., & Virkutyte, J., (2010). Comparative kinetic analysis of silent and ultrasound-assisted catalytic wet peroxide oxidation of phenol. *Ultrasonics Sonochemistry, 17*(3), 541–546.
45. Seshadri, R., Weiss, J., Hulbert, G. J., & Mount, J., (2003). Ultrasonic processing influences the rheological and optical properties of high-methoxyl pectin dispersions. *Food Hydrocolloids, 17*(2), 191–197.

46. Sheng, L., Wang, Y., Chen, J., Zou, J., Wang, Q., & Ma, M., (2018). Influence of high-intensity ultrasound on foaming and structural properties of egg white. *Food Res. Int., 108*, 604–610.
47. Theraja, B. L., (2008). *Modern Physics* (pp. 310–315). New Delhi–India: S. Chand Publishing.
48. Urbain, W., (1986). *Food Irradiation* (pp. 118–123). San Diego–CA, USA: Academic Press Inc.
49. Villamiel, M., & De Jong, P., (2000). Inactivation of *Pseudomonas fluorescens* and *Streptococcus thermophilus* in trypticase® soy broth and total bacteria in milk by continuous-flow ultrasonic treatment and conventional heating. *Journal of Food Engineering, 45*(3), 171–179.
50. Wang, D., Ma, X., Yan, L., Chantapakul, T., Wang, W., Ding, T., Ye, X., & Liu, D., (2017). Ultrasound assisted enzymatic hydrolysis of starch catalyzed by glucoamylase: Investigation on starch properties and degradation kinetics. *Carbohydrate Polymers, 175*, 47–54.
51. Weiss, J., Kristbergsson, K., & Kjartansson, G. T., (2011). Engineering food ingredients with high-intensity ultrasound; Chapter 10. In: Feng, H., & Barbosa-Cánovas, G. V., (eds.), *Ultrasound Technologies for Food and Bioprocessing* (pp. 239–285). New York–USA: Springer.
52. Wekhof, A., (2000). Disinfection with flash lamps. *PDA J. Pharm. Sci. Technol., 54*, 264–276.
53. Yasui, K., (2002). Influence of ultrasonic frequency on multi-bubble sonoluminescence. *Journal of the Acoustic Society of America, 112*(4), 1405–1413.
54. Yuste, J., Capellas, M., Pla, R., Fung, D. Y., & Murmur, M., (2007). High pressure processing for food safety and preservation: A review. *Journal of Rapid Methods & Automation in Microbiology, 9*(1), 1–10.
55. Zhang, Z., Bai, G., Xu, D., & Cao, Y., (2020). Effects of ultrasound on the kinetics and thermodynamics properties of papain entrapped in modified gelatin. *Food Hydrocolloids, 105*, E-article 105757.
56. Zheng, L., & Sun, D. W., (2005). Ultrasonic assistance of food freezing; Chapter 23. In: Sun, D. W., (ed.), *Emerging Technologies for Food Processing* (pp. 603–626). London–UK: Elsevier Ltd.
57. Sale, A. J. H., & Hamilton, W. A. (1967). Effects of high electric fields on microorganisms. Killing of bacteria and yeasts. *Biochimica et Biophysica Acta (BBA) General Subjects, 148*(3), 781–788.
58. Langevin, P. (1917). French patent 505703. Crystal oscillator.

PART II

MICROWAVE PROCESSING OF FOODS

CHAPTER 11

APPLICATIONS OF MICROWAVE TECHNOLOGY IN FOOD PROCESSING

ANIKET P. SARKATE, NILEEMA S. GORE, and BHAGWAN K. SAKHALE

ABSTRACT

Microwave (MW) technology is the most emerging and promising technology used in the food industry, and it is popular due to more benefits than the conventional methods. Applications of MW-aided technique is enormously increasing in food processing since last two decades with successful industrial application, such as drying, ultra-sonication, ohmic heating, electron irradiation, MW-assisted freezing, MW-assisted OD, MW-assisted infrared heating, and MW-vacuum processing. MW technique also paved the way for various extraction processes of essential compounds. This chapter focuses on various aspects such as working principal of MW, instrumentality of MW, knowledge of equation which helps in modeling of MW and its application; dielectric properties of agricultural food material which leads to an effective understanding of controlled effect of MW on food material, baking, roasting, etc., are potential food application. In general, overall greater success of MW takes account of high thermal efficiency, shorter processing time, reduced operational cost and improved product quality will lead to greater quality and safety of different processed products in the food industry for tradition or futuristic food processing technologies.

11.1 INTRODUCTION

Development of new methods leads to ease of using technique in the most recent way for efficient production in any industry. Hence, it is necessary for

Advanced Research Methods in Food Processing Technologies. Junaid Ahmad Malik, Megh R. Goyal, Preeti Birwal, & Ritesh B. Watharkar (Eds.)
© 2024 Apple Academic Press, Inc. Co-published with CRC Press (Taylor & Francis)

any industry to accept new changes for optimized production in a particular field and increase the yield accordingly. This will not only increase the profit but also reduces the time and capital invested. New technique overcomes the drawbacks of conventional methods. Various new techniques are employed in the food industry as well, such recent one is the application of microwave (MW) technology in the food sector. Utilization of MWs is traced back in the 1940s, since then commonly applied in food and chemical engineering processes [20]. MW is a type of electromagnetic energy, which covers a portion of the electromagnetic spectrum. It is identified by keeping the wavelength interval from 1 mm to 1 m and frequency from 300 MHz to 300 GHz. For industrial food processing, frequency used in between 915 and 2,450 MHz and for domestic up to 2,450 MHz. MW does not cause changes in molecular structure as is a non-ionizing radiation [21]. MW heating is a method that offers a technique of heating requiring neither conduction nor convection. Substance temperature can be elevated by directly or indirectly. Direct system produces heat within substance itself as in MW techniques. Three processes can take place in MW as reflection, transmission, and absorption [57]. From last two decades, use of MW has extensively implemented for food processing industry. MW technique is proved to be fast, effective, maintains quality and food nutrients. Numbers of MW applications are pasteurization, heating, sterilization, drying, tempering, and baking [8].

MW assisted extraction (MAE) has important consideration in case of proactive element as studied [23]. Over 10 million MW ovens are sold in the United States and Europe per year. These numbers along with the approximately 225 million of MWs applied in the industries attracts the significance of MW assisted heating [55]. Advances in MW oven design and narrowing the gap in cost between MW and thermal processing will provide an initiative for the development of newer MW processes. Modeling shows an important part in understanding the optimal design of heating system for MW research [25].

This chapter not only explores the importance of MW for food industries, but also includes the mechanism of heating, effect of dielectric properties, modeling, and various industrial applications.

11.2 HEATING MECHANISM

MW heating is much faster compared to traditional method as heat is produced in the material itself. This action resulted in rapid elevation in temperature due to higher fluxes rather than conduction, convection and/or radiation as in conventional heating [5]. Mechanism involves two modes viz. dielectric and ionic. Most of the food items contain some amount of water/moisture, hence, dipolar properties of water resulted into dielectric heating. As water molecule tends to realign itself along the oscillating electric field related to electromagnetic radiation, heat is generated due to rotation in high frequency electric field. Ions present in food generate the ionic heating. High frequency oscillating electric field is also accountable for generating the heat. Kinetic energy is actually imparted to ions by an electric field, so that the field alternates heat rapidly [10]. Energy transfer mechanism is a function of time, temperature, and frequency. Various other parameters also affect heating mechanism as di-electrical properties of food, etc. [41].

11.3 DIELECTRIC PROPERTIES

These properties are important for understanding the relationship between electromagnetic energy and material. They govern the efficiency of the heating process as well as quality. Rate of change of polarization is caused by oscillating electric field applied to a dielectric material (dipoles in the materials aligning themselves with the field representing displacement current in the dielectric material, hence power is transformed into heat) [31]. The capability to absorb, transmit, and reflect electromagnetic energy is observed. Food material is said to be 'lossy dielectric materials' they are neither good electric insulator nor good electric conductors. Dielectric properties can be prejudiced by many factors such as temperature, temperature, moisture content, salt content, frequency of the MWs and other ingredients and hence, establish a powerful relation between food materials and electric field [47]. Various parameters related to dielectric properties are discussed in subsections.

11.3.1 PERMEABILITY AND PERMITTIVITY

Permeability doesn't contribute to heating greatly as values for foodstuff are generally similar to that of free space [30]. Dielectric properties influencing reflection of electromagnetic waves at interfaces and attenuation of wave energy within materials are called permittivity. Complex relative permittivity ε* is expressed as:

$$\varepsilon^* = \varepsilon' - j\varepsilon'' \ (j = \sqrt{-1}) \qquad (1)$$

The real part ε' is referred to as dielectric constant, which is a dimensionless quantity and has ability to store energy when the material is exposed to an electric field; while the dielectric loss factor ε" is also dimensionless, which is the imaginary part, influences energy absorption and attenuation and also gives information regarding heat generation. Mechanisms contributing to dielectric loss in heterogeneous combination contains polar, electronic, atomic, and Maxwell–Wagner responses [67]. The variations with temperature of dielectric constant and dielectric loss factor lead to a comprehensive study of measuring these parameters in foods under a variety of temperature conditions and frequencies such as 300, 915, and 2,450 MHz [63].

11.3.2 LOSS TANGENT

Loss Tangent is defined as the tangent of dielectric loss angle or the ratio of ε" over ε':

$$\tan(\delta) = \frac{\varepsilon''}{\varepsilon'} = \frac{k''}{k'} \tag{2}$$

where; κ' (relative dielectric constant); and κ" (relative dielectric loss), which are given as $\kappa' = \varepsilon'/\varepsilon_0$ and $\kappa'' = \varepsilon''/\varepsilon_0$. Here, ε_0 is the permittivity of free space ($\varepsilon_0 = 8.854 \times 10^{-12}$ F/m) [4, 13]. It is also an index for material to generate heat [66].

11.3.3 PENETRATION DEPTH

It is noted as, the distance into a sample where, MW power reduced to 1^{-e} or 36.8% of its transmitted value. It is also defined as, the distance at which the MW power has been attenuated to 50% of P_{trans} [66]. Penetration depth is a function of ε' and ε" and given by following expression:

$$\text{dp} = \frac{\lambda_{0\sqrt{\varepsilon'}}}{2\pi\varepsilon''} \tag{3}$$

where; λ_0 = free space MW wavelength (λ_0 =12.2 cm for 2.45 GHz).

Other expanded expansion for dp is given by following equation:

$$\text{dp} = \frac{\text{c}}{2\pi f\sqrt{2\varepsilon'[\sqrt{1+(\varepsilon''/\varepsilon')^2}-1]}} \tag{4}$$

where; 'c' is the speed of light in free space (3×10^8 m/s); and 'f' is the frequency (Hz).

Usually, food products have ε"< 25, which implies a penetration depth of 0.6–1.0 cm. Penetration of MW is between 0.3 and 7 cm when food is with high moisture and at room temperature and frequency of 915 and 2,450 MHz. The dielectric properties predict the Penetration depth of material [58]. Penetration depth is inversely proportional to frequency and directly proportional to wavelength, i.e.:

$$\mathrm{dp} \propto \frac{\lambda}{f} \tag{5}$$

At frequencies near the MW heating regime, the penetration depths are smaller and often the size of the material to be treated, particularly when it is very wet, is many times larger than penetration depth and MW heating could result in unacceptable non-uniformities in the temperature distribution. Lower frequency provides greater penetration and while higher frequency causes surface heating [33]. Heating food effectively requires the sufficient knowledge of penetration depth which leads the path to optimum heating by MW and hence, should be considered by the food product developer as it will affect the overall product quality. Another importance of penetration depth is in the calculation of tube diameter for treating continuous flow with MW heating [53].

11.4 FACTORS AFFECTING DIELECTRIC PROPERTIES

The dielectric properties are affected by several factors. Generally, the amount of water is the dominant factor in hygroscopic materials like food, etc. Dielectric properties majorly depend on the frequency of applied alternating electric field, material temperature, density, composition, and structure of material (Table 11.1). The bulk density of air-particle mixture is a factor which affects dielectric properties in granular or particulate materials [39]. At frequencies above 1 GHz dipole polarization is noteworthy while below 1 GHz frequencies ionic losses are predominant. Due to electronic or atomic polarization, MW is not absorbed by the material and hence, dipole and ionic polarization comes into picture. At frequencies above 1 GHz dipole polarization is noteworthy while below 1 GHz frequencies ionic losses are predominant [52].

11.4.1 FREQUENCY

Materials that do not absorb radiofrequency (RF) and MW field are rare exceptions of some low-loss material, while for most materials, the dielectric properties vary with the frequency of the applied electric field. The orientation of the dipole with the applied electric field leads to polarization which contributes to the frequency dependency parameter of the dielectric properties of most material. The displacement of electrons of atoms with respect to the nucleus is called electronic polarization and the displacement of nuclei with respect to one another is called atomic polarization. As we have seen earlier, the dielectric properties varies inversely with frequency, i.e., decreases with an increase in frequency, hence, the dielectric constant of pure water decreases. The loss tangent, dielectric loss, changes in frequency for moist food has higher dielectric loss value. Dielectric properties, frequency, and energy absorbed are correlated to each other and change in any one parameter affects the rest considerably.

The dipole molecule within the food material (mostly water) has tendency to align themselves with the changes in the frequency of applied electric field but with constant increase in frequency this is not true and as a result, dielectric constant drops with increasing frequency in this region. The dielectric constant again levels off near the so-called optical value, and the loss factor again drops to a low value at higher frequency [17]. Some examples include: there was a considerable drop in dielectric properties of raw potatoes at frequency from 300 to 3,000 MHz and showed same when frequency increased further. The dielectric properties of apples were found to vary with maturity, dropping appreciably in the process of aging at a frequency of 300 to 900 MHz [65].

11.4.2 TEMPERATURE

Concentrations of mobile water and ions affect the dielectric properties of food at low temperature at MW frequencies. Frozen *food* products are considered as *a homogenous mixture* of solid (Ice) and aqueous ions. With the product of high water content, it was observed that in the melting zone, there was a sharp increase in ε' and ε" which was true till melting, but after melting it was seen that ε' decreases with a gradual further increase in temperature [52]. The increment in the ε' and ε" values is traced at the lower frequencies and vigorous rise with temperature above 60°C. Stunted values of the loss factor are seen at about 40 or 50°C [38]. In polar materials, relaxation time decreases with increasing temperature.

Elevated values of dielectric properties with temperature have been observed in area of dispersion or dielectric loss while it will decrease in absences of dielectric loss [35]. Viz. temperature dependency and frequency dependency contributes significantly in variation of dielectric properties and henceforth, values decreases or increases accordingly. At frequency 27 and 40 MHz, a gradual decrease in the dielectric constants was remarkable with elevated temperature for a food material with much water (66%) [40].

11.4.3 DENSITY

With variation in the electromagnetic field of the quantity of mass interacting, the dielectric properties are influenced. Density is defined by mass per unit volume, hence, will also affect dielectric properties which was particularly remarkable with the pulverized or granular materials, also referred to as particulate dielectric. Relation sandwiched between the dielectric properties of solid material and pulverized sample of such solid, were important in understanding the nature of density dependency. The dielectric properties are necessary when the sample is available in particulate form such cereal grains where kernels do not fit in measuring instruments and pure minerals in pulverized form for purification [34]. The linear regression of ε' and ε" was also studied. Correlation coefficients < 0.5, other than loss factor at 11.7 GHz for both dielectric properties at 2.45, 11.7, and 22.0 GHz frequencies (Table 11.1) [36].

TABLE 11.1 Microwave and Physical Properties for Selected Fruits and Vegetables (at 23°C) [36]

Fruits and Vegetables	Tissue Density (g/cm³)	Dielectric Properties					
		2.45 GHz		11.7 GHz		22.0 GHz	
		ε'	ε"	ε'	ε"	ε'	ε"
Potato	0.995	51.5	16.3	47.7	6.3	48.5	1.8
Sweet potato	0.949	49.8	15.2	49.9	4.8	49.4	1.5
Peach	0.944	58.3	12.7	49.4	5.2	49.9	1.8
Watermelon	0.847	58.9	15.2	50.4	2.6	49.9	1.3
Cantaloupe	0.912	55.2	13.6	50.4	3.7	49.1	1.2
Cucumber	0.906	63.5	13.0	54.6	3.0	49.5	0.9

Source: Reprinted with permission from Ref. [36]. © 1983 ASABE.

11.4.4 MOISTURE CONTENT

The heating of foodstuffs in a MW oven is mainly determined by the moisture content (also known as free water content) in it which makes it a crucial parameter. Most of the food contents 50–90% of water. Higher water content decreases the penetration depth and translates into greater MW absorption. When moisture content is high, there will be larger dielectric losses and product will heat efficiently as there is more dipole molecule that will align themselves and help in heating. When specific heat capacity is lower, low-moisture products can heat up as well and wave penetration depth will increase [15].

Several other factors (such as: storage, composition, geometry, etc.) also affect foodstuff dielectric properties used in the food processing industry. Hence, their consideration is also important the factors mentioned before. For example, dielectric constants of albumen and yolk at 10 MHz were lower than those of fresh albumen and yolk when eggs were stored at 15°C for 1 week. After 2 weeks in storage, it was observed that these dielectric constants rose and remained at higher levels for the rest of the 5-week period [14].

11.5 MEASUREMENT TECHNIQUES

Measurement of dielectric properties is essential in the food industry, as the texture of food components varies at various temperature and frequency conditions in MW heating. Physicochemical changes can be analyzed by these techniques. Based on the desired frequency and kind of foodstuff to be processed, the various techniques are dogged. Availability of equipment, resources of studies and extent of the research the choice of measurement is determined. The measurement methods are classified into four major groups, such as for free space, lumped circuit, transmission line, and resonator method [65].

In Lumped circuit, lower frequencies (<100 MHz) can only be used. Sample is a part of the insulator. With exceptions like gases and low-loss materials, measurement of all other material is possible. Measurement frequency range is from 50 MHz to above 100 GHz comes under the resonator in which material is filled partially or completely. The shift in resonant frequency is related to the dielectric properties.

The transmission line method is considered as sensitive, more specific than other methods and has a narrow range of frequencies. However, it is a difficult and time-consuming method and with the exception of gases materials in terms of sample preparation.

Transmit and receive horns are on the opposite side of the sample in the free space reflection method. Reflection and transmission coefficients from the sample are primary requirement of the free space S-parameter method.

Calibration standards in the free space are required. A resonant structure between two large reflective surfaces (flat or parabolic) is established in the free space cavity method. This technique is useful under extremely high temperatures; hence, they are non-contacting and non-destructive. Special calibration considerations and large flat, thin, the parallel-faced sample is required [18]. It is essential that the best technique and equipment should be preferred and also the proper understanding of the principles and techniques is crucial to get the reliable data. For proper measurement, the measurements of a few food samples for which dielectric properties should be well-known for standard calibration. However, such standards are only hand-full, and the dielectric constant and loss factors have values in the range. Hence, it is advisable that two different systems should be used and the final result should be given after comparison between two for appropriate values [37].

11.6 DIELECTRIC BEHAVIOR OF FOOD MATERIAL

The dielectric behavior is significant for the processing of food material, which contains some water percentage, hence it acts as a prime factor of consideration. Deviation in the dielectric properties by salt ions is noted but is adversely affected by the loss factor. Henceforth, salt solution is an important point of consideration to check behavioral changes in the food material. The salt concentration might show significant proportionality, as it is directly proportional to the dielectric loss and inversely to dielectric constant. The rise in temperature causes a decrease in the dielectric constant of salt solution while the loss factor depends on ranges of temperature.

For hard red winter wheat, the dielectric constant always decreases depending on the frequency and the moisture content of the wheat, but the loss factor may increase or decrease. The dielectric constants for corn, wheat, barley, oats, rye, and soybean have been determined by model studies and dielectric properties depends on the frequency, moisture content, bulk density over a wide range. The linear profile was observed between moisture content and constants at frequencies of 20 MHz, 300 MHz, and 2.45 GHz for rough rice. In all fruits and vegetables (FAV) under study, the dielectric constant showed a monotonic decrease with the frequency range from 0.2–20 GHz. The properties were increased at all frequencies with moisture content, but abnormal values were observed in nuts. At high moisture levels, the dielectric constant and loss factor were increased rapidly along with temperature in nuts. Dielectric properties of codfish showed abrupt changes in the dielectric constants in the region of freezing temperature. In case of meat, both the dielectric constant and loss factor increased with the increase in moisture content. The straight-line increment was seen in both dielectric constant and loss factor when the temperature was increased. However, opposite results were observed for beef and pork, for which there was a decrease in both values at freezing to temperature of 60°C. In baked food products, the dielectric constant and loss factor not only were decreased with the baking time; but both values were decreased also as the water content was reduced [9].

11.7 NON-UNIFORM TEMPERATURE DISTRIBUTION: A THREAT TO MICROWAVE (MW) PROCESSING

Though MW has several advantages over conventional heating, there are some setbacks associated with it. The temperature distribution profile varies during MW heating, and several researchers observed non-uniform temperature distribution in the MW heating in a variety of food materials. Some region was severely exposed to the heat while the other was exposed to the low heat. Although various solutions have been suggested, but the perfect solution has not been presented to this drawback till date. Hence, overcoming this drawback will lead to hurdle-free use of MW heating in the food processing industry [64].

11.8 MATHEMATICAL MODELLING

For optimizing the functions in MW heating, various mathematical modeling techniques have been used. In order to learn more about heat generation, many researchers modeled the MW heating by either Maxwell's equation or Lambert's law to evaluate electromagnetic field distribution. Maxwell's equation is implied in various numerical modeling. Electromagnetic field distribution inside MW space is governed by Maxwell's equation. For constant permittivity and permeability, Maxwell's equations can be presented as follows:

$$\nabla \times E = -j\omega\mu H$$
$$\nabla \times H = -j\omega\varepsilon_0\varepsilon^* E$$
$$\nabla \times E = 0$$
$$\nabla \times H = 0 \tag{6}$$

where; E is the electric field intensity (V/m); ω is the angular frequency (rad/s); μ is the permeability of free space (H/m); H is the magnetic field intensity

(A/m); ε_0 is the permittivity of free space (F/m); and ε^* is the complex relative permittivity [12].

Magnetic and Electric field intensities are time harmonics. The above equation also describes propagation of electric field, magnetic field and forced heat equation (absorption and diffusion of heat). Both the damped wave equation and transport equation are derived from the Maxwell's equation [16]. Lambert's law for exponential decay of MW absorption is given as:

$$P(x) = P_0 e^{-2\alpha x} \tag{7}$$

where; P_0 is the power at the surface; x is the distance measured from the surface; P(x) is the power dissipated at a distance x; and α is the attenuation constant (which is a function of wavelength of radiation (λ_0), relative dielectric constant (k') and loss tangent (tan δ)) and is given as:

$$\alpha = \frac{2\pi}{\lambda_0}\sqrt{\frac{k'\left(\left(1+tan^2\delta\right)^{\frac{1}{2}}-1\right)}{2}} \tag{8}$$

Framing simple models on the basis of basic geometry are useful applications of Lambert's law, and prediction of the changes in the product size and effect of change in product composition can be effectively done. Lambert's law showed promising experimental data for solving unsteady state heat transfer differential equations and can be applied for large 1-D systems [7]. However, comparison of both the laws revealed that Maxwell's equation could predict accurately the temperature profiles than Lambert's law. Errors were found with Lambert's law as it does not incorporate internal reflection values.

In theoretical models, it is important to include internal reflections as they cause variation in heating rates due to the internal standing wave patterns. Temperature profiling is benefitted when predicted with the use of Maxwell's equation [4]. The law is only valid for the semi-infinite specimens having dimensions wider than the wavelength [43]. Moreover, it is necessary that prediction should be more accurate and useful. Hence, better mathematical modeling will encourage the use of MW more vigorous in food industries.

11.9 APPLICATIONS OF MICROWAVE (MW) HEATING IN THE FOOD INDUSTRY

Looking at the applications rather than the food industry, MW is used in several ways in drying, blanching, sterilization, pasteurization, MAE, tempering, roasting, baking, thawing, packaging, etc. Other applications are also seen in medical waste treatment, wastewater remediation, sludge processing, contaminated soil remediation, integrated remedial methodology and activated carbon regeneration, etc. Besides this, various MW treatment and MW-assisted techniques have shown potential results not only in food but also in other industries [69]. As every coin has two sides; similarly, MW has some drawbacks, such as price, operating expenses; and safety concerns (to limit MW leakage) during scaling up limits the use of MW in industries [68]. However, neglecting all these factors, we can observe that MW is tremendously used in industries over the past 10 years.

11.9.1 MICROWAVE (MW) BLANCHING

Before the preservation procedures, blanching process helps to inactivate enzymes (such as peroxidase, ascorbic acid (AA) oxidase and lipoxygenase (LOX)) that are responsible for deterioration actions, i.e., effective loss in product quality, environmental impact and energy cost. Blanching also minimizes these negative effects [3]. This process also influences sensory characteristics, such as color, texture, and flavor. It was found that blanching of green beans for 30–90s was able to increase the negative a_L values and a_L/b_L ratio during frozen storage. Sensory qualities were superior in the blanched samples for 30–90s. Unblanched products are not organoleptically suitable at all storage stages. Nowadays, MW blanching studies are being conducted [6].

In a continuous conveyer MW oven, blanching was done with 4 magnetrons (power of 1.25 kW) and conveyer speed of 0.5 m/min). In polypropylene (PP) trays, 250 g of carrot slices per batch were blanched. Carrot nutritional quality was superior in MW blanching compared to water blanching. Even after frozen storage for three months, the nutritional quality was higher (vitamin, total sugar, etc.). There was a significant difference in the carotene content due to MW blanching [22]. During MW blanching, temperature, and power levels were not only influenced by vegetables and by its quantity, shape, and location in the MW oven. The ratio of surface area to volume was higher in leafy vegetables than that of other vegetables, and MW blanching was comparable to water blanching. Lower yields of products are the result of moisture evaporation. MW blanching was more superior to the traditional method [49].

Other such advantages were observed in nuts. The time and cost were significantly reduced by MW blanching of peanuts. The correlation between temperature, content of moisture and ability of blanching was established by continuous belt processing method. Effective blanching was possible at temperature up to 110°C and moisture content of 5.5% or less. Peanuts showed significantly better results [54].

11.9.2 MICROWAVE (MW) BAKING

Microwave (MW) is used in baking to accelerate the baking process. The MW baking results also in higher drying rate and uniform moisture profiles [44]. For short instance of time under favorable oven conditions in dough or a batter, there is heating due to the interaction of MW with certain molecules. Browning reaction does not occur due to lack of time for specified changes, which render the product quality in terms of volume, texture, color, and crust. The pre-gelatinized starches are helpful for the issues related to low volume. Hydrocolloids can tackle the high moisture loss.

Texture can be improved with the use of some additives. The browning effect can be developed by using susceptors to increase surface temperature. Changes in the products during MW-aided baking techniques can be examined by water mobility, protein relaxation, starch gelatinization, etc. Problems in MW-baked products can be minimized by modeling variations in electrical, thermal, and functional properties of the material. Improvement of MW baking is still a challenge for quality improvement of bakery products. If all the drawbacks are improved, then it will an efficient technique [59].

11.9.3 MICROWAVE (MW) THAWING AND TEMPERING

Thawing is a wonderful application of MW techniques. Advantages of MW thawing compared with the conventional process are reduction in time, space, and other food properties (microbial deterioration, drip loss, etc.). However, localized overheating is an issue. This overheating is due to the fact that water absorbs more radiation thus leading to overheated loss of water and chemical deterioration with temperature. For an efficient MW thawing, temperature uniformity is a key step, for which feedback controller can be used along with variable continuous power. Material properties, dimensions, magnitude, and frequency of electromagnetic radiation are the main parameters that affect the thawing rates. The variable thermal property and irregular shapes should be avoided. For optimum thawing rates, modeling is helpful [26].

Tempering at home is done by leaving out the product in air at ambient temperature. Due to several drawbacks of this technique, MW tempering or RF tempering was introduced, because these are fast methods compared with traditional methods. MW tempering allows the temperature uniformly thus allowing proper mechanical handling or processing (such as: cutting, slicing, etc.). There was no compromise in the food quality even at refrigeration temperature. There was a reduction in drip loss and space requirement for MW tempering [48].

While selecting thawing or tempering process, one must consider the time required for thawing, bacteriological condition and appearance of the product, capital, and operating costs of a system, and effluent disposal, etc. Uniformity in heating and control of end temperature are essential requirements in MW tempering [56].

11.9.4 MICROWAVE (MW) DRYING

The MW drying results in supreme quality of product, optimum utilization of energy along with cut-down time of drying, etc. However, uneven drying and non-uniform temperature distribution are consistent issues related to MW drying [28]. Besides these disadvantages, MW drying is a fourth-generation technology with all its benefits. The MW drying is categorized into the following three stages:

1. **Heating up stage:** There is an increase in temperature, as there is energy transfer from radiation to thermal in the moisture containing material.
2. **Quick Drying Stage:** A uniform temperature is maintained, and vaporization of moisture takes place due to the generated thermal energy. In porous material, rate of vaporization depends on the local rate of thermal energy transformation from MW.
3. **Decreased Drying Stage:** The energy of vaporization is less than thermal energy transformed from MW energy.

Among all these phases, the heating phase is relatively small. Moisture distribution is determined by the second phase, as most of the moisture loss takes place here. Another application of drying process is the dielectric heating with MW energy. Other issues are textural damages and limited product penetration of the MW radiation into the product.

Various MW-assisted techniques are MW-assisted air drying (MWAD), MW-assisted drying in the whole air drying process (MDWAD), MW-assisted drying as final stage of air drying (MDFSAD), MW-assisted vacuum drying (MWVD), microwave-enhanced spouted bed drying (MWSD), MW-assisted freeze-drying (MWFD), MW-assisted finish drying following osmotic dehydration (MDOD) [70]. MW-assisted drying (MWD) is an important application. Materials, such as food, wood, ceramics, oils, etc., are dried by this method because the effectiveness is twice. Regularity of power absorbed with constant motions avoids hotspot formation [50].

11.9.5 MICROWAVE (MW) PASTEURIZATION AND STERILIZATION

In pasteurization, not only the pathogens are destroyed, but also the other bacteria and enzymes are inactivated for safe intake of food products. Pasteurization is an important process in the milk and fruit juices to reduce health hazards. The product is unstable at room temperature as the bacterial spores are not killed. In the ultra-high temperature (UHT) process, extreme high temperature is employed for an instance of time. Depending on the destruction of pathogens, heat-resistant, disease-causing microorganisms, temperature, and time duration are determined [1]. Results of MW pasteurization have shown an effective and successful application when compared to the controlled parameters. The results were accurate and took less time than the standard time for the pasteurization process alone [19].

The sterilization of any food product improves its shelf life and safety. Various sterilization methods are targeted heating, electroporation, magnetic field coupling and rupture of cell membrane. MW selective heating results in faster microbial destruction as the temperature within the microbial bodies are higher than the ambient temperature. The leakage of cellular material occurs due to the electroporation mechanism, in which the electrical potential across the cell membrane can generate pores in cells. The mechanism for proteins and nucleic acid in combination with the magnetic field leads to phagocytosis, and destruction of the microbial cell is termed as magnetic field coupling mechanism [13]. Inhomogeneous food (such as: fish in gravy packed in pouches; and chicken meat, etc.) showed promising results when it was treated with MW sterilization [61].

11.9.6 MICROWAVE-ASSISTED EXTRACTION (MWE)

Among several available extraction technologies, the biocomponents can be extracted using MW-assisted method. The MW-extraction of polyphenols and caffeine from green tea leaves showed high extraction, high extraction selectivity, shorter time of extraction at low cost, and less labor intensive. MAE is also speedy, harmless, and more eco-friendly for food and medicinal industries [46]. The similar observations for MAE of glycyrrhizic acid (GA) from licorice root were recorded. It was observed that the MAE resulted in equal, but not significantly greater than the conventional method [45]. MAE has also been used for extraction of flavonoids [51]. A maximum pectin yield (19.24%) of the product was obtained with MAE at optimum conditions [29].

Alternative laboratory-scale MAE method has been developed. Being a speedy process, the MAE has decreased requirements of solvents and higher extraction efficiency, etc. MAE of dietary polyphenols, including flavonoids (flavones, flavonols, isoflavones, catechins, flavanones, and cinnamic acid derivatives), anthocyanins, and carotenoids (lycopene, lutein, carotenes, etc.) is successful in the food industry [27].

11.9.7 MICROWAVE (MW) ROASTING

The use of MW can be effective in roasting of various food materials in comparison with conventional roasting [60]. Negative aspects associated with conventional roasting can be overcome by MW roasting. In the study related to the roasting of cocoa beans, there was a reduction in the process time and temperature required for roasting.

The lowermost possible values of volatile and total acidity and the topmost values of the fat content of the roasted beans are physiochemical property assets when MW roasting is applied to any process within less time than the conventional methods [24]. The negative impacts of MW roasting included variation in physiochemical property in sunflower seeds. There was a difference between the refractive indices of the control oils, which were 1.4610 and 1.4640, respectively. With increase in roasting time, refractive index was decreased. Browning reaction was also observed from light yellow to brown color of control oil when density was increased [2]. In the case of peanuts, some unwanted and toxic components were found even though it took lower time of roasting. The darkening of peanut oil was also observed [32].

Various other MW technologies are: ultrasonication, ohmic heating, electron irradiation, freezing, OD [22], infrared heating, and MW-vacuum processing [11].

11.10 EFFECTS OF MICROWAVE (MW) ENERGY

Various adverse effects have also been seen, such as thermal and nonthermal effects. Thermal effect is dependent on penetration depth and frequency. For frequency of 1,000 MHz, heating can occur in the deep tissue; while for 3,000 MHz, heating occurs superficially. At a frequency of 2,450 MHz, a combinational effect occurs. Nonthermal effect is associated with studies in the laboratory with use of MWs. Excessive radiation can lead to cataracts, sterility or cancer [42]. The parts of the human body are rapidly damaged, such as lens in eye ball and testicles. Damage of cataract is irreversible. Inactivation of certain enzymes, lining up of particulate matter, cellular differentiation was inhibited in chick embryos; and clotting time of blood has been affected [62].

11.11 SUMMARY

MW has been successfully used in food industries for blanching, baking, tempering, thawing, drying, extraction, roasting, etc. In this chapter, various MW-assisted techniques and modeling law responsible for optimization of MW working have been described. Brief introduction of dielectric properties is remarkable with the use of MWs. Dielectric properties of food material add up proper indication for food behavioral patterns. The obstacles in MW applications can be resolved if a proper understanding of the mathematical model is implemented. The proficiency in modeling methods may result in cost effectiveness in the use of MWs. MW technology leads to better product quality in less time.

KEYWORDS

- **dielectric properties**
- **drying**
- **extraction**
- **food processing**
- **glycyrrhizic acid**
- **microwave**
- **ultrasonication**

REFERENCES

1. Ahmed, J., & Ramaswamy, H., (2007). Microwave pasteurization and sterilization of foods. In: *Food Science and Technology Handbook of Food Preservation* (2nd edn., pp. 691–711). Boca Raton–FL: CRC Press.
2. Anjum, F., Anwar, F., Jamil, A., & Iqbal, M., (2006). Microwave roasting effects on the physicochemical composition and oxidative stability of sunflower seed oil. *Journal of the American Oil Chemists' Society, 83*(9), 777–784.
3. Arroqui, C., Rumsey, T., Lopez, A., & Virseda, P., (2002). Losses by diffusion of ascorbic acid during recycled water blanching of potato tissue. *Journal of Food Engineering, 52*, 25–30.
4. Barringer, S. A., Davis, E. A., Gordon, J., Ayappa, K. G., & Davis, H. T., (1995). Microwave-heating temperature profiles for thin slabs compared to maxwell and lambert law predictions. *Journal of Food Science, 60*(5), 1137–1142.
5. Bengtsson, N. E., & Ohlsson, T., (1974). Microwave heating in the food industry. *Proceedings of the IEEE, 62*, 44–55.
6. Brewer, M. S., Klein, B. P., Rastogi, B. K., & Perry, A. K., (1994). Microwave blanching effects on chemical, sensory and color characteristics of frozen green beans. *Journal of Food Quality, 17*, 245–259.
7. Chandrasekaran, S., Ramanathan, S., & Basak, T., (2011). Microwave material processing: A review. *AIChE Journal, 58*(2), 330–363.
8. Chandrasekaran, S., Ramanathan, S., & Basak, T., (2013). Microwave food processing: A review. *Food Research International, 52*, 243–261.
9. Datta, A. K., & Anantheswaran, R. C., (2001). *Handbook of Microwave Technology for Food Applications* (pp. 69–107). New York: M. Dekker Inc.
10. Datta, A. K., & Davidson, P. M., (2000). Microwave and radio frequency processing. *Journal of Food Science, 65*, 32–41.
11. Ekezie, F. C., Sun, D., Han, Z., & Cheng, J., (2017). Microwave-assisted food processing technologies for enhancing product quality and process efficiency: A review of recent developments. *Trends in Food Science & Technology, 67*, 58–69.
12. Geedipalli, S., Rakesh, V., & Datta, A., (2007). Modeling the heating uniformity contributed by a rotating turntable in microwave ovens. *Journal of Food Engineering, 82*, 359–368.
13. Guo, Q., Sun, D., Cheng, J., & Han, Z., (2017). Microwave processing techniques and their recent applications in the food industry. *Trends in Food Science & Technology, 67*, 236–247.
14. Guo, W., Trabelsi, S., Nelson, S., & Jones, D., (2007). Storage effects on dielectric properties of eggs from 10 to 1800MHz. *Journal of Food Science, 72*(5), 335–340.
15. Heddleson, R. A., & Doores, S., (1994). Factors affecting microwave heating of foods and microwave induced destruction of foodborne pathogens: A review. *Journal of Food Protection, 57*, 1025–1037.
16. Hill, J. M., & Marchant, T. R., (1996). Modelling of microwave heating. *Applied Mathematical Modeling, 20*(1), 3–15.
17. Içier, F., & Baysal, T., (2004). Dielectrical properties of food material; Part-1: Factors affecting and industrial uses. *Critical Reviews in Food Science and Nutrition, 44*(6), 465–471.

18. Içier, F., & Baysal, T., (2004). Dielectrical properties of food materials, Part-2: Measurement techniques. *Critical Reviews in Food Science and Nutrition, 44*, 473–478.
19. Jaynes, H. O., (1975). Microwave pasteurization of milk. *Journal of Milk and Food Technology, 38*(7), 386–387.
20. Joseph, O. I., (2017). Microwave heating in food processing. *BAOJ Nutrition, 3*, 1–9.
21. Kalla, A. M., & Devaraju, R., (2017). Microwave energy and its application in food industry: A review. *Asian Journal of Dairy and Food Research, 36*(1), 37–44.
22. Kidmose, U., & Martens, H. J., (1999). Changes in texture, microstructure and nutritional quality of carrot slices during blanching and freezing. *Journal of the Science of Food and Agriculture, 79*, 1747–1753.
23. Krishnaswamy, K., Orsat, V., Gariépy, Y., & Thangavel, K., (2013). Optimization of microwave-assisted extraction of phenolic antioxidants from grape seeds (*Vitis vinifera*). *Food and Bioprocess Technology, 6*, 441–455.
24. Krysiak, W., (2011). Effects of convective and microwave roasting on the physicochemical properties of cocoa beans and cocoa butter extracted from this material. *Grasas Y Aceites (Fats & Oils), 62*(4), 467–478.
25. Ku, H. S., Siores, E., Taube, A., & Ball, J. A. R., (2002). Productivity improvement through the use of industrial microwave technologies. *Computers and Industrial Engineering, 42*, 281–290.
26. Li, B., & Sun, D., (2002). Novel methods for rapid freezing and thawing of foods–A review. *Journal of Food Engineering, 54*, 175–182.
27. Li, Y., Fabiano-Tixier, A., Abert-Vian, M., & Chemat, F., (2012). Microwave-assisted extraction of antioxidants and food colors. *Food Engineering Series Microwave-Assisted Extraction for Bioactive Compounds, 2012*, 103–125.
28. Li, Z. Y., Wang, R. F., & Kudra, T., (2011). Uniformity issue in microwave drying. *Drying Technology, 29*(6), 652–660.
29. Maran, J. P., Sivakumar, V., Thirugnanasambandham, K., & Sridhar, R., (2013). Optimization of microwave assisted extraction of pectin from orange peel. *Carbohydrate Polymers, 97*, 703–709.
30. Marra, F., Zhang, L., & Lyng, J. G., (2009). Radio frequency treatment of foods: Review of recent advances. *Journal of Food Engineering, 91*, 497–508.
31. Meda, V., Orsat, V., & Raghavan, V., (2017). Microwave heating and the dielectric properties of foods. *The Microwave Processing of Foods, 2017*, 23–43.
32. Megahed, M. G., (2001). Microwave roasting of peanuts: Effects on oil characteristics and composition. *Food / Nahrung, 45*(4), 255–257.
33. Metaxas, A. C., & Meredith, R. J., (1988). *Industrial Microwave Heating* (2nd edn., pp. 78–82). London: Peter Peregrinus Ltd.
34. Nelson, S. O., (1991). Dielectric properties of agricultural products-measurements and applications. *IEEE Transactions on Electrical Insulation, 26*(5), 845–869.
35. Nelson, S. O., (1973). Electrical properties of agricultural products: A critical review. *Transactions of the ASAE, 16*, 384–400.
36. Nelson, S. O., (1983). Dielectric properties of some fresh fruits and vegetables at frequencies of 2.45 to 22 GHz. *Transactions of the ASAE, 26*, 613–616.
37. Nelson, S. O., (1999). Dielectric properties measurement techniques and applications. *Transactions of the ASAE, 42*(2), 523–529.
38. Nelson, S. O., & Trabelsi, S., (2006). Dielectric spectroscopy of wheat from 10 MHz to 1.8 GHz. *Measurement Science and Technology, 17*, 2294–2298.

39. Nelson, S. O., & Trabelsi, S., (2012). Factors influencing the dielectric properties of agricultural and food products. *Journal of Microwave Power and Electromagnetic Energy, 46*(2), 93–107.
40. Nelson, S. O., & Bartley, P. G., (2002). Frequency and temperature dependence of the dielectric properties of food materials. *Transactions of the ASAE, 45*(4), 1223–1227.
41. Ohlsson, T., & Bengtsson, N., (2001). Microwave technology and foods. *Advances in Food and Nutrition Research, 43*, 65–140.
42. Okeke, C., Abioye, A. E., & Omosun, Y., (2014). Microwave heating applications in food processing. *IOSR Journal of Electrical and Electronics Engineering, 9*(4), 29–34.
43. Oliveira, M., & Franca, A., (2002). Microwave heating of foodstuffs. *Journal of Food Engineering, 53*(4), 347–359.
44. Orsat, V., Raghavan, G. S. V., & Krishnaswamy, K., (2017). Microwave technology for food processing. *The Microwave Processing of Foods, 2017*, 100–116.
45. Pan, X., Liu, H., Jia, G., & Shu, Y. Y., (2000). Microwave-assisted extraction of glycyrrhizic acid from licorice root. *Biochemical Engineering Journal, 5*, 173–177.
46. Pan, X., Niu, G., & Liu, H., (2003). Microwave-assisted extraction of tea polyphenols and tea caffeine from green tea leaves. *Chemical Engineering and Processing: Process Intensification, 42*, 129–133.
47. Puligundla, P., Abdullah, S. A., Choi, W., Jun, S., Oh, S., & Ko, S., (2013). Potentials of microwave heating technology for select food processing applications – A brief overview. *Journal of Food Processing & Technology, 4*(11), 210–218.
48. Raghavan, V., Orsat, V., & Meda, V., (2005). Microwave processing of foods. *Stewart Postharvest Review, 3*(2), 1–7.
49. Ramesh, M., Wolf, W., Tevini, D., & Bognar, A., (2002). Microwave blanching of vegetables. *Journal of Food Science, 67*(1), 390–398.
50. Rattanadecho, P., & Makul, N., (2016). Microwave-assisted drying: A review of the state-of-the-art. *Drying Technology, 34*(1), 1–38.
51. Routray, W., & Orsat, V., (2011). Microwave-assisted extraction of flavonoids: A review. *Food and Bioprocess Technology, 5*, 409–424.
52. Ryynänen, S., (1995). The electromagnetic properties of food materials: A review of the basic principles. *Journal of Food Engineering, 26*, 409–429.
53. Salazar-González, C., Martín-González, M. F. S., López-Malo, A., & Sosa-Morales, M. E., (2012). Recent studies related to microwave processing of fluid foods. *Food and Bioprocess Technology, 5*, 31–46.
54. Schirack, A., Sanders, T., & Sandeep, K., (2007). Effect of processing parameters on the temperature and moisture content of microwave-blanched peanuts. *Journal of Food Process Engineering, 30*, 225–240.
55. Schubert, H., & Regie, M., (2006). Novel and traditional microwave applications in the food industry. *Advances in Microwave and Radio Frequency Processing, 2006*, 259–270.
56. Shaheen, M. S., El-Massry, K. F., El-Massry, A. H., & Anjum, F. M., (2012). Microwave applications in thermal food processing; Chapter 1. In: Cao, W., (ed.), *The Development and Application of Microwave Heating*. In-Tech [online]. https://www.intechopen.com/chapters/40744 (accessed on 09 February 2023).
57. Smith, J. S., & Hui, Y. H., (2008). *Food Processing: Principles and Applications* (p. 524). London: John Wiley & Blackwell.

58. Sosa-Morales, M. E., Valerio-Junco, L., López-Malo, A., & García, H. S., (2010). Dielectric properties of foods: Reported data in the 21st century and their potential applications. *LWT–Food Science and Technology, 43*, 1169–1179.
59. Sumnu, G. A., (2001). Review on microwave baking of foods. *International Journal of Food Science & Technology, 36*, 117–127.
60. Takagi, S., Ienaga, H., Tsuchiya, C., & Yoshida, H., (1999). Microwave roasting effects on the composition of tocopherols and acyl lipids within each structural part and section of a soybean. *Journal of the Science of Food and Agriculture, 79*, 1155–1162.
61. Tang, Z., Mikhaylenko, G., Liu, F., Mah, J., Pandit, R., Younce, F., & Tang, J., (2008). Microwave sterilization of sliced beef in gravy in 7-oz trays. *Journal of Food Engineering, 89*, 375–383.
62. Tape, N. W., (1970). Application of microwave energy in food manufacture. *Canadian Institute of Food Technology Journal, 3*(2), 39–43.
63. To, E. C., Mudgett, R. E., Wang, D. I. C., Goldblith, S. A., & Decareau, R. V., (1974). Dielectric properties of food materials. *Journal of Microwave Power, 9*, 303–315.
64. Vadivambal, R., & Jayas, D. S., (2010). Non-uniform temperature distribution during microwave heating of food materials—A review. *Food and Bioprocess Technology, 3*, 161–171.
65. Venkatesh, M. S., & Raghavan, G. S. V., (2005). An overview of dielectric properties measuring techniques. *Canadian Biosystems Engineering, 47*, 7.15–7.30.
66. Venkatesh, M. S., & Raghavan, G. S. V., (2004). An overview of microwave processing and dielectric properties of agri-food materials. *Biosystems Engineering, 88*(1), 1–18.
67. Wang, S., Tang, J., Johnson, J. A., Mitcham, E., Hansen, J. D., Hallman, G., Drake, S. R., & Wang, Y., (2003). Dielectric properties of fruits and insect pests as related to radio frequency and microwave treatments. *Biosystems Engineering, 85*, 201–212.
68. Wray, D., & Ramaswamy, H. S., (2015). Novel concepts in microwave drying of foods. *Drying Technology, 33*, 769–783.
69. Wu, T., (2008). Environmental perspectives of microwave applications as remedial alternatives: Review. *Practice Periodical of Hazardous, Toxic, and Radioactive Waste Management, 12*, 102–115.
70. Zhang, M., Tang, J., Mujumdar, A., & Wang, S., (2006). Trends in microwave-related drying of fruits and vegetables. *Trends in Food Science & Technology, 17*(10), 524–534.

CHAPTER 12

POTENTIAL OF MICROWAVE HEATING IN FOOD PROCESSING

MOHAN NAIK,
YASHASWINI JAPANNANAYAKANAKOTE PUTTALINGAPPA, and
N. VENKATACHALAPATHY

ABSTRACT

Microwave (MW) heating is an efficient thermal processing method in food processing with several advantages, such as uniform heating (when combined with other techniques), timesaving, energy-efficient, simple, and versatile design of equipment. MW processing has certain limitations, such as non-uniform heating or development of "hotspots" and unsuitable for large or thick food materials. The combination of the MW with other heating methods (such as freezing, convective drying, infrared drying, vacuum drying, ultrasound, plasma, and pulsed spout drying) is gaining special interest. Still in food processing, there is a need for an efficient method without damaging the quality characteristics. The chapter aims to highlight the applications of MW heating in various unit operations involved in food processing.

12.1 INTRODUCTION

During the past five decades, microwaves (MWs) heating has established a wide array of applications in food processing operations, such as drying, thawing, concentration, dehydration, extraction, pasteurization, sterilization, tenderization, tempering, baking, blanching, puffing, and cooking. It extends numerous advantages over the traditional method of heating because higher rate of heat transfer, lesser time of product exposure, uniform heating,

Advanced Research Methods in Food Processing Technologies. Junaid Ahmad Malik, Megh R. Goyal, Preeti Birwal, & Ritesh B. Watharkar (Eds.)
© 2024 Apple Academic Press, Inc. Co-published with CRC Press (Taylor & Francis)

lower maintenance, and safe handling procedures [37]. Furthermore, it retains most of the quality characteristics than the conventional methods. MW processing has some industrial non-compliances, such as: non-uniform temperature distribution and overheating. The efforts have been made by various researchers to overcome these limitations to combine with other technologies like imparting motion/rotation, ohmic heating, vacuum drying, convective heating, infra-red heating, ultra-sonication, etc. [12].

In the electromagnetic spectrum, the frequency of MWs categorized with wavelength ranges from 1.00 mm–1.00 m at frequencies between 300 MHz and 300 GHz. MW frequencies of 915 MHz and 2,450 MHz are most pertinent and have wide range of application in the field of scientific study [31].

Thermal energy efficiency is a predominant factor that has to be considered for the successful operation of food processing plant as it influences the cost economics, quality of the final product and degree of environmental impact. When energy efficiency is considered then MW holds an upper hand because in conduction and convection modes of heating, the heat transfer takes place from surface of the food to interior resulting in increased energy requirement and processing time. Volumetric heating of the food product takes place when exposed to MW energy, greater penetration power along with volumetric heating will result in a reduction of the temperature gradient between the outer surface and inner side of the substance [35].

The recent advancements in MW processing include MW-freeze drying, MW-vacuum drying, MW-convective drying, MW-frying, MW-freezing, MW-infrared drying, MW-ohmic drying, MW assisted-osmotic drying and MW assisted-ultrasonication. Potentiality of these hybrid techniques has been tested by various investigators. However, for industrial applications, it has to be tested and to optimize the process parameters and operating conditions. Collectively, MW heating serves as an energy-efficient technology, which has added several advantages over conventional methods of processing. Hence, MW has become popular even at the domestic level.

This chapter is focused on the potential applications of MW in various food processing unit operations, such as pasteurization and sterilization, baking, blanching, cooking, and MW drying. Furthermore, the effects of MW heating on quality characteristics of various categories of foodstuffs have been discussed.

12.2 MECHANISM OF MICROWAVE (MW) HEATING

MW is transmitted in the form of electromagnetic waves, when MWs interact with a material, part of the energy is transmitted/reflected/absorbed by the food material, and heat dissipation occurs (Figure 12.1). These characteristics phenomenon leads generation of heat within in the food material. Dipolar interactions occur within the food substance because of molecular motion and intermolecular friction [37], which results in changes in temperature gradient within the material and release of heat in all directions [19]. In liquids with higher ionic concentrations may be get heated faster with increasing frequency of the process [41].

MW heating mechanism is fundamentally dependent on numerous factors, such as dielectric properties of food material and penetration depth, nature of food material, equipment design, MW frequency, amount of water present in foods, mass, volume, product composition and load [21]. Generally, food substances are poor electrical insulators and conductors; hence, it has been categorized as "lossy dielectric materials." Dielectric properties of foods are key influential factors that are responsible for the interaction between the food components and electric field and there are many factors, such as temperature, amount of water present, ionic concentration, frequency of the MW and other elements that can influence the dielectric properties of the food material [6].

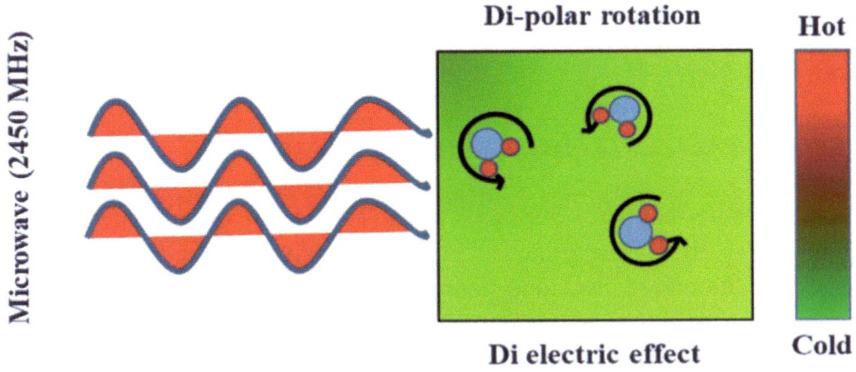

FIGURE 12.1 Mechanism of microwave heating.

Dielectric properties can be defined as follows:

$$E = E' - jE'' \quad (1)$$

where; E is the dielectric property; E′ is the dielectric constant; E″ is the dielectric loss factor; and δ is the dielectric loss angle ($\tan\delta = E''/E'$).

The relative permittivity/dielectric constant is the real part of the dielectric properties which implies the ability to store electric energy and the imaginary part of the dielectric property, termed as dielectric loss, which indicates the ability and convert electric energy into heat.

The power dissipated in a material is proportional to the frequency of the source, the dielectric loss of the material, and the square of the field strength within it. The power dissipated per unit volume can be expressed as:

$$P = \omega E_0 E' (\tan\delta) M^2_{rms} = 2\pi f E_0 E'' E^2_{rms} \quad (2)$$

where; E_0 is the permittivity of free space (8.85310212 F/m); f is the MW frequency (Hz); and M is the electrical field strength (V/m) in the material.

The power penetration depth (P_D) is defined as the distance at which the power density drops to a value of *1/e* from its value at the surface [32]. Eqn. (3) is applicable for food materials, which are non-magnetic ($\mu r = 1$) in nature.

$$Penetration\,depth = \frac{v}{\sqrt{2\pi f}[k'\{\sqrt{1+(k''/k')^2} - 1\}]^{1/2}} \quad (3)$$

where; v is the velocity of light given by $v = (\mu_0 E_0)^{-1/2}$;ω is the angular frequency; and μ_0 is the permeability of free space ($\mu_0 = 4\pi \times 10^{-7}$ H/m).

12.3 APPLICATIONS OF MICROWAVE (MW) HEATING IN FOOD PROCESSING

12.3.1 PASTEURIZATION AND STERILIZATION

Pasteurization and sterilization are the major step in the area of food processing and preservation. The principal aim is to reduce the active, pathogenic, spoilage microorganisms and potent enzyme mainly in liquid foods. The presence of pathogenic microorganisms and enzymes accelerates spoilage and destroys nutritional composition and organoleptic characteristics of food-stuffs. Traditional thermal methods have been commonly adopted in food and milk processing industry till date; It is found that high temperatures employed to reduce the microbial load and also result in thermal degradation of volatile and thermo-sensitive substances. Which results in loss of sensory appeal and nutritive value [4]. The preservation effects of MW heating are generally attributed to thermal effects, various works of literature have been reported that MW heating technique for the inactivation of bacteria, virus, yeast, fungi, and mold follows first-order kinetics. This novel heating method results in selective heating of the microbial cells to higher temperature level than the surrounding fluid leads rapid devastation of microbial cells and electroporation of cellular membrane occurs thereby eliminates the microbial load [18].

Siguemoto et al. [40] determined the impact of conventional and MW pasteurization on flavor and nutritional quality of cloudy apple juice. The authors have been reported that MW pasteurized samples well retained most of the volatile compounds and little changes in sugar-acid ratio and phenolic compounds. Similarly, González-Monroy et al. [17] has reported 2,450 MHz, 490 W MW treatment with a target temperature of 90°C and holding time of 12s could effectively inactivate the pectin methylesterase (PME) activity in green and tamarind beverages. Furthermore, significant changes in sensorial attributes, color, and no much changes in total soluble solids (TSS), pH and titratable acidity (TA) were reported.

MW sterilization has been recognized as a validated processing method to produce low-acid, shelf-stable packaged foods by FDA and USDA-FSIS [45]. The MW sterilization process is a high-temperature-short-time (HTST) process which can cause minimal quality deterioration and rapid microbial destruction. MW sterilization process (target temperature: 128°C; time: 3 min) results superior quality products than conventional canning (retort temperature: 120°C; time: 45 min) and retorting foil pouching (temperature: 125°C; time 13 min) [36]. Other authors have been investigated the consequence of package barrier characteristics on storage life of vitamin-C-fortified sweet potato puree (SPP) processed with MW sterilization. The decrease in vitamin-C, color, and increase in β-carotene level after treatment and storage was reported [53].

To date, there are a limited number of industrial pasteurization/sterilization devices that are available for packed foods and or food ingredients. Joel Haamer developed an industrial-scale in-pack meal pasteurization system at atmospheric pressure for meal size of 150 g up to 1 kg. In this developed system, the MW energy is used to create steam. During processing, steam is introduced to create over-pressure-temperature within the pack. The main function of vent is to release excess energy as steam and to remove the oxygen. When MW energy is removed, the products fall below the boiling point and the steam seals aseptically and creates a vacuum. The sealed platter/dish is cooled in the spiral chiller. Furthermore, valve can be used for reheating of packed food tray [14].

The main problem linked with the MW method of heating is uneven temperature distribution with the formation of cold spots [7]. To overcome this liability combination conventional and MW heating can be employed [48].

12.3.2 DRYING

Drying is one of the basic methods employed for the preservation of agro-horticultural commodities. Drying extends the shelf life of the food substance, thereby there is a scope of technological intervention in this process. Drying mainly aims at reducing the moisture content within the substance, this process is considerable consumes more time and energy in the food industry. New novel methods of drying are intended to minimize drying time and energy requirement with minimum loss in quality aspects of the food substances [22]. In traditional heating methods, thermal energy is passed to the surface of the material whereas, in MW drying/heating, heat is generated inside the material, thereby higher heat transfer can be achieved. It has been reported that in MW heating temperature would rise much faster than in traditional heating [35]. MW energy is used as a single or supplementary heat source in drying processes like, air drying, vacuum drying, and freeze-drying. Several researchers have reported that there is a significant reduction in drying time of carrot slices (25–90%) [34], kiwifruit (40–89%) [30], garlic [39], oyster mushrooms (75%) [3].

Vacuum drying is another important method of drying under reduced pressure. Under this condition, the evaporation of water occurs at lower temperatures with minimum changes in chemical, sensorial, and nutritional attributes of foods. MW heating is a volumetric process hence, it serves as a potent way to transmit energy in the system. This helps again in falling rate drying periods and reduces the time of drying process. In MW-vacuum drying effective moisture diffusivity would increases. MW vacuum drying results porous products, which rehydrates more speedily and thoroughly than the air-dried product [16]. MW freeze drying is one of the choice to overcome the prevailing disadvantage of the conventional drying process. It can produce the similar quality products as that of traditional freeze-drying and can reduce the drying time effectively. The quality of the freeze-dried product is best to compare with other conventional techniques, but it is a long time processing and brings high energy consumption issue [50].

The advancement MW drying has led the development of hybrid techniques like pulsed fluidized bed MW freeze-drying, low-frequency MW (915 MHz) assisted pulse-spouted bed freeze-drying, MW-assisted hot air fluidized bed drying (FBD), MW-assisted FBD, vacuum-assisted MW drying, Ultrasound-assisted hybrid intermittent MW-hot air drying and MW-assisted foam-mat drying. These technologies have shown greater potential against conventional drying methods. The operating parameter and process conditions are needed to optimize for industrial application for different types of commodities.

12.3.3 BAKING

Baking is the process of cooking food stuffs by using dry heat. Where heat is conveyed to the product surface mainly by convection from the heating medium and by radiation from oven walls followed by conduction to the center [46]. Wherein MW baking, food material is surrounded by air at ambient temperature, which is not heated by MWs. MWs interacts with charged particles and polar molecules. As a result, heat would produce interior of the food substance and it is conducted through the food material. Considerably MW takes lower time for heating due to the rapid heat generation. MW processing parameters like temperature and baking time have a strong influence on kinetic reactions of starch gelatinization and browning. Hence, it is prerequisite to confirm the gelatinization and browning reactions are completely accomplished when baking time is reduced, or else, the quality of the product can be reduced [47].

MW baking significantly reduces the baking time (up to five times) compared to the traditional convective baking process. Similar quality characteristics were observed in both the cases with slight changes in color development on the pastry surface. Furthermore, different structural formation in terms of pore size distribution was reported [15]. Accelerated stalling, low volume, color defects, crust formation, firm, tough texture and higher moisture loss are the major drawback of the MW baking to convection baking. Moisture loss can be minimized by the addition of instant starches or hydrocolloids. Type and nature of starches present in raw materials also significantly contribute to the baking quality of the baked foods. A study conducted by Seyhun et al. [38] using corn, potato, waxy corn, amylo-maize, and pre-gelatinized starches to hinder the staling in MW baked cake (1.5 min at 100% power). Authors have been reported that pre-gelatinized starches most effectively hinder the stalling or moisture during baking and storage. Cake prepared by adding amylo-maize starch produces more firm texture and it has been noticed that amylose content was leached in control samples baked by MW. To alleviate the MW baking problems, efforts have been done by combining MW with other methods. Conjoining of MW energy with halogen lamp is effectively solves the MW baking problems and adds advantages like lesser time of baking, browning, and crispiness to the baked products.

Sumnu et al. [44] evaluated the quality attributes of cakes baked by MW, infrared, and infrared-MW combination of techniques. MW resulted in a poor quality of baked cakes whereas infrared-MW baked cakes found good color formation. Halogen lamp-combined with MW produces a similar quality of cakes obtained from the convection baking process. The combination of IR-MW (5 min at 70% halogen lamp and 50% MW power level) produces a superior quality of cakes. Under these optimum conditions, about 75% of decline in baking time was observed by the authors. In halogen lamp-MW combination, MW plays a dominant role in terms of affecting quality characteristics. Furthermore, by the increase in power reduction in specific volume, the total color difference increased weight loss and firmness as reported. Loaves of bread backed by halogen-MW combination have shown acceptable specific volume and color development. Furthermore, higher firmness and weightless was recorded [23].

Dielectric properties of the foods affected by the numerous factors like MW frequency, moisture content, temperature, salt concentration and nutritional composition. The thorough knowledge of MW frequency and dielectric properties helps to determine the optimum frequency for desirable application. Alaa et al. [1] experimented on dielectric properties of MW (0.915–2.450 GHz) baked cake and its constituents. It is driven fact that the dielectric characteristics of food materials increased with increase in frequency. The dielectric constant for cake batter and flour was independent of temperature. The dielectric constant of Madeira cake constituent, i.e., sugar would be increased with an increase in product temperature. At higher frequency level (2,450 MHz), the dielectric loss factor for cake batter and flour relatively decreased with increase in temperature. Whereas at a lower frequency level (915 MHz) which was significantly not affected. During the initial period of MW baking, the dielectric properties of the cake batter was increased up to the limit, and the latter decreased gradually until the completion of the baking process.

12.3.4 BLANCHING

Blanching is one of the common pre-processing practices used to preserve the quality and improve the storage life of fruits and vegetables (FAV). The main aim of this process is to inactivate active enzymes, reduce the volume of the material by expelling intracellular trapped air, reduce the microbial population, and removes detrimental odors and flavors [5]. Conventional blanching is practiced by hot water and steam. MW-assisted blanching reduces the time and energy required to achieve the inactivation of enzymes. Since it is a process that allows HTST treatment of solid foods through volumetric heating. It is essential to preserve thermo-sensitive nutrients, vitamins, and other active compounds present in foods [10].

Evidence have shown that fruit peel (orange, mango, banana, and apple) treated at 720 W MW power level up to 3 min of treatments shown changes in the total phenolic compound, antioxidant capacity, and browning index. Increase in free and bound phenolic fractions and antioxidant activity increased [9]. MW power level higher than 900 W greatly alter the cellular microstructure and disintegrate the cell-matrix (middle lamella). As a result, changes in cell wall polymers was noticed by the authors [25]. MW blanching effectively inactivates ascorbic acid (AA) oxidase and retains L-ascorbic acid and dehydroascorbic acid.

Low-temperature long time water blanching (holding time: 12 min; temperature: 70°C) shown vitamin C loss, which is mainly due to mass transfer phenomena rather than temperature degradation, while after high-temperature short treatments (holding time: 5 min; temperature: 90°C) the decrease of total vitamin C content seemed to be mainly related to thermal degradation than due to the leakage of the nutrients in the blanching medium [51]. MW hot-air flow rolling dry-blanching modify the microstructure and affects the drying characteristics, water content and migration. And also improves the drying features and quality of FAV. The MW power of 3 W/g and 9 min of treatment time increased polysaccharide, formaldehyde, and phenolic content of oyster mushroom. The active enzymes like polyphenol oxidase (PPO) and peroxidase (POD) activities were completely inactivated. Blanching more than 12 min of treatment results decrease in brightness, polysaccharides, and phenolic content [43].

The effectiveness of MW blanching mainly depends on process conditions, material characteristics. Liburdi et al. [27] have investigated the effect of MW blanching and geometric shapes of vegetables on quality attributes. The vegetables like potatoes, savoy cabbage and white cabbage were cut into different shapes, namely cylindrical, large parallelepiped, small parallelepiped and cuboid-shape. The inactivation of POD firstly accomplished at the core of the sample and steadily increases towards the periphery of the samples with an increase of power. The MW exposure of 120 s at 160 W, 75 s at 160 W and 45 s at 350 W effectively inactivates POD activity in cylinder, large parallelepiped and small parallelepiped samples, respectively. The geometric shape plays a selective role in the optimization of process conditions for MW blanching of FAV. Collectively MW blanching serves as an efficient, time-energy saving method for preservation of biological materials.

12.3.5 COOKING

Cooking is one of the thermal processing methods of foods at a higher temperature in different heating medium (hot water, steam, hot oil, hot sand, radiation, etc.). The main purpose of cooking is to alter a foods in an appetizing, safe, and nutritious, which poses various changes in nutritional composition, texture, and quality characteristics (color, aroma, softness, hardness, and other texture associated characteristics). The thermal cooking methods like autoclaving, microwaving, sous vide, stewing, baking, frying, steaming, boiling, roasting, grilling, etc., are in practice at household and industrial level of food processing. These methods have been proven a wide array of changes in sensorial, digestion, and quality characteristics aspect.

In general, cooking using boiling water has the disadvantages like consumes relatively more time, nutrients are lost by leaching, needs intermittent mixing, consumes more fuel and needs relatively more human intervention. An alternative to the conventional cooking method, novel methods like MW cooking, ohmic cooking, infrared cooking, radiofrequency (RF) cooking and electromagnetic induction cooking has been shown a promising effect on food quality and safety. Among all the treatments MW processing widely accepted throughout the globe. The various scientific investigations have been reported the potential pros and cons of MW cooking on various aspect of foodstuffs.

Liu et al. [29] evaluated the effect of a different method of cooking on the quality and nutrient aspect of edible fungi. Its different degree of quality aspect in different cooking methods was noticed. More importantly, changes in physical, chemical, enzymatic modification and digestion aspects were observed. MW cooked fungi samples released more amount of protein, fat, and minerals during digestion was reported. Losses of protein bio-accessibility were more in fried samples. MW cooking (20 min, 750 W) better retains nutrients and save energy than the conventional cooking method. However, around 70% loss of Vitamin-C and 27% loss of total phenolic compound was reported in case of MW cooked courgetti, which would be lower than that of the conventional cooking [11].

Xu et al. [52] experimented on the effect of MW and steam cooking on quality of purple sweet potatoes. It was reported that combination of MW-steam cooking (MW power: 500 W; steam: 1,700 W for 12 min) improves the total phenolic content, flavonoids, phenolic acids and anthocyanins. Synergistic effect of method would reduce the cooking time and retains most of the nutrients compared with individual steaming.

Nowadays, the demand for low oil low calorie fried foods has increased due to an increase in nutritional awareness among the consumers. The fried foods are more popular because of their characteristics aroma, taste, and sensory attributes. Reducing oil uptake during frying is one the challengeable task of preparation of fried foods. Many pre-frying treatments like blanching, coating, and dehydration are in practice to reduce the oil absorption. In this concern, nowadays vacuum frying and MW frying gained special interest. These methods result in high quality fried products with low oil content. The MW power level and frying time have a great influence on moisture content hence, moisture (%) decreases with increase in frying time and MW power. The optimum condition for preparation of potato chips using a MW is 550 W power level for 2.5 min. Under these conditions, the oil uptake ratio was relatively lower than that of conventional frying method [33].

Devi et al. [8] have studied the consequence of ultrasound on MW-vacuum fried mushroom chips. Increase in temperature and MW power level increase in moisture loss was reported. It was observed that ultrasound-MW-vacuum frying could reduce oil content by 16–20% and produces a better quality of fried chips. Ultrasound-MW-vacuum frying (MW power of 1,000 W at 90°C) accelerates the frying rate. By these hurdle technologies, the energy consumption was decreased to 28–39% with lower oil uptake, and enhance in color, crispness, chlorophyll, and vitamin C content was noticed.

From the literature, it can be concluded that MW frying combined with ultra-sonication, vacuum frying produces high quality fried products with improved textural and national quality products. Similarly reduces the oil uptake (%) and considerably consumes lower energy.

12.3.6 EXTRACTION

Microwave-assisted extraction (MWE) is a profound method widely used for extraction and recovery of valuable compounds from biological materials. The methods offer several advantages like high extraction rate, high-quality products, low cost, lower extraction time, lower energy consumption and less solvent use over the conventional extraction process. It mainly works on the mechanism of ionic conductance and dipole rotation. Other than extractions MW also used for purification, concentration, and drying. MW induces cell wall damage by forced superheating of water molecules entrapped and uninterrupted collisions within the matrix. These changes cause sudden diffusion of compounds within cells into the surrounding solvent [49].

Al-Dhabi et al. [2] assessed the quality of MW-assisted polysaccharide from *Syzygium cumini* seeds waste. The MW power of 515 W, pH of 3.2, solid to liquid ratio of 1:15 g/ml and MW heating time of 3.1 min resulted in the highest yield of 4.71% of polysaccharide. The morphological and physicochemical changes were observed in the extracted polysaccharide. The impact of MWE of soymilk from soybean was investigated. The MWE resulted in the highest yield of 24% soymilk with a protein content of 44.4%. Compared to protein extracted from the conventional method, MWE resulted in high quality of protein in terms of protein solubility and protein digestibility. The optimum extraction condition for soymilk using MWE was power level of 675 W, a target temperature of 80°C and a stirring speed of 160 rpm [49]. Rapid cellular destruction and heating are the key features of MWE.

Fardhyanti et al. [13] analyzed the quality of mace essential oil extracted by MW-assisted hydro-distillation process. Under optimum extraction condition: 800 W MW power and 20 min of extraction time, the highest yield of 8.63% was obtained. The quality characteristics were comparable with the oil extracted from the conventional hydro-distillation method.

Ibrahim et al. [20] illustrated that oil yield increases with an increase in MW power. MWE resulted in a maximum yield of 37% at 330 MW and 20 min. MW accelerates the oil yield with a shorter duration of time than the conventional process. The changes or degradation in oil quality was noticed even at a lower power level. The degradation of oil quality may be attributed to the hydrolysis of the oil component at a higher temperature. Which resulted in changes in the acid value of extracted oil.

Extraction of pectin from conventional hot water for several hours (Temperature: 60–100°C; pH: 1.5–3.0) is tedious-time consuming process and results in lower pectin yield. Extraction pectin from biological material using MW shown the greater potential of extraction. Su et al. [42] evaluated the characteristics of pectin from orange peel by MWE with added surfactant. It was reported that pectin extracted by MWE with added surfactant results maximum yield of 32.8% under optimum extraction condition. The addition of a surfactant (Tween-80 (8 g/L, w/v)) would increase the pectin yield up to 17%. Changes in pectin yield and quality characteristics were reported in the case of pectin extracted from different methods.

Collectively by reviewing all the literature in this section, it can be concluded that the destructive nature of MW enhances the release of compounds from cellular materials. The degree of effects can be measured by evaluating the quality of products. Under optimized condition, extraction yield would be higher and expect lower degradation of compounds.

12.4 RECENT RESEARCH DEVELOPMENTS IN MICROWAVE (MW) HEATING

The major drawback of MW is non-uniform heating or formation of "hotspot," which leads degradation of the product. To overcome this problem, from the last five decades various developments in MW heating method was notices. To alleviate the problem of "hotspot" MW heating is combined with other methodologies like rotation, mixing, sample moving, forced air, vacuum, and modifying the composition of the surrounding media.

The advancement in the application of MW heating led the development of various hybrid techniques like ultrasonic (US)-MWE, pulsed fluidized bed-MW freeze-drying, low-frequency MW-assisted pulse-spouted bed freeze-drying, MW-assisted hot air FBD, low-pressure MW plasma sterilization, MW-assisted FBD, vacuum-assisted MW drying, US-assisted hybrid intermittent MW-hot air drying and MW-assisted foam-mat drying.

In recent years, MW freeze-drying is known as a potential method for obtaining a high-quality dried product. Liu et al. [28] evaluated the effect of spray drying and MW freeze-drying on quality of hydrolyzed/glycosylated ovalbumin. The results showed that product dried under spray drying resulted in higher protein solubility, emulsifying activity, foaming capacities and gel hardness than MW freeze-drying. However, MW freeze-dried products better in color, emulsion stability, foam stability, water/oil absorption capacity, and thermal stability.

Similarly, Li et al. [26] studied the consequence of US-assisted osmotic dehydration (OD) pretreatment before pulsed fluidized bed MW freeze-drying of Chinese yam. US treatment increases the mass transfer, loss factor and makes the water more mobile. Higher ultrasound power level increases the cell size. US pretreated-pulsed fluidized bed MW freeze-dried samples superior in quality in terms of color, texture, reduced astringency, sourness, and antioxidant capacity. Pulsed fluidized bed MW freeze-drying recognized for high quality dried products. The evidence is shown that it reduces the rehydration ration time, color degradation. Furthermore, it retains crispness, volatile compounds and antioxidant activities of freeze-dried products. More importantly, it reduces approximately 71.9% drying time compared to freeze-drying and pulsed fluidized bed MW freeze-drying accelerates the drying process effectively.

Kim et al. [24] conducted a study on the impact of MW-cold plasma (CP) treatments on decontamination of onion powder. The authors used helium as a plasma-forming gas, with simultaneous exposure to low MW density at 170 $mW \cdot m^{-2}$ or high MW density at 250 $mW \cdot m^{-2}$. The high MW density plasma

(400 W for 40 min) effectively reduces the activity of *Bacillus cereus* spores more than low MW density plasma. Quality characteristics like color, antioxidant capacity, and quercetin concentrations were not changed much during storage.

Overall, the diversified application of MW heating led to the investigation of novel techniques combined with MW heating. The MW-integrated technologies have shown potential for dying, dehydration, decontamination, extraction, purification, cooking, baking, and other wider applications.

12.5 SUMMARY

This chapter mainly highlights the fundamental of MW heating, mechanism, and its application in food processing. Furthermore, the chapter gives an example of MW processing of various products and the effect of MW on quality characteristics. It also overviewed the recent trends, developments, and limitations in MW preservation of foods. Still, there is a need for such technology, which brings desirable changes in food products by preserving native quality characteristics. For industrial applications, continuous MW system is usually needed with high-quality output.

KEYWORDS

- **conventional methods**
- **dipole rotation**
- **ionic conduction**
- **microwave heating**
- **microwaves**
- **polyphenol oxidase**
- **sweet potato puree**
- **total soluble solids**

REFERENCES

1. Ala'a, H., Hararah, M. A., Megahey, E. K., McMinn, W. A. M., & Magee, T. R. A., (2010). Dielectric properties of microwave-baked cake and its constituents over a frequency range of 0.915–2.450 GHz. *Journal of Food Engineering, 98*(1), 84–92.

2. Al-Dhabi, N. A., & Ponmurugan, K., (2020). Microwave assisted extraction and characterization of polysaccharide from waste jamun fruit seeds. *International Journal of Biological Macromolecules, 152*, 1157–1163.
3. Bhattacharya, M., Srivastav, P. P., & Mishra, H. N., (2015). Thin-layer modeling of convective and microwave-convective drying of oyster mushroom (*Pleurotus ostreatus*). *Journal of Food Science and Technology, 52*(4), 2013–2022.
4. Bhushand, D. M., Vyawarea, A. N., Wasnik, P. G., Agrawal, A. K., & Sandey, K. K., (2017). *Processing Technologies for Milk and Milk Products: Methods, Applications, and Energy Usage* (pp. 219–251). Burlington – ON: Apple Academic Press.
5. Binsi, P. K., Ninan, G., Zynudheen, A. A., Neethu, R., Ronda, V., & Ravishankar, C. N., (2014). Compositional and chill storage characteristics of microwave-blanched sutchi catfish (*Pangasianodon Hypophthalmus*) fillets. *International Journal of Food Science & Technology, 49*(2), 364–372.
6. Buffler, C. R., (1993). *Microwave Cooking and Processing: Engineering Fundamentals for the Food Scientist* (pp. 18–113). Berlin, Germany: Springer.
7. Chandrasekaran, S., Ramanathan, S., & Basak, T., (2013). Microwave food processing: A review. *Food Research International, 52*(1), 243–261.
8. Devi, S., Zhang, M., & Law, C. L., (2018). Effect of ultrasound and microwave assisted vacuum frying on mushroom (*Agaricus bisporus*) chips quality. *Food Bioscience, 25*, 111–117.
9. Dibanda, R. F., Akdowa, E. P., & Tongwa, Q. M., (2020). Effect of microwave blanching on antioxidant activity, phenolic compounds and browning behavior of some fruit peelings. *Food Chemistry, 302*, E-article 125308.
10. Dorantes-Alvarez, L., Jaramillo-Flores, E., González, K., Martinez, R., & Parada, L., (2011). Blanching peppers using microwaves. *Procedia Food Science, 1*, 178–183.
11. Douiri-Bedoui, I., Abdellaoui, H., Alexa, R., Jacolot, P., Druon, C., Tessier, F. J., & Laguerre, J. C., (2011). Optimization of microwave cooking of courgette in terms of nutrient preservation and energy consumption. *Procedia Food Science, 1*, 805–813.
12. Ekezie, F. G. C., Sun, D. W., Han, Z., & Cheng, J. H., (2017). Microwave-assisted food processing technologies for enhancing product quality and process efficiency: A review of recent developments. *Trends in Food Science and Technology, 67*, 58–69.
13. Fardhyanti, D. S., Sediawan, W. B., & Hisyam, A., (2019). Kinetics of mace (*Myristicae Arillus*) essential oil extraction using microwave-assisted hydrodistillation: Effect of microwave power. *Industrial Crops and Products, 131*, 315–322.
14. Gander, R. E., McClements, J. D., Sanderson, L. K., Rostad, B. A., Josephson, K. E., & Pratt, A. J., (1994). Sprint start instrumentation. *IEEE Transactions on Instrumentation and Measurement, 43*(4), 637–643.
15. Garg, A., Malafronte, L., & Windhab, E. J., (2019). Baking kinetics of laminated dough using convective and microwave heating. *Food and Bioproducts Processing, 115*, 59–67.
16. Giri, S. K., & Prasad, S., (2007). Drying kinetics and rehydration characteristics of microwave-vacuum and convective hot-air dried mushrooms. *Journal of Food Engineering, 78*(2), 512–521.
17. González-Monroy, A. D., Rodríguez-Hernández, G., Ozuna, C., & Sosa-Morales, M. E., (2018). Microwave-assisted pasteurization of beverages (tamarind and green) and their quality during refrigerated storage. *Innovative Food Science and Emerging Technologies, 49*, 51–57.

18. Guo, Q., Sun, D. W., Cheng, J. H., & Han, Z., (2017). Microwave processing techniques and their recent applications in the food industry. *Trends in Food Science & Technology, 67*, 236–247.
19. Hossan, M. R., Byun, D., & Dutta, P., (2010). Analysis of microwave heating for cylindrical shaped objects. *International Journal of Heat and Mass Transfer, 53*(23, 24), 5129–5138.
20. Ibrahim, N. A., & Zaini, M. A. A., (2018). Microwave-assisted solvent extraction of castor oil from seeds. *Chinese Journal of Chemical Engineering, 26*(12), 2516–2522.
21. Icier, F., & Baysal, T., (2004). Dielectric properties of food materials, part I: Factors affecting and industrial uses. *Critical Reviews in Food Science and Nutrition, 44*(6), 465–471.
22. Kassem, A. S., Shokr, A. Z., El-Mahdy, A. R., Aboukarima, A. M., & Hamed, E. Y., (2011). Comparison of drying characteristics of Thompson seedless grapes using combined microwave oven and hot air drying. *Journal of the Saudi Society of Agricultural Sciences, 10*(1), 33–40.
23. Keskin, S. O., Sumnu, G., & Sahin, S., (2004). Bread baking in halogen lamp–microwave combination oven. *Food Research International, 37*(5), 489–495.
24. Kim, J. E., Oh, Y. J., Won, M. Y., Lee, K. S., & Min, S. C., (2017). Microbial decontamination of onion powder using microwave powered cold plasma treatments. *Food Microbiology, 62*, 112–123.
25. Latorre, M. E., De Escalada, P. M. F., Rojas, A. M., & Gerschenson, L. N., (2013). Blanching of red beet (*Beta Vulgaris* L.) roots: Effect of hot water or microwave radiation on cell-wall characteristics. *LWT–Food Science and Technology, 50*(1), 193–203.
26. Li, L., Zhang, M., & Wang, W., (2020). Ultrasound-assisted osmotic dehydration pretreatment before pulsed fluidized bed microwave freeze-drying (PFBMFD) of Chinese yam. *Food Bioscience, 2020*, E-article 100548.
27. Liburdi, K., Benucci, I., & Esti, M., (2019). Effect of microwave power and blanching time in relation to different geometric shapes of vegetables. *LWT – Food Science and Technology, 99*, 497–504.
28. Liu, L., Dai, X., Kang, H., Xu, Y., & Hao, W., (2020). Structural and functional properties of hydrolyzed/glycosylated ovalbumin under spray drying and microwave freeze drying. *Food Science and Human Wellness, 9*(1), 80–87.
29. Liu, Y., Duan, X., Zhang, M., Li, C., Zhang, Z., Liu, A., & Wu, W., (2020). Cooking methods effect on the nutrients, bio-accessibility and antioxidant activity of *Craterellus cornucopioides*. *LWT – Food Science and Technology, 131*, E-article 109768.
30. Maskan, M., (2001). Kinetics of color change of kiwifruits during hot air and microwave drying. *Journal of Food Engineering, 48*(2), 169–175.
31. Meredith, & Roger J., (1998). *Engineers' Handbook of Industrial Microwave Heating* (pp. 19–50). London, UK: The Institution of Electrical Engineers.
32. Metaxas, A. A., & Meredith, R. J., (1983). *Industrial Microwave Heating* (pp. 112–178). London: UK: The Institution of Engineering and Technology.
33. Oztop, M. H., Sahin, S., & Sumnu, G., (2007). Optimization of microwave frying of potato slices by using Taguchi technique. *Journal of Food Engineering, 79*(1), 83–91.
34. Prabhanjan, D. G., Ramaswamy, H. S., & Raghavan, G. V., (1995). Microwave-assisted convective air drying of thin layer carrots. *Journal of Food Engineering, 25*(2), 283–293.

35. Pradeep, P., Abdullah, S. A., Choi, W., Jun, S., Oh, S., & Ko, S., (2013). Potentials of microwave heating technology for select food processing applications–brief overview and update. *Journal of Food Processing and Technology, 4*(11), E-article 1000278.
36. Risman, P. O., Ohlsson, T., & Wass, (1987). Principles and models of power density distribution in microwave oven loads. *Journal of Microwave Power and Electromagnetic Energy, 22*(4), 193–198.
37. Salazar-González, C., San Martín-González, M. F., López-Malo, A., & Sosa-Morales, M. E., (2012). Recent studies related to microwave processing of fluid foods. *Food and Bioprocess Technology, 5*(1), 31–46.
38. Seyhun, N. A. D. İ. D. E., Sumnu, G., & Sahin, S. E. R. P. İ. L., (2005). Effects of different starch types on retardation of staling of microwave-baked cakes. *Food and Bioproducts Processing, 83*(1), 1–5.
39. Sharma, G. P., & Prasad, S., (2006). Optimization of process parameters for microwave drying of garlic cloves. *Journal of Food Engineering, 75*(4), 441–446.
40. Siguemoto, É. S., Purgatto, E., Hassimotto, N. M., & Gut, J. A., (2019). Comparative evaluation of flavor and nutritional quality after conventional and microwave-assisted pasteurization of cloudy apple juice. *LWT–Food Science & Technology, 111*, 853–860.
41. Singh, R. P., & Heldman, D. R., (2014). *Introduction to Food Engineering* (pp. 389–396). San Diego – USA: Gulf Professional Publishing.
42. Su, D. L., Li, P. J., Quek, S. Y., Huang, Z. Q., Yuan, Y. J., Li, G. Y., & Shan, Y., (2019). Efficient extraction and characterization of pectin from orange peel by a combined surfactant and microwave assisted process. *Food Chemistry, 286*, 1–7.
43. Su, D., Lv, W., Wang, Y., Wang, L., & Li, D., (2020). Influence of microwave hot-air flow rolling dry-blanching on microstructure, water migration and quality of *Pleurotus eryngii* during hot-air drying. *Food Control, 2020*, E-article 107228.
44. Sumnu, G., Sahin, S., & Sevimli, M., (2005). Microwave, infrared and infrared-microwave combination baking of cakes. *Journal of Food Engineering, 71*(2), 150–155.
45. Tang, J., Feng, H., & Lau, M., (2002). Microwave heating in food processing. *Advances in Bioprocessing Engineering, 1*, 1–43.
46. Therdthai, N., & Zhou, W., (2003). Recent advances in the studies of bread baking process and their impacts on the bread baking technology. *Food Science and Technology Research, 9*(3), 219–226.
47. Therdthai, N., Zhou, W., & Adamczak, T., (2002). Optimization of the temperature profile in bread baking. *Journal of Food Engineering, 55*(1), 41–48.
48. Vadivambal, R., & Jayas, D. S., (2010). Non-uniform temperature distribution during microwave heating of food materials: A review. *Food and Bioprocess Technology, 3*(2), 161–171.
49. Varghese, T., & Pare, A., (2019). Effect of microwave assisted extraction on yield and protein characteristics of soymilk. *Journal of Food Engineering, 262*, 92–99.
50. Wang, J., Xiong, Y. S., & Yu, Y., (2004). Microwave drying characteristics of potato and the effect of different microwave powers on the dried quality of potato. *European Food Research and Technology, 219*(5), 500–506.
51. Xanthakis, E., Gogou, E., Taoukis, P., & Ahrné, L., (2018). Effect of microwave assisted blanching on the ascorbic acid oxidase inactivation and vitamin C degradation in frozen mangoes. *Innovative Food Science & Emerging Technologies, 48*, 248–257.

52. Xu, Y., Chen, Y., Cao, Y., Xia, W., & Jiang, Q., (2016). Application of simultaneous combination of microwave and steam cooking to improve nutritional quality of cooked purple sweet potatoes and saving time. *Innovative Food Science & Emerging Technologies, 36*, 303–310.
53. Zhang, H., Patel, J., Bhunia, K., Al-Ghamdi, S., Sonar, C. R., Ross, C. F., & Sablani, S. S., (2019). Color, vitamin C, B-carotene and sensory quality retention in microwave-assisted thermally sterilized sweet potato puree: Effects of polymeric package gas barrier during storage. *Food Packaging and Shelf Life, 21*, E-article 100324.

CHAPTER 13

INTERMITTENT MICROWAVE DRYING OF STRAWBERRY FRUITS

LEANDRO LEVATE MACEDO, JEFFERSON LUIZ GOMES CORRÊA, IRINEU PETRI JÚNIOR, WALLAF COSTA VIMERCATI, and CINTIA DA SILVA ARAÚJO

ABSTRACT

Strawberries are nutritionally rich, with appreciated sensory characteristics. However, it is a highly perishable fruit, and the processing of the strawberry is recommended. Convective drying is the most traditional method for reducing the moisture content of a food. However, this process can be very time-consuming, consumes a large amount of electrical energy and can cause undesirable changes in the fruit properties, such as color change and degradation of vitamins and bioactive compounds.

13.1 INTRODUCTION

Strawberry is a pseudo-fruit, belonging to the group of non-climacteric fruits, of great sensory appreciation, mainly in olfactory, taste, and visual terms, which makes strawberry one of the most produced and consumed fruits in the world. Moreover, strawberry has low caloric and high content of antioxidant compounds, such as vitamin C and anthocyanins, in addition to dietary fibers and various minerals. However, it has low physical resistance, high respiratory rate and high water activity, which makes the strawberry highly perishable. Strawberry processing is an important and interesting strategy to add value, develop other products and increase shelf life, reducing post-harvest losses, making it necessary to use a complementary drying method, such as convective and microwave (MW) drying [2, 7, 9].

Advanced Research Methods in Food Processing Technologies. Junaid Ahmad Malik, Megh R. Goyal, Preeti Birwal, & Ritesh B. Watharkar (Eds.)
© 2024 Apple Academic Press, Inc. Co-published with CRC Press (Taylor & Francis)

Convective drying is the most used method for food drying. It consists of exposing the material to a heated airflow [37]. However, this method has some limitations. As a result, some procedures, such as ethanol pretreatment (ET), have been implemented for convective drying and alternative methods, such as MW drying, have been developed to streamline the drying process and reduce food damage [30, 33, 34].

Ethanol can be added to foods by spraying, dipping, or immersion. ET is effective in reducing the drying time of various foods. Thus, the food is exposed to the drying air for a shorter time, minimizing undesirable changes to the dried product, such as reducing the color change, preserving heat-sensitive compounds, such as vitamin C, betalains, carotenoids, anthocyanins, phenolic compounds, and antioxidants [5, 33, 34, 42].

MW drying is gaining prominence nowadays, given the significant reduction in drying time caused by the rapid heating and volumetric heating of the product. Such characteristics are associated with improving the quality of the dried product, making MW drying a promising food technology [22, 32]. However, the continuous power supply may overheat the product. Thus, an alternative solution is to use the MW discontinuously, that is, intermittently, in which the magnetron is turned on and off during the drying process, causing the uniform distribution of moisture and heat [28].

This chapter focuses on drying of strawberry fruits using ethanol pretreatment and intermittent MW food processing technologies.

13.2 STRAWBERRY

Strawberry is one of the most produced and consumed fruits in the world. Although it is often referred to as a fruit, the strawberry is a pseudofruit since the pericarp is developed from an accessory organ [13]. In the case of strawberries, fertilization stimulates the growth of the receptacle, which, in turn, becomes fleshy, constituting a pseudofruit [4].

Strawberries fall into the group of non-climacteric fruits, as well as citrus fruits, cherry, grape, and pineapple. This characteristic is due to the gradual reduction of respiration and there is no production of endogenous ethylene. Ripening is still complete in the plant, before harvesting. Thus, strawberries should be harvested already mature, which reduces their lifetime [4, 13].

The color and flavor of the strawberry are attractive in sensory terms and, together with the nutritional properties, leads to it arousing interest in its consumption [19, 25]. The sensory appreciation of the strawberry is the result of multiple interactions between the constituents of the strawberry, such as sugars, acids, and pigments, and the human senses, such as taste, sight, touch, and smell [48]. This makes the strawberry the fifth most consumed fruit in the United States, behind only bananas, apples, oranges, and grapes [7, 20]. The nutritional composition of fresh strawberries is rich and diversified, with several minerals and vitamins (Table 13.1).

Strawberry has an attractive activity antioxidant. The antioxidant compounds contribute to the strawberry being classified as a functional food with various preventive and therapeutic health benefits, with positive effects on various diseases. The antioxidant power of strawberries is attributed to their concentration of polyphenols. The high concentration of these compounds in the strawberry makes it included among the 100 dietary sources richest in polyphenols and is classified among the foods that provide more than one mg of polyphenols per serving [7, 25].

Anthocyanins belong to the group of flavonoids and are among the main polyphenols found in strawberries. Anthocyanins aid in the protection of the brain, liver, and kidney, and act in the prevention of cardiovascular diseases, obesity control and the treatment of cancer. In addition to these beneficial health properties, anthocyanins are water-soluble pigments that give blue, purple, and red colors, giving the product an attractive color [7, 8, 25].

TABLE 13.1 Centesimal Composition, Minerals, and Vitamins Present in 100 g of Fresh Strawberry

Centesimal Composition		**Mineral Content**		**Vitamin Content**	
Water	90.95 g	Calcium	16.00 mg	Vitamin C	58.80 mg
Energy	32.00 kcal	Iron	0.41 mg	Thiamin	0.02 mg
Protein	0.67 g	Magnesium	13.00 mg	Riboflavin	0.02 mg
Ash	0.40 g	Phosphorus	24.00 mg	Niacin	0.39 mg
Total lipid	0.30 g	Potassium	153.00 mg	Pantothenic acid	0.13 mg
Carbohydrate	7.68 g	Sodium	1.00 mg	Vitamin B_6	0.05 mg
Sugars	4.89 g	Zinc	0.14 mg	Folate	24.00 μg
Sucrose	0.47 g	Copper	0.048 mg	Choline	5.70 mg
Glucose	1.99 g	Manganese	0.39 mg	Vitamin A	1.00 μg
Fructose	2.44 g	Selenium	0.40 μg	Lutein + zeaxanthin	26.00 μg
Dietary fiber	2.00 g	–	–	Vitamin E	0.29 mg
–	–	–	–	Vitamin K	2.20 μg

Source: Adapted from Ref. [25]

Like the vast majority of fruits, strawberries have a high moisture content, high water activity and high physical fragility. These significantly contribute to increased perishability, by favoring the development of microorganisms and oxidative reactions, especially when subjected to physical damage during harvesting, transportation, and marketing, causing nutritional, sensory, and economic losses [13]. Strawberries, when stored at room temperature, have a very short shelf life of one to two days [39].

Several strawberry processing techniques have been studied, such as modified atmosphere packaging (MAP), active packaging, edible coatings and films and ultrasound [6]. In addition to these methods, strawberry drying is a good alternative for strawberry processing. Several methods have been used, foam mat drying [26, 51], convective drying [34, 38], freeze-drying [14], spray drying [41], MW drying [24] and infrared drying [1].

13.3 DRYING OF FOODS

Drying is one of the most traditional methods in food preservation, to increase the shelf life of products, going from days or weeks to months or even years. Simultaneously, it contributes to the reduction of demands on packaging, transport, and storage, implying cost reduction. Drying refers to a thermal process applied to remove moisture present in the food, converting solid, semi-solid, or liquid materials into solids with low moisture content. Removal is based on the difference in vapor pressure between the food and the surrounding air [31, 34–36, 40, 52].

Many drying methods have been developed over the years, where each method has been developed to meet the specific needs of each food. However, inevitably, regardless of the method, from the simplest and oldest, to the most sophisticated and current method, drying not only changes the moisture content of food but can also change physical, sensory, and nutritional properties. In most cases, these changes are undesirable and irreversible, even after rehydration [31, 40, 45]. However, many studies have been developed to minimize these changes, to maintain the beneficial properties of the pre-drying product [5, 31, 34–36].

13.3.1 DRYING WITH ETHANOL PRETREATMENT

Pretreatments applied to fruits before drying generally aim to improve the drying process and with the preservation of compounds and characteristics of the food [53]. Among these pretreatments, the addition of low-boiling organic solvents, such as ethanol, isopropanol, and ethyl acetate, has been carried out on the wet material by immersion in a solution with the organic solvent [43, 44, 53] or dripped/spread under the surface of the food [5, 16, 33, 49] or even spraying the organic solvent during drying, changing the composition of the atmospheric air [10, 11, 46].

Ethyl alcohol or ethanol is an organic compound of molecular formula CH_3CH_2OH, belonging to the family of alcohols, presenting a boiling temperature of 78°C. This compound has been mostly used solvent as a pretreatment to convective drying, because of its low toxicity and low cost [33]. The U.S. Food and Drug Administration (FDA) considers ethanol a GRAS (generally recognized as safe) substance [21].

The addition of ethanol acts on four main aspects: reducing the heat of water vaporization; the opening of channels; increasing permeability of cellular membrane; and the Marangoni effect [16, 23, 33, 34, 43, 46]:

- The addition of ethanol reduces the heat of vaporization of the water present in the food, due to the mixture of the water+ethanol formed. As a result, water evaporation is increased;
- The rapid evaporation of ethanol and the ethanol+water mixture forms flow channels and pores in the food, facilitating the displacement of pure water from inside the product to the drying air;
- Ethanol increases the permeability of the cell wall of the food, both in the sense of dehydration and for rehydration, since the use of ethanol causes structural changes in the tissue, promoting the dissolution and disorganization of the cell wall compounds;
- The Marangoni effect promotes the transfer of mass at an interface between two fluids of different surface tension. Water has a higher surface tension than ethanol. In food pre-treated with ethanol, the surface tension gradient causes the ethanol, when it evaporates first, to increase the displacement of the water.

Studies have shown several beneficial effects of pretreatment with ethanol, both in the process and in the characteristics of the dried product. The reduction of drying time is the main contribution of ethanol in the process. Studies have observed reductions between 10% and 60% in drying time with the inclusion of this pretreatment [33, 43, 44, 46]. Reducing process time decreases energy consumption, reducing equipment usage time and labor time.

Reducing the drying time reduces the exposure of the material to drying conditions, such as high temperatures. As a result, less degradation of thermosensitive compounds has been observed, such as minimizing the color change [16], decreased shrinkage [49], reduced degradation of ascorbic acid (AA) [47, 53], cumarin [49], volatile compounds [10, 11], higher aromas retention [53], increased preservation of antioxidant compounds [5], and better rehydration [53].

13.3.2 INTERMITTENT MICROWAVE DRYING (IMD)

The term "microwave" refers to electromagnetic waves in the frequency range 300 MHz to 300 GHz with a wavelength from 1 mm to 1 m. MW drying is gaining increasing interest in food drying in recent years. This drying method is based on a unique volumetric heating mode facilitated by electromagnetic radiation at 915 or 2,450 MHz. The waves penetrate the material, causing agitation of the molecules and, consequently, heating is generated throughout the product, producing steam, which moves due to the internal pressure gradient [22, 32].

The heating generated by MWs occurs mainly by two mechanisms: dipolar reorientation and ionic conduction. In the first, the dipolar molecule, like water, tries to orient itself through the electric field. However, the electric field generated by the MW changes direction 2.45 billion times per second, causing agitation of the molecules, generating friction and heating inside the product, especially in materials with high moisture content, such as fruits. In the second mechanism, the ions present in the food move according to the electric field, generating heating. However, in food, the heating generated by the agitation of the molecules is more expressive than that generated by the ions [22, 32].

MW drying features reduced processing time, low energy consumption, precise process control and uniform heating [17, 22, 30, 32]. Furthermore, because MWs mainly heat the wet areas of the food, this method can preserve quality characteristics, such as color, density, rehydration capacity, texture, and nutritional parameters [15, 22, 27, 29]. Another highlight of this drying method is that bound water molecules which are difficult to remove can also be excited by MWs, facilitating their removal [32].

Continuous exposure to MWs can cause the product to overheat, leading to the degradation of various compounds or even the burning of the product. Thus, the batch process is called intermittent microwave drying (IMD) and has been an interesting alternative for drying food. In this process, the magnetron is turned on and off periodically, allowing moisture and temperature to be evenly distributed during drying, preventing overheating of the food. IMD has less energy consumption than convective drying and its use is associated with the production of dehydrated foods of good physical, nutritional, and chemical qualities [12, 27, 28, 30, 50].

Cao et al. [12] observed that the inclusion of intermittent MWs during vacuum drying of litchi fruits saved at least 31% of the energy needed, reduced the product browning, and increased the sensory acceptance of the dry product. Guclu et al. [50] observed that IMD was more effective in creating new aroma compounds in sweet red peppers than infrared and hot air drying.

13.3.3 DRYING OF STRAWBERRY FRUITS

The application of ethanol treatment (ET) in strawberry drying was studied by Macedo et al. [34]. In this study, ET was enhanced by dipping strawberries cut into 10 mm edge cubes in 95% ethanol. Afterwards, the samples were submitted to heated air drying. It was observed that ET reduced drying time by up to 30%, which reduced electrical energy consumption. Furthermore, ET reduced the hygroscopicity of the dried product and contributed to the preservation of anthocyanins, phenolic compounds and antioxidant activity of the strawberry. Therefore, the application of ET is an interesting procedure in strawberry drying. However, more studies can be developed to improve the strawberry drying process with ET.

Several MW drying studies of strawberry have been carried out. However, studies that assess the process operating intermittently are still rare. Amami et al. [3] studied the drying of strawberry slices using the intermittent system. The use of this system resulted in dried strawberries with higher antioxidant activity and total phenolic content, in addition to being a process with lower total cost. Therefore, like drying with ET, IMD is a good choice for strawberry drying.

13.3.4 GENERAL CONSIDERATIONS

The high perishability of strawberries requires processing methods to be used, aiming to extend the shelf life and also preserve the qualitative parameters of the product. One of the ways to process the strawberry is subjecting it to drying. Drying, especially the method using heated air, stands out among the various food processing techniques. Ethanol pretreatment can help reduce drying time and retain antioxidant compounds in fruits, simply, and inexpensively. In this same context, MW drying has stood out for its short process time, low energy consumption, and good preservation of the qualitative parameters of the product.

13.4 SUMMARY

Ethanol pretreatment of the strawberry consists of adding ethanol to the wet material, either by spraying, dipping, or immersion. This procedure is an efficient procedure in reducing drying time and minimizing the degradation caused during convective drying. In turn, the MW method is an alternative for drying food, capable of promoting volumetric heating generated by the exposure of the material to electromagnetic waves. However, continuous exposure to electromagnetic waves can cause the material to overheat, which can even lead to burning of the product. Thus, the intermittent system has stood out successful in recent years. In this system, the equipment magnetron is switched on and off continuously during the entire drying process. It has been associated with preservation of thermosensitive compounds.

KEYWORDS

- **anthocyanins**
- **antioxidant compounds**
- **emerging technology**
- **ethanol treatment**
- **food preservation**
- **intermittent microwave drying**
- **microwave drying**
- **strawberry**

REFERENCES

1. Adak, N., Heybeli, N., & Ertekin, C., (2017). Infrared drying of strawberry. *Food Chemistry, 219*, 109–116.
2. Alvarez-Suarez, J. M., Mazzoni, L., Forbes-Hernandez, T. Y., Gasparrini, M., Sabbadini, S., & Giampieri, F., (2014). The effects of pre-harvest and post-harvest factors on the nutritional quality of strawberry fruits: A review. *Journal of Berry Research, 4*(1), 1–10.

3. Amami, E., Besombes, C., Kechaou, N., & Allaf, K., (2018). Strawberries hybrid drying combining airflow, DIC technology and intermittent microwaves. In: *Proceedings of 21st International Drying Symposium.* Valencia: Universitat Politècnica València. Available from: http://ocs.editorial.upv.es/index.php/IDS/IDS2018/paper/view/7556 (accessed on 09 February 2023).
4. Antunes, L. E. C., Júnior, C. R., & Schwengber, J. E., (2016). Strawberry. In: *Plant Physiology Manual: Crop Physiology (Manuel on Vegetative Physiology: Crop Physiology)* (p. 589). Brazil: Embrapa.
5. Araújo, C. Da. S., Corrêa, J. L. G., Dev, S., Macedo, L. L., Vimercati, W. C., Rodrigues De, O. C., & Pio, L. A. S., (2020). Influence of pretreatment with ethanol and drying temperature on physicochemical and antioxidant properties of white and red pulp pitayas dried in foam mat. *Drying Technology, 38*, 1–10.
6. Baicu, A. A., & Popa, M. E., (2018). Trends in prolonging the post-harvest life of strawberries: A review. *Annals of the University Dunarea de Jos of Galati, Fascicle VI: Food Technology, 42*(1), 9–16.
7. Basu, A., Nguyen, A., Betts, N. M., & Lyons, T. J., (2014). Strawberry as a functional food: An evidence-based review. *Critical Reviews in Food Science and Nutrition, 54*(6), 790–806.
8. Bendokas, V., Skemiene, K., Trumbeckaite, S., Stanys, V., Passamonti, S., Borutaite, V., & Liobikas, J., (2019). Anthocyanins: From plant pigments to health benefits at mitochondrial level. *Critical Reviews in Food Science and Nutrition, 60*(19), 1–14.
9. Bhat, R., & Stamminger, R., (2015). Preserving strawberry quality by employing novel food preservation and processing techniques–Recent updates and future scope–an overview. *Journal of Food Process Engineering, 38*(6), 536–554.
10. Braga, A. M. P., Pedroso, M. P., Augusto, F., & Silva, M. A., (2009). Volatiles identification in pineapple submitted to drying in an ethanolic atmosphere. *Drying Technology, 27*(2), 248–257.
11. Braga, A. M. P., Silva, M. A., Pedroso, M. P., Augusto, F., & Barata, L. E. S., (2010). Volatile composition changes of pineapple during drying in modified and controlled atmosphere. *International Journal of Food Engineering, 6*(1), https://doi.org/10.2202/1556-3758.1835.
12. Cao, X., Chen, J., Islam, M. N., Xu, W., & Zhong, S., (2019). Effect of intermittent microwave volumetric heating on dehydration, energy consumption, antioxidant substances, and sensory qualities of litchi fruit during vacuum drying. *Molecules, 24*(23), E-article 24234291; [online]. https://doi.org/10.3390/molecules24234291.
13. Chitarra, M. I. F., & Chitarra, A. B., (2005). *Pós-colheita de Frutas e Hortaliças: Fisiologia e Manuseio (Postharvest of Fruits & Vegetables Physiology & Handling)* (2nd edn., p. 785) Saulo Paulo – Lavras, Brazil: UFLA.
14. Ciurzyńska, A., Bajno, J., Olsiński, I., Pisarska, A., Ostrowska-Ligęza, E., Pałacha, Z., & Lenart, A., (2018). Sorption properties and phase transitions temperature of freeze-dried strawberry model based on hydrocolloids with a tailored structure. *Drying Technology, 36*(10), 1209–1223.
15. Corrêa, J. L. G., Dev, S. R. S., Gariepy, Y., & Raghavan, G. S. V., (2011). Drying of pineapple by microwave-vacuum with osmotic pretreatment. *Drying Technology, 29*(13), 1556–1561.
16. Corrêa, J. L. G., Braga, A. M. P., Hochheim, M., & Silva, M. A., (2012). The influence of ethanol on the convective drying of unripe, ripe, and overripe bananas. *Drying Technology, 30*(8), 817–826.

17. Da Costa, F. O., Alvarenga, T. F., De Mesquita, T. V. C., & Petri, Jr. I., (2021). Hybrid drying of pulped arabica coffee cherry beans (*Coffea arabica* L.) using a hexagonal microwave dryer designed by numerical simulations. *Journal of Food Process Engineering, 44*(5), E-article 13666; https://doi.org/10.1111/jfpe.13666.
18. Da Cunha, R. M. C., Brandão, S. C. R., De Medeiros, R. A. B., Da Silva, Jr. E. V., Fernandes Da, S. J. H., & Azoubel, P. M., (2020). Effect of ethanol pretreatment on melon convective drying. *Food Chemistry, 333*, E-article 127502, https://doi.org/10.1016/j.foodchem.2020.127502.
19. Du, X., Plotto, A., Baldwin, E., & Rouseff, R., (2011). Evaluation of volatiles from two subtropical strawberry cultivars using GC-olfactometry, GC-MS odor activity values, and sensory analysis. *Journal of Agricultural and Food Chemistry, 59*(23), 12569–12577.
20. FAO (Food and Agriculture Organization of the United Nations). *FAOSTAT*. http://www.fao.org/faostat (accessed on 09 February 2023).
21. FDA (US Food & Drug Administration), (2019). *Food Additive Status List*. https://www.fda.gov/food/food-additives-petitions/food-additive-status-list (accessed on 09 February 2023).
22. Feng, H., Yin, Y., & Tang, J., (2012). Microwave drying of food and agricultural materials: Basics and heat and mass transfer modeling. *Food Engineering Reviews, 4*(2), 89–106.
23. Funebo, T., Ahrné, L., Prothon, F., Kidman, S., Langton, M., & Skjöldebrand, C., (2002). Microwave and convective dehydration of ethanol treated and frozen apple – physical properties and drying kinetics. *International Journal of Food Science and Technology, 37*(6), 603–614.
24. Gamboa-Santos, J., & Campañone, L. A., (2019). Application of osmotic dehydration and microwave drying to strawberries coated with edible films. *Drying Technology, 37*(8), 1002–1012.
25. Giampieri, F., Alvarez-Suarez, J. M., & Battino, M., (2014). Strawberry and human health: Effects beyond antioxidant activity. *Journal of Agricultural and Food Chemistry, 62*(18), 3867–3876.
26. Guazi, J. S., Lago-Vanzela, E. S., & Conti-Silva, A. C., (2019). Development of smoothies from dehydrated products of strawberry and banana pulps obtained through foam-mat drying. *International Journal of Food Science and Technology, 54*(1), 54–61.
27. Guclu, G., Keser, D., Kelebek, H., Keskin, M., Emre, S. Y., Soysal, Y., & Selli, S., (2021). Impact of production and drying methods on the volatile and phenolic characteristics of fresh and powdered sweet red peppers. *Food Chemistry, 338*, E-article 128129; https://doi.org/10.1016/j.foodchem.2020.128129.
28. Huang, D., Men, K., Tang, X., Li, W., & Sherif, S. A., (2021). Microwave intermittent drying characteristics of *Camellia oleifera* seeds. *Journal of Food Process Engineering, 44*(1), E-article 13608; https://doi.org/10.1111/jfpe.13608.
29. İzli, N., & Polat, A., (2020). Intermittent microwave drying of apple slices: Drying kinetics, modeling, rehydration ratio and effective moisture diffusivity. *Tarim Bilimleri Dergisi (J. of Agricultural Sciences), 26*(1), 32–41.
30. Junqueira, J. R. De. J., Corrêa, J. L. G., & Ernesto, D. B., (2017). Microwave, convective, and intermittent microwave-convective drying of pulsed vacuum osmodehydrated pumpkin slices. *Journal of Food Processing and Preservation, 41*(6), E-article 13250; https://doi.org/10.1111/jfpp.13250.

31. Karam, M. C., Petit, J., Zimmer, D., Baudelaire, D. E., & Scher, J., (2016). Effects of drying and grinding in production of fruit and vegetable powders: A review. *Journal of Food Engineering, 188*, 32–49.
32. Kumar, C., & Karim, M. A., (2019). Microwave-convective drying of food materials: A critical review. *Critical Reviews in Food Science and Nutrition, 59*(3), 379–394.
33. Macedo, L. L., Corrêa, J. L. G., Araújo, C. Da. S., Vimercati, W. C., & Pio, L. A. S., (2021). Process optimization and ethanol use for obtaining white and red dragon fruit powder by foam mat drying. *Journal of Food Science, 86*(2), 426–433.
34. Macedo, L. L., Corrêa, J. L. G., Da Silva, A. C., Vimercati, W. C., & Júnior, I. P., (2021). Convective drying with ethanol pretreatment of strawberry enriched with isomaltulose. *Food and Bioprocess Technology, 2021*, E-article 0123456789; https://doi.org/10.1007/s11947-021-02710-2.
35. Macedo, L. L., Da Silva, A. C., Vimercati, W. C., Saraiva, S. H., & Teixeira, L. J. Q., (2021). Influence of yacon syrup concentration and drying air temperature on properties of osmotically pre-dehydrated dried banana. *Heat and Mass Transfer, 57*(3), 441–451.
36. Macedo, L. L., Vimercati, W. C., Araújo, C., Saraiva, S. H., & Teixeira, L. J. Q., (2020). Effect of drying air temperature on drying kinetics and physicochemical characteristics of dried banana. *Journal of Food Process Engineering, 43* (April), 1–10.
37. Maisnam, D., Rasane, P., Dey, A., Kaur, S., & Sarma, C., (2015). Recent advances in conventional frying of foods. *Journal of Food Technology and Preservation, 1*(1), 25–34.
38. Méndez-Lagunas, L., Rodríguez-Ramírez, J., Cruz-Gracida, M., Sandoval-Torres, S., & Barriada-Bernal, G., (2017). Convective drying kinetics of strawberry (*Fragaria ananassa*): Effects on antioxidant activity, anthocyanins and total phenolic content. *Food Chemistry, 230*, 174–181.
39. Muley, A. B., & Singhal, R. S., (2020). Extension of postharvest shelf life of strawberries (*Fragaria ananassa*) using a coating of chitosan-whey protein isolate conjugate. *Food Chemistry, 329*, E-article 127213; https://doi.org/10.1016/j.foodchem.2020.127213.
40. Omolola, A. O., Jideani, A. I. O., & Kapila, P. F., (2017). Quality properties of fruits as affected by drying operation. *Critical Reviews in Food Science and Nutrition, 57*(1), 95–108.
41. Pellicer, J. A., Fortea, M. I., Trabal, J., Rodríguez-López, M. I., Gabaldón, J. A., & Núñez-Delicado, E., (2019). Stability of microencapsulated strawberry flavor by spray drying, freeze drying and fluid bed. *Powder Technology, 347*, 179–185.
42. Rojas, M. L., Silveira, I., & Augusto, P. E. D., (2020). Ultrasound and ethanol pretreatments to improve convective drying: Drying, rehydration and carotenoid content of pumpkin. *Food and Bioproducts Processing, 119*, 20–30.
43. Rojas, M. L., & Augusto, P. E. D., (2018a). Ethanol and ultrasound pretreatments to improve infrared drying of potato slices. *Innovative Food Science and Emerging Technologies, 49* (April), 65–75.
44. Rojas, M. L., & Augusto, P. E. D., (2018b). Ethanol pretreatment improves vegetable drying and rehydration: Kinetics, mechanisms and impact on viscoelastic properties. *Journal of Food Engineering, 233*, 17–27.
45. Sagar, V. R., & Suresh, K. P., (2010). Recent advances in drying and dehydration of fruits and vegetables: A review. *Journal of Food Science and Technology, 47*(1), 15–26.
46. Santos, K. C., Guedes, J. S., Rojas, M. L., Carvalho, G. R., & Augusto, P. E. D., (2021). Enhancing carrot convective drying by combining ethanol and ultrasound as pretreatments: Effect on product structure, quality, energy consumption, drying and

rehydration kinetics. *Ultrasonics Sonochemistry, 70*, E-article 105304; https://doi.org/10.1016/j.ultsonch. 2020.105304.
47. Santos, P. H. S., & Silva, M. A., (2009). Kinetics of L-ascorbic acid degradation in pineapple drying under ethanolic atmosphere. *Drying Technology, 27*(9), 947–954.
48. Schwieterman, M. L., Colquhoun, T. A., Jaworski, E. A., Bartoshuk, L. M., Gilbert, J. L., Tieman, D. M., Odabasi, A. Z., et al., (2014). Strawberry flavor: Diverse chemical compositions, a seasonal influence, and effects on sensory perception. *PLoS One*, *9*(2); [online]. https://doi.org/10.1371/journal.pone.0088446.
49. Silva, M. G., Celeghini, R. M. S., & Silva, M. A., (2018). Effect of ethanol on the drying characteristics and on the coumarin yield of dried guaco leaves (*Mikania laevigata* Schultz BIP. Ex Baker). *Brazilian Journal of Chemical Engineering, 35*(3), 1095–1104.
50. Tepe, T. K., & Tepe, B., (2020). The comparison of drying and rehydration characteristics of intermittent-microwave and hot-air dried-apple slices. *Heat and Mass Transfer, 56*(11), 3047–3057.
51. Vimercati, W. C., Macedo, L. L., Araújo, C. Da. S., Teixeira, L. J. Q., & Saraiva, S. H., (2019). Efeito da temperatura na cinética de secagem em leito de espuma e na degradação de antocianina em morango (effect of temperature on foam bed drying kinetics and anthocyanin degradation in strawberry). *Brazilian Journal of Food Technology, 22*, E-article 22118; https://doi.org/10.1590/1981-6723.22118.
52. Vimercati, W. C., Macedo, L. L., Da Silva, A. C., Maradini, F. A. M., Saraiva, S. H., Teixeira, L. J. Q., Correa, J. L. G., & Minozzo, M. G., (2021). Effect of blanching and drying methods of spinach on the physicochemical properties and cooking quality of enriched pasta. *Journal of Food Measurement and Characterization, 2021*. [online]; https://doi.org/10.1007/s11694-021-01155-7.
53. Wang, X., Feng, Y., Zhou, C., Sun, Y., Wu, B., Yagoub, A. E. A., & Aboagarib, E. A. A., (2019). Effect of vacuum and ethanol pretreatment on infrared-hot air drying of scallion (*Allium fistulosum*). *Food Chemistry, 295*, 432–440.

PART III

RESEARCH ADVANCES IN THE PROCESSING OF FOODS

CHAPTER 14

HIGH-PRESSURE PROCESSING OF FOODS AND FOOD PRODUCTS: AN OVERVIEW

CINU VARGHESE, V. HARSHA, K. L. CLAUDIA, and ALPHY NISA STEAPHEN

ABSTRACT

High-pressure processing (HPP) or ultra-high-pressure or high-hydrostatic-pressure processing is a novel non-thermal processing method, an alternative to conventional thermal processes. HPP ensures product safety and quality, retaining color, flavor, and fresh appearance. HPP holds a promise for providing consumers with high quality, delicious, and fresh foods. This chapter focuses on the concept, principles, components, and types of HPP and its application in the food processing industry.

14.1 INTRODUCTION

The organic nature and moisture content of foods make them susceptible to microbial, physical, and chemical spoilage. Therefore, preservation is crucial for keeping food edible, proper texture, nutritious, and appealing for a long time. Traditional methods of preservation (such as freezing, drying, pasteurization, and chilling) are still available; however, today's consumers are increasingly demanding natural, tasty convenience foods without additives or preservatives [34]. Due to this demand, several non-thermal food processing technologies have been developed, and high-pressure technology has proved to be successful [35]. Processes of high hydrostatic pressure

Advanced Research Methods in Food Processing Technologies. Junaid Ahmad Malik, Megh R. Goyal, Preeti Birwal, & Ritesh B. Watharkar (Eds.)
© 2024 Apple Academic Press, Inc. Co-published with CRC Press (Taylor & Francis)

(HHP) or high-pressure processing (HPP) is a method of preservation and sterilization that involves processing of foods under high pressure [7]. The process of HPP uses high pressure, between 100 and 1,000 MPa, using heat or without, thus preserving most foods without affecting their flavor, nutritional value, or texture [5].

Microbial inactivation by pressure was demonstrated experimentally in 1899 by B. H. Hite [19], who showed that high pressure treatment cannot only destroy microorganisms but can also prevent milk from spoiling and keep its flavor intact. Compared to untreated milk, treated milk stayed sweeter for 24–60 hours longer. Linton & Patterson [24] concluded that pressure does not necessarily kill bacteria spores, but it does normally kill vegetative bacteria, and Carbon dioxide gas pressure inactivates bacteria more effectively than the hydrogen or nitrogen gas pressures. In the late 1900s, researchers focused on killing spores by HPP and found that biological spores can germinate under moderate pressures, and they can be easily killed with heat, pressure, or ionizing irradiation methods. Nevertheless, if the pressure is too high at the beginning, germination will not occur. Using moderate pressure, however, is not always effective since some bacteria are more resistant to germination under pressure, though some of them may survive [3, 16, 37].

High pressure processing (HPP) is suitable for processing of liquid foods as well as solid foods with high moisture content. Although pressure processing kills microorganisms, yet covalent bonds are almost untouched by low temperatures of 0–40°C. This specific feature of HPP (at moderate temperatures) makes this innovation unique since the method has less effect on foods [27]. The method preserves food quality without the use of harsh thermal processes or chemical preservatives [4]. HPP is most effective for acidic foods, since pressure-tolerant spores cannot live in low pH environment [3, 7]. High pressure processed foods have more commercial success than traditional methods despite their high cost, since the demand for minimally processed foods is now higher than it was before [13].

HPP affect the macromolecules (nucleic acids, proteins, polysaccharides), cellular components, and functional structure by destroying non-covalent bonds, which lead to compaction of food volume and fatal results in bacteria. The inactivation of enzymes by this bond destruction will further prevent the enzymatic degradation of foods [36, 38]. However, the covalent bonds have no effect on HP processing. Therefore, the sensory properties of treated foods are similar to those of fresh foods [32]. The HPP can make a food product last longer while protecting its sensory (color, texture, or taste) and nutritional properties by deactivating microorganisms (and destroy spores at elevated temperatures) and enzymes that cause spoilage. The HP pasteurized food products use less energy and generate fewer wastes [12]. During the treatment, low temperatures are generally used and processed product is free from preservatives and additives; and food is treated uniformly throughout and can be processed in-package. All these advantageous properties are attractive to consumers [11, 12]. This chapter focuses on the technology of HPP of foods and food products.

14.2 SALIENT FEATURES OF HPP

HPP is a non-thermal technology, which is carried out at ultra-high pressure (100–1,000 MPa) [9, 40]. High pressure processing of foods is mainly intended for preservation and modification of a food material [15, 39]. This processing technology is designed to inactivate microorganisms, spores, viruses, and enzyme with minimal degradation of nutritional properties, texture, and flavors [15]. This can be used for both liquid and solid food materials as a cleaner and greener process to achieve the preservation effects. HPP is carried out at room temperature with minimal energy consumption during heating and subsequent cooling process [18]. Also, the food products are packaged in such a manner that prevents further contamination after pasteurization [35]. The important advantages of HPP are:

- Best retention of the essential oils compared to conventional heat sterilization, freezing, and drying [28].
- Compared to heat treatment methods, HPP significantly reduces the leeching out of cell constituents (example: potato blanching).
- Environmentally friendly and no waste technology.
- Extend the shelf life of the treated food products.
- Food can be treated at ambient temperature or lower temperatures [21].
- High pressure treatment (400–600 MPa) retains more color, phenols, anthocyanins, and ascorbic acid from most of the fruit purees compared to thermal treatment [33].
- Inactivation of microbials leads to the quality as well as the safety of food products.
- It permits the transmittance of pressure all over the food material, without considering the geometry and size of the material, thus making size reduction optional.
- Maintain the freshness, sensory characteristics, and nutritional properties of food products.

- Minimal or no use of food preservatives, these aids authentic or additive free products.
- No chance of contamination and cross contamination during the processing, ensures the safety aspects of processed commodities.
- Non thermal innovative processing technique.
- Pressure-time combinations required for several species of pathogenic microorganisms can be established using high pressure and it will help for advanced predictions [17].
- This can be used to develop ingredients with novel functional properties [35].
- This is a preservation technique for most of the value-added fruit and vegetable products.

14.3 PRINCIPLE OF OPERATION OF HPP

14.3.1 LE CHATELIER'S PRINCIPLE

This principle claims that *"if a change in conditions is applied on a system which is in equilibrium, then the system will try to counter to that change and bring back the equilibrium"* [37]. This implies that high pressure favors the reaction that results in a drop in volume while opposing those that involve volume increase [25, 29].

14.3.2 ISO-STATIC PRINCIPLE

According to iso-static principle, application of pressure is instantaneous and uniform throughout the sample (Figure 14.1). When a food material is compressed by uniform pressure from all directions, the material returns to its original shape during decompression [40]. Regardless of size and shape of the food material, the pressure effects are instantaneously and uniformly distributed within the food material.

14.3.3 PRINCIPLE OF MICROSCOPIC REORDERING

According to this principle, *"for a given temperature, increase in pressure increases the degree of ordering of molecules of a particular commodity"* [41]. High pressure alters the interatomic distance between the molecules that result in changes in physical or chemical reactions [40]. According to this principle of microscopic reordering, pressure limits vibration, rotation, and transitional motion, which results in an increase in the molecular order.

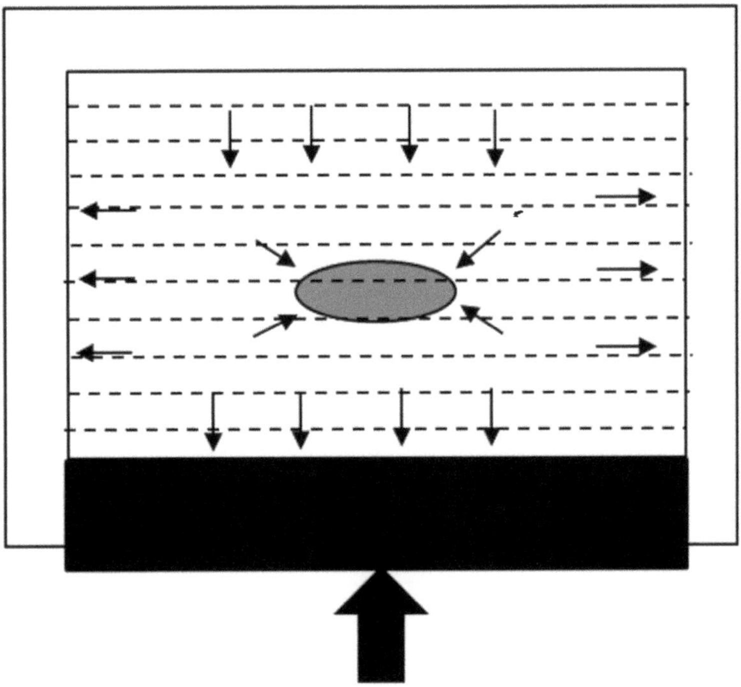

FIGURE 14.1 The isostatic principle in HPP.
Source: Adapted from Ref. [40]

14.4 COMPONENTS OF HPP SYSTEM

An industrial HPP system comprise of a pressure' vessel, vessel closures, yoke, high pressure pump and intensifier (generating target pressure), control system and a product handling system (loading and removing the materials) (Figure 14.2) [9, 30]. Pressure vessel is the most important part of the HPP system. High pressure vessel can be framed with two or more concentric cylinders [10]. The outer layer compresses the inner cylindrical layers, and the pressure chamber walls are always under some residual pressure. The food material, which is to be treated, is placed in the pressure vessel capable of maintaining the required pressure. The food material is immersed in a liquid which acts as pressure transmitting medium. These liquids can be sodium benzoate, ethanol, silicon oil, glycol or castor oil. In addition to this, water can also act as transmitting media. Design of pressure vessel is a critical factor, and it depends on the operating pressure and vessel diameter.

The pressure vessel has an opening and closing system to allow rapid loading and unloading of the vessel. Two closures (also named as plugs) are present at the ends of the pressure vessel. While designing material handling efficiency, sealing system and safety of pressure vessel should be considered [25]. At higher pressure range, closures subjected to wear and proper maintenance and periodic checking are necessary. Yoke-frame is a structure that holds all the HP vessels and closures. This is fabricated by laminating a few steel sheets or winding steel frames. The operations of the pressure vessel is controlled by a computer system.

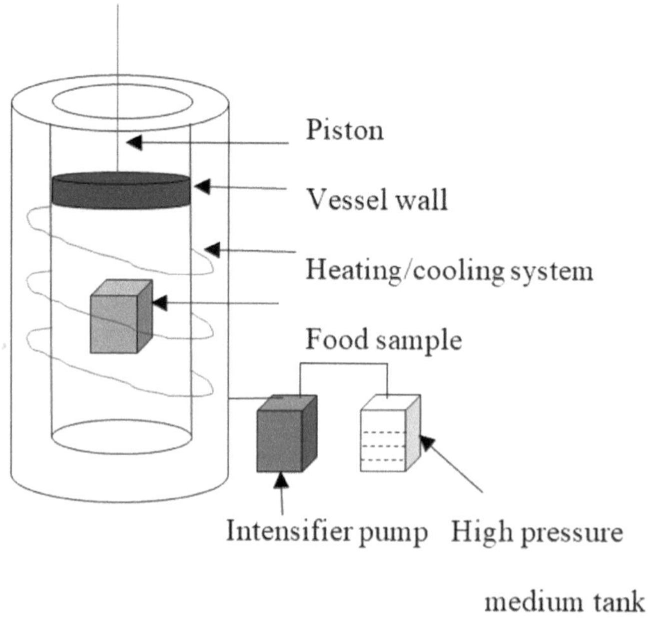

FIGURE 14.2 Basic components of HPP system.

Source: Adapted from Ref. [31]

14.5 TYPES OF HPP SYSTEM

Based on the operation, there are three types of HPP system: batch, semi-continuous, and continuous systems [9]. Batch and continuous types are mainly suitable for high pressure pasteurization. Although batch system can process both liquid and solid food products that need to be pre-packed before processing; yet continuous and semi-continuous systems can only process liquid food products.

14.5.1 BATCH SYSTEM

High pressure batch system mainly consists of a cylindrical pressure vessel, closures, an intensifier pump, and a process control system. Packaged food product is loaded into the pressure vessel and then the vessel is closed. The pressure transmitting fluid is filled into the vessel either by means of a pumping system or by using piston-displacement system until the required pressure has been achieved. The food material is held for the required time and then the vessel is relaxed by freeing out transmitting media [10]. After this, the food materials are unloaded and stored conventionally [9]. Usually, food materials are subjected to 3–5 min at 600 MPa for processing. Around, 5–6 cycles/h are enough to complete all stages from loading to unloading. There is no contamination or cross-contamination during the process [9].

14.5.2 SEMI-CONTINUOUS HPP SYSTEM

It consists of two or more pressure vessels, low pressure pump, high pressure transfer pump, holding tank, sterilized tanks and controlling valves (Figure 14.3). A free moving piston is present in the pressure vessel to avoid contact with food products from the pressurizing fluid. Control valves are provided to avoid cross contamination when new products are being processed. In semi-continuous HPP, a low-pressure pump is used to pump the liquid food product into the pressure vessel. During the filling, a free piston is displaced. Once the vessel is filled, the inlet valve is closed and then pressurizing liquid is added behind the piston and flowing food material is compressed. After the appropriate holding time, the pressure is released from the transmitting fluid. And the processed food material can be stored in a sterile tank.

14.5.3 CONTINUOUS HPP SYSTEM

Continuous system can only process liquid food products (Figure 14.4). The food materials flow through a tube system, which pressurizes the food material at a pressure of 100 MPa or more by using high-pressure intensifiers. To avoid high heating and shearing, a unit to depressurize the product is applied. Following the decompression, the treated product then goes to a sterilized storage tank, and it is packed finally [6].

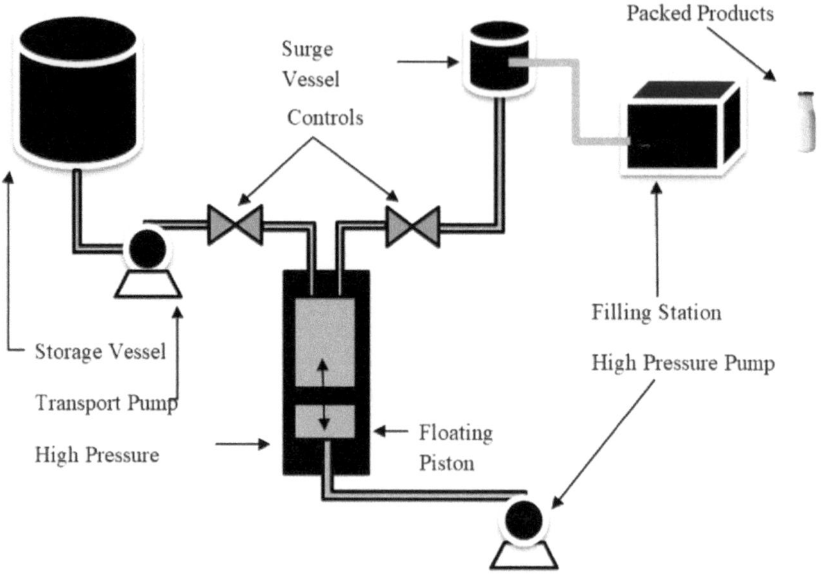

FIGURE 14.3 Semi-continuous HPP system.
Source: Adapted from Ref. [42]

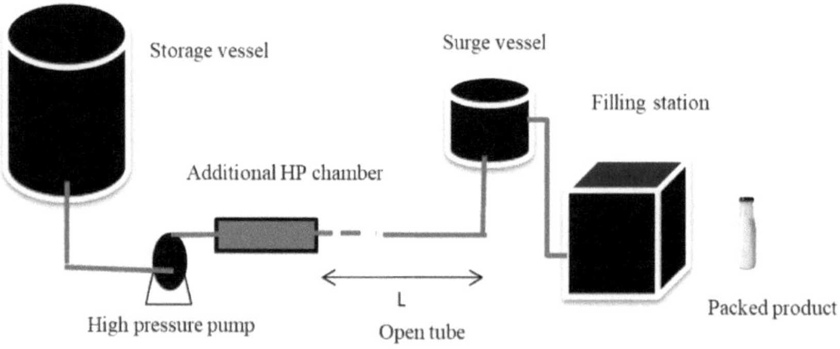

FIGURE 14.4 Continuous high-pressure processing system.
Source: Adapted from Ref. [42]

14.6 EFFECTS OF HPP ON MICROORGANISMS

When food items are processed under high pressure to extend the shelf-life, the microorganisms and enzymes get killed or inactivated. Inactivation of microorganisms under HP is accompanied by membrane destruction due to the generated shear forces. This leads to disruption of cell function, local heat damage, and protein deformation. When pressure is applied from all sides, compression of microbial cells occurs gradually; and after decompression the cells may return to its initial size. The cells then become physiologically inactive as the cellular material is oozed out. Also, high pressure affects the cell morphology and biochemical reactions, genetic mechanism and denaturation of protein, denaturation of some major enzymes and ribosomes [24].

Increasing the pressure during treatment will depend on the level of inactivation or effect on microorganism. A pressure of 50 MPa can suppress protein synthesis in microorganisms and can reduce the number of ribosomes. The 100 MPa pressure can result in partial protein denaturation. And 200 MPa may induce damage to the cell membrane and structure. When the pressure is raised above 300 MPa, enzymes, and proteins are irreversibly denatured, which will cause cell membranes rupture, internal substances are excreted, and it results in the death of bacteria [1, 17].

14.6.1 BACTERIA

High pressure destroys spoilage pathogenic microorganisms. The microscopic organisms that cause food decay are: *Listeria monocytogenes, Staphylococcus aureus, E coli* and species of *Salmonella, Vibrio,* and *Campylobacter*. The cell walls of Gram-negative bacteria are weaker than that of the Gram-positive bacteria. Hence, Gram-negative bacteria are more pressure sensitive than the gram-positive bacteria. Acid foods, ready-to-eat meats and dairy products contain *Listeria monocytogenes* (gram-positive), a major pathogenic bacteria.

14.6.2 FUNGI

Fungi are of two types based on the molecular structure, yeasts, and molds. In general, yeast, molds, and vegetative cells are more susceptible to pressure; therefore, inactivation of this microorganism is achieved at lower pressures. Yeasts are generally nonfood pathogens, whereas the growth of toxic mold is a food safety concern. Low pressure treatment (< 400 MPa) for a couple of minutes is enough to kill the pathogens. Most of the molds can be inactivated at pressure ranging from 300 to 600 MPa (e.g., *Penicillium roqueforti* present in cheese) [37].

14.6.3 BACTERIAL SPORES

Bacterial spore inactivation is a serious challenge in the food industry. These spores are pressure resistant. High pressure of >800 MPa may be required for the inactivation of bacterial spores at ambient temperature conditions. Bacterial spores in low acid food products need high-pressure range (1,400 or 1,500 MPa) for inactivation of particular species at ambient temperature. Species of *Clostridium* and *Bacillus* (such as: *Clostridium botulinum, Clostridium perfringens* and *Bacillus cereus*) are foodborne spore-forming bacteria. *Clostridium botulinum* is a major concern since it can grow, germinate, and produce a paralytic neurological toxin mainly in low acid foods. Therefore, heat treatment (at temperature >80°C) along with HPP can effectively kill the spores, especially in low acid food products [2].

14.6.4 VIRUSES

Depending on the structural diversity, viruses show resistant to pressure. Compared to naked virus, enveloped virus is more sensitive to pressure. Complete inactivation can be achieved for the suspensions of *feline calicivirus* (at 275 MPa for 5 minutes), adenovirus (at 400 MPa for 15 minutes), and hepatitis A (at 450 MPa for 5 minutes). Inactivation of microorganisms by high pressure processing follows a given trend, such as gram-negative > yeast/mold > gram-positive > spores. Various factors influence the response of microorganism including pathogens to high pressure treatments; and clearly, these factors must be considered while optimizing HPP parameters in microbiological safety. The major factors affecting the sensitivity of microorganisms to high pressure include temperature, water activity (aw), pH, pressure, and holding time.

14.7 HPP PACKAGING OF FOODS

Food packaging material should be selected properly to maintain the physical integrity during the processing. It should be flexible to withstand the compressive forces and should be heat sealable. Good maintenance of packages throughout the storage period is necessary to avoid recontamination of foods. Metal cans, glass bottles and paperboard-based packages are not used for HPP packaging. These may deform or fracture when subjected to compression [26]. HPP can be done in foods that are in bulk or pre-packaged with flexible or semirigid materials. For commercial products, flexible packaging materials are suitable for HPP packaging. Polypropylene, polyester tubes, polyethylene pouches and nylon-cast polypropylene pouches are the most common materials used for HPP packaging. Plastic materials are also suited for the packaging of HPP application. When exposed to mechanical stresses in HPP, these flexible food packaging structures responded effectively, reversible after compression and showed resilience [8]. Structures with inorganic coatings like Al2O3 and SiO2 have performed well in terms of mechanical and barrier properties [31]. Packaging materials based on biopolymers are possible alternatives in the associated environmental challenges. Other types of materials need further research and evaluation for HPP packaging [23].

The light barrier qualities of the packaging materials were unaffected by HPP at different temperatures. The combination of high pressure and high temperature (600 MPa, 90°C and 10 min) enhanced the water vapor transmission rate of LDPE/EVOH/LDPE, PET/LDPE/PA/EVOH/PA/LDPE, and PET printing/PET_{met}/LDPEcoex packaging, as well as the oxygen transmission rate (OTR) of LDPE/PA/LDPE, LDPE/EVOH/LDPE, PET/LDPE/PA/, EVOH/PA/LDPE in terms of water vapor and oxygen barrier properties [26]. Requirements for HPP packaging materials include visual integrity, gas permeability, head space, seal, and physical strength properties, printability, and global migration of packaging components into the food either refrigerated or shelf-stable products [20].

14.8 APPLICATIONS OF HPP IN FOOD PRESERVATION

HPP has proved a promising technology in food preservation against pathogenic and spoilage microorganisms in food products. The Food and Drug Administration (FDA) and the United States Department of Agriculture (USDA) have recommended HPP as an alternative to traditional high-temperature pasteurization. The advancements of HPP developments in the last decade have made its use cost-effective in food handling for preserving quality and pathogen inactivation in variety of seafood products, such as shellfish, mullet, salmon, cod, squid, and shrimp [14]. Commercial applications for food products [30]; pasteurization and sterilization of fruits, salad dressings, yogurts, and sauces; pasteurization of meats and vegetables; and decontamination of high-risk and high-value heat-sensitive components (such as: flavorings and vitamins). Among the alternate innovative food processing techniques available, HPP appears to be one of the most promising for dairy products, because it allows for the development of new shelf-stable foods while retaining their natural nutritional values with excellent organoleptic characteristics [31].

Combining HPP with other processing treatments may help to lower the processing costs. Suggestions for lowering process severity and cost during high-pressure pasteurization and sterilizing have been proposed by researchers. Continuous juice pasteurization has been proposed using HPP in combination with Carbon dioxide or argon. Combining pressure treatment with mild heat treatment and non-thermal preservation technologies (such as: irradiation, PEF processing and ultrasound) could be an another option. Antimicrobial preservatives can be added to the process [5].

14.9 CHALLENGES

The primary issue for HPP mainly for well-established industrial scale procedures is to determine the exact mechanism that can result in the inactivation of the bacterial spores [14]. For explaining bacterial inactivation and evaluating microbial processes, kinetic models are required. Process validation studies are also needed. The high levels of HPP needs could be reduced by combining non-thermal technology with HPP. Processing equipment can be improved. Further research is needed to establish HPP on a commercial level with proper packaging and storage. The support from public research funds and novel ideas of partnership formation are required for the wide acceptance of HPP products [5].

14.10 SUMMARY

Nowadays, consumers demand natural flavor, taste, aromas, and fresh appearance of food materials. Traditional preservation methods (such as: physical and chemical methods) are extensively used to reduce microbial growth and to improve the food quality. The pasteurization and sterilization are the most widely used preservation techniques. Although the thermal treatment reduces the growth of microorganism effectively, yet there are high chances of alteration of vitamins, fresh appearance, flavors, and other sensory attributes. HPP is one of the most innovative approaches for preserving food freshness, nutrition, taste, and flavor, thus resulting in high-quality food products without the use of chemical preservatives. HPP is an eco-friendly method. The risk of recontamination is limited since the products are packed.

It removes harmful bacteria, thus ensuring safety. Therefore, HPP results in superior quality, a safe food product with a longer shelf-life.

KEYWORDS

- **Food and Drug Administration**
- **high hydrostatic pressure**
- **high-pressure processing**
- **iso-static principle**
- **Le Chatelier's principle**
- **microscopic reordering**
- **oxygen transmission rate**

REFERENCES

1. Abe, F., (2007). Exploration of the effects of high hydrostatic pressure on microbial growth, physiology and survival: Perspectives from piezophysiology. *Bioscience, Biotechnology and Biochemistry, 71*(10), 2347–2357.
2. Abee, T., & Wouters, J. A., (1999). Microbial stress response in minimal processing. *International Journal of Food Microbiology, 50*(1, 2), 65–91.
3. Adams, M. R., & Moss, M. O., (2007). *Food Microbiology* (p. 307). New Delhi – India: New Age International Pvt. Ltd.
4. Aymerich, T., Picouet, P. A., & Monfort, J. M., (2007). Decontamination technologies for meat products. *Meat Science, 78*(1, 2), 114–129.
5. Balasubramaniam, V. M., & Farkas, D., (2008). High-pressure food processing. *Food Science and Technology International, 14*(5), 413–418.
6. Bermúdez-Aguirre, D., Maria, G., Corradini, M. G., Kezban, C. K., & Barbosa-Cánovas, G. V., (2016). High pressure processing in combination with high temperature and other preservation factors. In: Balasubramaniam, V. M., Barbosa-Cánovas, G. V., & Lelieveld, H. L. M., (eds.), *High Pressure Processing of Food* (pp. 193–215). Cham: Springer.
7. Brown, A. C., (2007). *Understanding Food: Principles and Preparation* (3rd edn., p. 547). Boston–MA: Cengage Learning.
8. Caner, C., Hernandez, R. J., & Harte, B. R., (2004). High-pressure processing effects on the mechanical, barrier and mass transfer properties of food packaging flexible structures: A critical review. *Packaging Technology and Science, 17*(1), 23–29.
9. Elamin, W. M., Endan, J. B., Yosuf, Y. A., Shamsudin, R., & Ahmedov, A., (2015). High pressure processing technology and equipment evolution: A review. *Journal of Engineering Science and Technology Review, 8*(5), 75–83.

10. Farkas, D. F., & Hovver, D. G., (2011). High pressure processing. *J. of Food Science, 65*(8), 47–64.
11. Fellows, P. J., (2007). *Food Technology: Principles and Practice* (p. 609). Oxford: Woodhead Publishing.
12. Fellows, P. J., (2017). *Food Processing and Technology: Principles and Practices* (p. 12733). Oxford: Woodhead Publishing.
13. Fellows, P. J., (2000). *Food Processing Technology: Principles and Practice* (2nd edn., p. 591). Boca Raton – FL: CRC Press.
14. Ginsau, M. A., (2015). High pressure processing: A novel food preservation technique. *IOSR Journal of Environmental Science, 9*(5), 2319–2399.
15. Heinz, V., & Buckow, R., (2010). Food preservation by high pressure. *Journal Fur Verbraucherschutz Und Lebensmittelsicherheit (Journal of Consumer Protection and Food Safety), 5*(1), 73–81.
16. Hendrickx, M. E. G., & Knorr, D. W., (2002). *Ultra-High-Pressure Treatments of Foods* (p. 333). Food Engineering Series; Cham: Springer + Business Media.
17. Huang, H. W., Lung, H. M., Yang, B. B., & Wang, C. Y., (2014). Responses of microorganisms to high hydrostatic pressure processing. *Food Control, 40*(1), 250–259.
18. Huang, H. W., Wu, S. J., Lu, J. K., Shyu, Y. T., & Wang, C. Y., (2017). Current status and future trends of high-pressure processing in food industry. *Food Control, 72*(12), 1–8.
19. Jay, J. M., Loessner, M. J., & Golden, D. A., (2008). *Modern Food Microbiology* (7th edn., p. 790). Cham: Springer Science + Business Media.
20. Juliano, P., Koutchma, T., Sui, Q., Barbosa-Cánovas, G. V., & Sadler, G., (2010). Polymeric-based food packaging for high-pressure processing. *Food Engineering Reviews, 2*(4), 274–297.
21. Knorr, D., (1996). Advantages, opportunities and challenges of high hydrostatic pressure application to food systems. *Progress in Biotechnology, 13*(C), 279–287.
22. Krebbers, B., Matser, A. M., Koets, M., & Van, D. B. R. W., (2002). Quality and storage-stability of high-pressure preserved green beans. *Journal of Food Engineering, 54*(1), 27–33.
23. Li, T., Zhao, L., Wang, Y., Wu, X., & Liao, X., (2021). Effect of high-pressure processing on the preparation and characteristic changes of biopolymer-based films in food packaging applications. *Food Engineering Reviews, 13*(3), 454–464.
24. Linton, M., & Patterson, M. F., (2000). High pressure processing of foods for microbiological safety and quality. *Acta Microbiologica et Immunologica Hungarica, 47*(2, 3), 175–182.
25. Mahadevan, S., & Karwe, M. V., (2016). Effect of high-pressure processing on bioactive compounds. In: *High Pressure Processing of Food (Food Engineering Series)* (pp. 479–507). New York – UK: Springer International Publishing.
26. Marangoni, L., Alves, R. M. V., Moreira, C. Q., Cristianini, M., Padula, M., & Anjos, C. A. R., (2020). High-pressure processing effects on the barrier properties of flexible packaging materials. *Journal of Food Processing and Preservation, 44*(11), 1–12.
27. Masson, P., (1993). Effects of high pressure on proteins. *Food Reviews International, 9*(4), 611–628.
28. Matser, A. M., Krebbers, B., Van, D. B. R. W., & Bartels, P. V., (2004). Advantages of high-pressure sterilization on quality of food products. *Trends in Food Science and Technology, 15*(2), 79–85.

29. Medina-Meza, I. G., Barnaba, C., & Barbosa-Cánovas, G. V., (2014). Effects of high-pressure processing on lipid oxidation: A review. *Innovative Food Science and Emerging Technologies, 22*, 1–10.
30. Muntean, M. V., Marian, O., Barbieru, V., Cătunescu, G. M., Ranta, O., Drocas, I., & Terhes, S., (2016). High pressure processing in food industry: Characteristics and applications. *Agriculture and Agricultural Science Procedia, 10*, 377–383.
31. Naik, L., Sharma, R., & Gaare, R., (2013). Application of high-pressure processing technology for dairy food preservation – future perspective: A review. *Journal of Animal Production Advances, 3*(8), 232–238.
32. Oey, I., Van, D. P. I., Van, L. A., & Hendrickx, M., (2008). Does high pressure processing influence the nutritional aspects of plant-based food systems? *Trends in Food Science and Technology, 19*(6), 300–308.
33. Patras, A., Brunton, N. P., Da Pieve, S., & Butler, F., (2009). Impact of high-pressure processing on total antioxidant activity, phenolic, ascorbic acid, anthocyanin content and color of strawberry and blackberry purées. *Innovative Food Science and Emerging Technologies, 10*(3), 308–313.
34. Rahman, M. S., (2007). *Handbook of Food Preservation* (2nd edn., p. 1088). Food Science and Technology Book Series; Boca Raton – FL: CRC Press.
35. Rastogi, N. K., Raghavarao, K. S. M. S., Balasubramaniam, V. M., Niranjan, K., & Knorr, D., (2007). Opportunities and challenges in high pressure processing of foods. *Critical Reviews in Food Science and Nutrition, 47*(1), 69–112.
36. San-Martin, M. F., Barbosa-Cánovas, G. V., & Swanson, B. G., (2002). Food processing by high hydrostatic pressure. *Critical Reviews in Food Science and Nutrition, 42*(6), 627–645.
37. Smelt, J. P., (1998). Recent advances in the microbiology of high-pressure processing. *Trends in Food Science and Technology, 9*(4), 152–158.
38. Smelt, J. P., Hellemons, J. C., & Patterson, M., (2001). Effects of high pressure on vegetative microorganisms. In: Hendrickx, M. E. G., & Knorr, D., (eds.), *Ultra High-Pressure Treatments of Foods* (pp. 55–76). New York, US: Kluwer Academic/Plenum Publishers.
39. Wang, S., (2009). Book review. *Trends in Food Science & Technology, 20*(2), 105–106.
40. Yordanov, D. G., & Angelova, G. V., (2010). High pressure processing for food preserving. *Biotechnology and Biotechnological Equipment, 24*(3), 1940–1945.
41. Zulkurnain, M., Maleky, F., & Balasubramaniam, V. M., (2016). High pressure processing effects on lipids thermophysical properties and crystallization kinetics. *Food Engineering Reviews, 8*(4), 393–413.
42. Berg, R. W. Van Den, Hoogland, H., Lelieve, H. L. M., & Schepdael, L. V. (2001). High Pressure Equipment Designs. 297–313.

CHAPTER 15

APPLICATIONS OF ELECTROSPINNING TECHNOLOGY IN FOODS: AN OVERVIEW

HARINDERJEET KAUR BHULLAR, BARINDERJEET SINGH TOOR, and AMARJEET KAUR

ABSTRACT

Electrospinning is an effective tool for fabricating submicron to nano-scale fibers. Since the method avoids the application of any toxic chemical and severe temperature or pressure conditions, thus it ensures high efficacy. This technique uses biopolymers (such as proteins and carbohydrates) and some synthetic polymers due to their non-toxic, biodegradable, and biocompatible properties. Parameters including the solution parameters (viscosity, molecular weight of solution, conductivity of solution, and surface tension), processing conditions (solution flow rate, applied voltage, tip to collector distance) and ambient conditions (humidity and temperature) have significant effect on the quality of fibers formed thus all such parameters must be precisely controlled in order to produced desired results. The fibers fabricated by electrospinning have many merits as high surface to volume ratio, submicron to nano-scale diameter, desirable porosity, and tailored fiber morphology. The stabilization of nutraceuticals, involvement in active packaging and edible films, improving the viability of probiotic substances, immobilization of enzymes and construction of delivery systems are the few potential application of electrospinning. In spite of much research at laboratory scale, efforts are required to scale up this process for utilization by food industries.

Advanced Research Methods in Food Processing Technologies. Junaid Ahmad Malik, Megh R. Goyal, Preeti Birwal, & Ritesh B. Watharkar (Eds.)
© 2024 Apple Academic Press, Inc. Co-published with CRC Press (Taylor & Francis)

15.1 INTRODUCTION

Electrospinning is an innovative approach for producing submicron to nano-scale fibers from a range of polymers or biopolymers. The technique involved the application of an external electric field in order the induce the movement of polymer solution in the form of fine jets. There are various constraints that are important in this technique including concentration of polymer, surface tension, viscosity, molecular weight, applied voltage flow rate, etc. All such constraints are carefully optimized in order to obtain desired results [10, 22]. A range of synthetic polymers have been used for producing electrospun fibers, such as polyvinyl alcohol (PVA), polyethylene oxide (PEO) however the trend is sifting successively to biopolymers or food grade polymers owing to the concern of consumers towards the adverse effects of synthetic polymers on the human body [12, 67]. There are a number of electrospinning variants which can be executed to obtain electrospun fibers, i.e., blend, coaxial, and emulsion electrospinning. The availability of such variants provides more flexibility to this technique.

Electrospinning has its roots in the 19th century, but it has been broadly applied in the field of agriculture and food from the last two decades. Although there have been other techniques, like extrusion, the electrospinning has certain structural advantages including desired porosity, ultrafine fibers, and high surface area to volume ratio. Recent investigations have shown the potential of electrospun nanofibers for the development of functional foods, attributed to its ability to maintain the integrity of bioactive compounds as well as corresponding beneficial properties. Moreover, electrospun fibers have potential utilization in the development of active/intelligent packaging, edible films, texturized proteins, etc. [40, 71].

Encapsulation of bioactive components is the major application of electrospinning. Encapsulated bioactive compounds tend to maintain their antioxidant and nutritional properties which is ascribed to their enhanced thermal and light stability as well as resistance to oxidative reactions [41]. Also, encapsulation enhances the sensorial attributes of the food products by masking the off-flavors without affecting the other quality attributes [64]. Techniques like spray drying and solvent evaporation have also been utilized for long times for the encapsulation of bioactive agent; however, the high temperature conditions of these technologies adversely affect the structural integrity, stability, and health benefits of compounds. Moreover, the encapsulation efficiencies recorded in electrospinning are better as compared to other technologies which make this process efficient and effective. Non-food applications of the electrospinning include wound repairing and dressing, clothing, and textile, tissue engineering, air filtration, military protective clothing, energy conservation and storage, etc. [13, 54].

Overall, this chapter provides brief knowledge on the mechanism of electrospinning, the factors affecting the process and the various types of polymers/biopolymers being utilized for electrospinning. Further, the emphasis has also been given to the application of electrospun fibers in the field of food science and technology.

15.2 MECHANISM OF ELECTROSPINNING PROCESS

It is essential to have brief knowledge about the electrospinning process in order to optimize the experimental conditions as well as to maximize the output. The basic setup of electrospinning includes syringe pump with a metal needle, collector (rotating drum or plate) and high voltage supply (Figure 15.1). Before executing the actual process, the polymers are dissolved in certain solvents. The polymer solution thus formed is extruded through a syringe pump resulting in the formation of droplet at the tip of the metal needle. Thereafter, a specific electric field is applied between the collector plate and the tip of the needle. Electrotactic forces such as external coulombic force and internal electrostatic repulsion cause the distortion of droplet from hemispherical to conical shape (also known as Taylor cone). With the intensification of the electric field, the surface of the droplet tends to accumulate more and more electric charge. When a critical value of electric field intensity is attained, the surface tension of the polymer solution is overcome by the electrostatic repulsion. This leads to the ejection of an electrically charged jet of solution from the Taylor cone's tip towards the oppositely charged collector [10, 22]. Thereafter, solvent evaporation happens due to rapid whipping of jet between collector and tip of needle. Finally, sub-micron or nano-scale fibers of polymer solution deposit on the collector [11].

15.3 POLYMERS USED IN THE ELECTROSPINNING TECHNOLOGY

Broad range of natural as well as synthetic polymers have the ability to form electrospun fibers within submicron to nano-range that can be used for the immobilization of enzymes, encapsulation of bioactive compounds, fabrication of active packaging material, preparation of biosensors as well as clarification of beverages [65]. Depending upon the characteristics and their applicability, a blend of natural as well as synthetic polymers is also used.

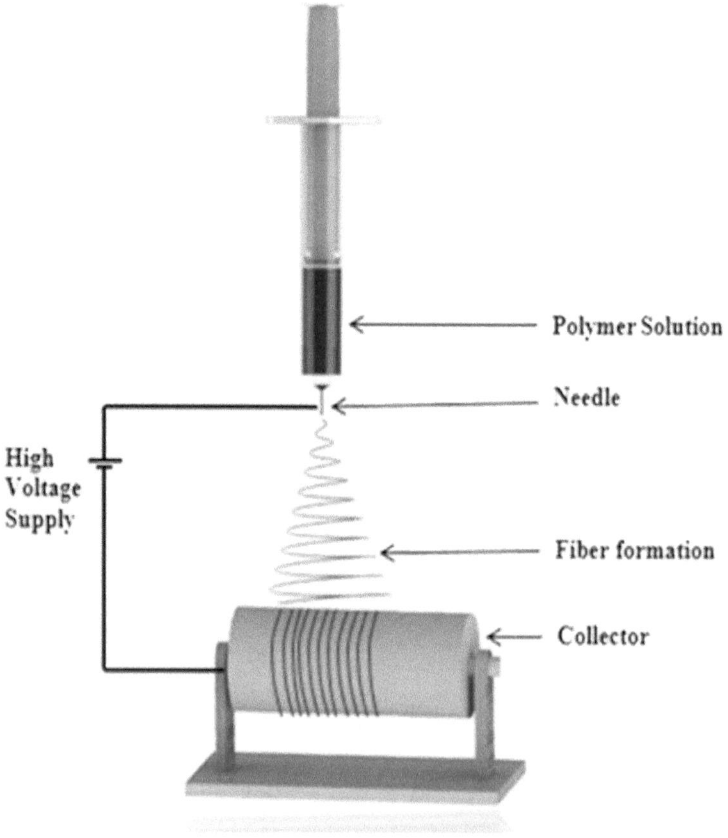

FIGURE 15.1 Schematic representation of electrospinning process.

15.3.1 NATURAL POLYMERS

Natural biomaterials are the proteins and polysaccharides obtained from natural sources such as plants, animals, and microorganisms. Naturally occurring biopolymers have gained immense interest due to their intrinsic ability to bind cells, low immunogenicity, non-toxicity, biocompatibility, renewability, biodegradability, antibacterial activity, and specific physical and chemical nature of biopolymers [64].

15.3.1.1 PROTEIN-BASED FIBERS

Protein-based electrospun fibers are amphiphilic in nature and have multiple binding sites which strongly binds the compounds by forming hydrogen and covalent bonds and also by electrostatic as well as hydrophobic interactions. But the secondary and tertiary structure of protein and less interaction among globular proteins hinders the electrospinning of proteins. To improve spinnability, random coil conformation is required. In order to achieve this, denaturation, change in pH, heating, breakage of cysteine bridges, usage of suitable solvents and blending of polymers is done [64]. Denaturation of proteins transforms globular albumin molecule into fibrillar form that improve the fiber production [17].

Various proteins used for electrospinning are whey protein isolate and concentrate, collagen, gelatin, caseinate, silk fibroin, egg albumen, soy protein isolate, amaranth protein isolate and zein (Figure 15.2). Of all the plant-based protein fibers, electrospun zein fibers exhibits excellent gas barrier properties, heat, and light stability, maintains antioxidant activity and antimicrobial activity of encapsulated material, enhances oxidative stability of fish oil, exerts gastro-mucoadhesive property of α-tocopherol, high encapsulation efficiency, enhanced bioaccessibility of resveratrol and fiber hydrophobicity. Thereof, zein fibers could be utilized for the encapsulation of several bioactive compounds such as gallic acid, curcumin, β-carotene, fish oil, α-tocopherol, rose hip seed oil and also for the preparation of food packaging materials. Whereas, other plant protein fibers as amaranth protein isolates and soy protein isolates are used in combination with other polymers.

Usually, amaranth protein isolates are spinned with pullulan to encapsulate quercetin, ferulic acid [2] and curcumin [12]. Amaranth protein isolates/pullulan electrospun fibers retain antioxidant activity and exhibit controlled release of core material. However, for the electrospinning of soy protein isolates, synthetic polymers are blended to it. In addition to this, denaturation of soy protein isolate could also improve its spinnability, encapsulation efficiency and retention of antioxidants of bioactive compounds. For example, red raspberry extract encapsulated in electrospun denatured soy protein isolate fibers retained a large amount of anthocyanin, exerted high antimicrobial activity in comparison to native soy protein isolates fibers [58].

Furthermore, amphiphilic nature of whey protein isolates nanofiber results in greater stability of encapsulated *Bifidobacterium strains* against relative humidity [39]. Besides ambient conditions, the solvent used for the electrospinning of gelatin greatly influence the fiber formation. For example, increase in the viscosity and conductivity during the electrospinning of vitamin A and E loaded gelatin led to the formation of stable fibers with controlled release ability [36]. Recently, organic solvents are replaced with surfactants for the fabrication of gelatin nanofibers.

Deng & co-workers [16] found that addition of Tween 80 and cationic cetyltrimethylammonium bromide to curcumin encapsulated gelatin nanofibers improve free radical scavenging activity, antibacterial activity, sustained, and controlled release of encapsulated curcumin. In contrast to this, anionic surfactants need to be incorporated for the fabrication of fish sarcoplasmic protein fibers as cationic surfactants produce dense fibers which hinders the release of encapsulated material [53]. While, for the fabrication of pure collagen highly volatile fluoro-alcohols as 1,1,1,3,3,3-hexafluoro-2-propanol fibers or 2,2,2-trifluoroethanol are used as solvents [72]. These solvents lower the denaturation temperature, dissolve collagen at adequate concentration to achieve electrospinning and maintain more than 50% of their native structure. Similarly, hexafluoro-2-propanol is used for the electrospinning of silk fibroin [17] But partial denaturation of native structure of these polymers, non-spinnability demands concern. Nevertheless, electrospun protein fibers provide high nutritional value, specific surface area, thermal resistance, moisture resistance, light stability, oxidative stability to the bioactive compound and can be used for the active packaging of food material and as a nutraceutical carrier in food industry.

15.3.1.2 CARBOHYDRATE-BASED FIBERS

Carbohydrates are thermostable and hydrophilic in nature but possess the ability to entrap hydrophobic compounds which makes them suitable for encapsulating thermo-degradable compounds. In contrast, strong hydrogen bonding of polysaccharides increases the solution viscosity and forms gel thereby, hinders the fiber formation. To overcome this, polysaccharides are blended with other biopolymers or synthetic polymers and solvents, hence decreasing the concentration of polysaccharides [22]. Carbohydrates, especially polysaccharides used for electrospinning, are classified on the basis of their source as plant origin, animal origin, from seaweeds and from microorganisms (Figure 15.2).

Electrospinning of chitosan is quite difficult because of its molar mass and polycationic nature, thus it requires the addition of other polymers (e.g., PVA) and solvents (tetrahydrofolate, acetic acid, trifluoroacetic acid, trifluoroacetic acid/dichloromethane) [23]. Similarly, to enhance the spinnability of alginate, other polymers such as PEO or solvents such as glycerol have been used. Alborzi & co-workers [3] found that folic acid encapsulated in sodium alginate/pectin/PEO electrospun nanofibers remained 100% stable at pH 3 even after 41 days of storage. This study shows that alginate electrospun fibers impart stability to photosensitive compounds. In another study, colon specific delivery of bovine serum albumin (BSA) was reported upon encapsulation in alginate/chitosan electrospun nanofibers [61].

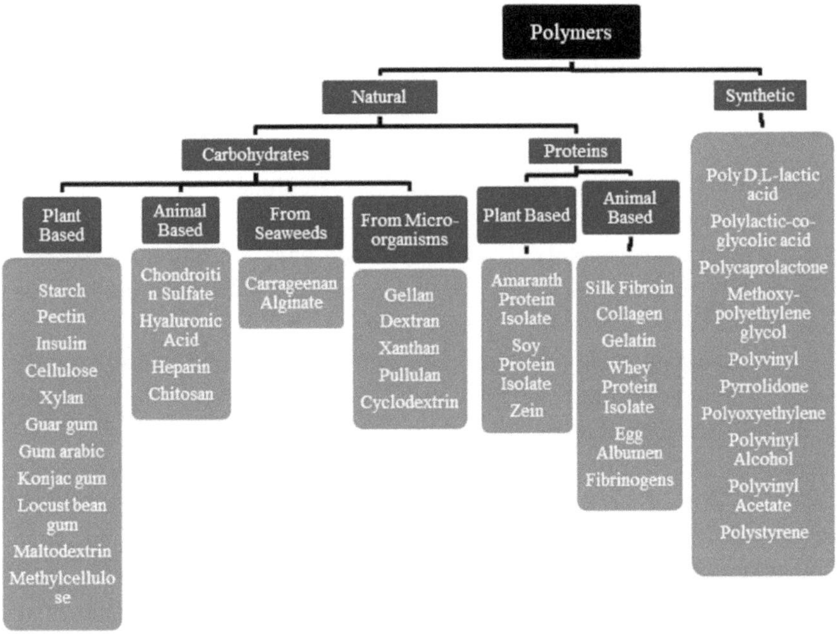

FIGURE 15.2 Classification of polymers.

Furthermore, cyclodextrin (CD) form an inclusion complex with bioactive compound and maintain their functionality. The addition of β-CD to polycaprolactone (PCL)/α-tocopherol enhanced UV stability and free radical scavenging activity of α-tocopherol [8]. In addition, nanofilm developed by encapsulating perillaldehyde (aroma compound) in pullulan/CD electrospun nanofibers, remain stable at ambient conditions as well as at high temperature but release the compound at high relative humidity thereby, such nanofilms can be potentially used for active packaging of food [41]. In addition, physicochemical properties of hyaluronic acid as shock absorber, molecular filter, biodegradability, biocompatibility also makes it suitable for nutraceutical carrier in food industry.

Polysaccharide-based nanofibers as chitosan nanofibers, alginate, and cellulose nanofibers have been used for the immobilization of enzymes to enhance their stability, bioactivity, and durability. For instance, immobilized hen egg-white lysozyme through cross-linked enzyme aggregates in chitosan nanofibers retains 75% of its activity even after 80 days of storage [48]. Similarly, naringinase immobilized on electrospun cellulose acetate nanofibers using layer by layer technique significantly removed 22.7% naringin and 60.7% limonene from grapefruit; thus, this technique can be used to remove bitterness from citrus fruit juices [26]. Cellulose acetate nanofibers have also been utilized for the controlled release of sea holly extract, to purify immunoglobulin G and also used as biosensors to detect the presence of photochromic compounds in solution.

Typically, proteins and carbohydrates are blended to obtain fibers of desired characteristics. For instance, egg albumen and cellulose acetate are blended to reduce surface tension and facilitates the fabrication of egg albumen/cellulose acetate fibers [66]. Similarly, pullulan/whey protein concentrate electrospun fibers increased cell viability and survival rate of *Bifidobacterium* strains during storage at 4° and 20°C [39].

Fibers from carbohydrates are biodegradable, biocompatible, mask undesirable odor, impart oxidative stability, thermal stability, retain antioxidant and antimicrobial activity of compounds and have high tensile strength. Therefore, polysaccharides based electrospun fibers can potentially be used for the designing of active packaging material, as drug delivery vehicle for control and sustained release of compounds, development of biosensor in beverage industry and for the immobilization of enzymes.

15.3.2 SYNTHETIC POLYMERS

Synthetic polymers are man tailored to accomplish desired mechanical strength, viscoelasticity, and degradation at specific area and rate. Most commonly used synthetic polymers are polyglycolide, polylactide, PVA, PCL, PEO, and polystyrene. In the food industry, these polymers are usually blended with natural polymers to improve spinnability, barrier properties, mechanical properties, and thermal stability of electrospun fibers by entrapping biopolymers in intertwined network. For instance, encapsulation of geraniol in PVA and γ-CD matrix increased the stability and shelf life of essential oil [35]. Similarly, coaxial electrospinning of ferulic acid/polyvinyl pyrrolidone/cellulose acetate improves sustained release of ferulic acid [68].

Also, PVA/β-CD/retinyl acetate electrospun nanofibers have better thermal and oxidative stability than PVA/retinyl acetate fibers. Thereof, the incorporation of synthetic polymer to natural polymers adds a greater advantage to use fibers for nutraceutical delivery, food packaging, preparation of edible films and encapsulation of bioactive compounds.

15.4 IMPORTANT FACTORS IN ELECTROSPINNING

During electrospinning, fabrication of fibers largely depends upon three major parameters as solution parameters, processing, and ambient conditions. The inherent attributes of polymer solution are polymer concentration, solution viscosity, surface tension, molecular weight of polymer, electrical conductivity, solvent used. These attributes are interrelated to each other, and little modifications in them produce fibers of desired characteristics. Furthermore, processing conditions include the distance of the electrode from the collector, flow rate and strength of applied voltage and also ambient temperature and relative humidity greatly affect the fiber morphology and diameter (Figure 15.3) [56].

15.4.1 SOLUTION PARAMETERS

15.4.1.1 VISCOSITY/CONCENTRATION OF SOLUTION

In electrospinning, specific solution concentration is desired to facilitate the formation of fibers. At low solution concentration, beads are formed and increase in concentration leads to the formation of beaded fiber, whereas at the optimum concentration nanofibers of desired diameters are formed. However, if the concentration of the solution is too high, then flow at the tip of the needle cannot be maintained [22].

Ki & co-workers [30] found power law relationship between diameter of fiber and concentration of solution, which states that with the increase in concentration there is an increase in diameter of fiber. Concentration and viscosity of the solution has direct relation as with the increase in the concentration, viscosity of solution also increases thereby increases the fiber diameter and uniformity which could be due to increased interchain association. As, when concentration of β-CD was increased during fabrication of β-CD/PVA fibers, solution's viscosity also increased this in turn led to the production of uniform fibers of larger average diameter. The same results were obtained when PVA/chitosan blend nanofibers were electro spun [27]. However, increase in the viscosity impedes the ejaculation of polymer solution from the needle. Thereby, specific concentration and viscosity is required to attain optimum polymer mobility, higher molecular orientation and to fabricate fibers with higher strength, stiffness, and crystallinity.

15.4.1.2 MOLECULAR WEIGHT OF THE POLYMERS

The molecular weight of the polymers depicts the entanglement of polymer chains in solution. Beads and fiber formation depend upon the entanglement number of polymer solution (η_e). Entanglement number is determined by the molecular weight of the polymer upon molecular weight of entangled polymer. For the fiber formation entanglement number of solution must be above 3.5. This number not only depends upon the molecular weight of the polymer but also on the solution concentration. As entanglement number of 5% polyvinyl pyrrolidone solution was 2 and beads were formed, but the increase in the concentration up to 10% corresponds to η_e above 3.5 and formation of pure fibers [43]. Furthermore, at particular polymer concentration, low molecular weight polymers form beads and high molecular weight polymers form nanofibers. But too high molecular weight polymers are also not required as they form fibers of larger average diameter due to extensive interchain linkages [21].

15.4.1.3 SURFACE TENSION

Surface tension is the counteracting force against the charge generated by the voltage applied [47]. During electrospinning, bead-less fiber can be obtained if surface tension is kept lower, whereas high surface tension unstable the jet and form droplets. Usually, surfactants are added to improve spinnability, to decrease the diameter of electrospun fiber and to reduce the occurrence of beaded fiber [62]. Increase in the concentration of non-ionic surfactant and TrionX-100 has significantly reduced the average diameter of fiber. In addition, surfactant Tween 20 was used in the blend of chitosan/PEO solutions at different acidity level, and upon electrospinning, nanofibers of different diameters were attained as high acidic solution produced less beaded fibers. It was observed that beadless fiber formation is not due to the addition of surfactant but due to increase in acetic acid concentration that increases viscosity and thereby lowers the surface tension [73].

15.4.1.4 CONDUCTIVITY OF SOLUTION

Electrical conductivity of solution is associated with the charge density on the surface of jet, as under applied electric field it increases the length of jet along its axis. It has been observed that solution with high electrical conductivity yield fiber of smaller diameter, whereas if conductivity is not sufficient, then beads will be formed, and fibers will not be uniform as electric charge will not be enough for elongation.

Usually, biopolymers are polyelectrolytic in nature, which is polymer dissociates into polycations and polyanions, thus increases the charge on the jet resulting in higher tension under applied electric field. Therefore, even high conductive solutions as natural polymers are unstable and have poor fiber forming ability than synthetic polymers. However, to enhance the conductivity, the salt solution or organic acids can be added. Zong et al. [74] observed the significant effect of ionic salts (NaH_2PO_4, KH_2PO_4 and NaCl) on morphology and diameter of electro spun fiber that is beadless fiber with smaller diameter, ranging from 200 to 1,000 nm were produced. Conductivity of many other polymers like polylactic acid (PLA), PVA, and PEO was also improved by adding ionic salts. This resulted in lessening the bead formation and increase in uniformity of fiber.

15.4.1.5 SOLVENT USED

Polymer solution is prepared by dissolving 10–20% of polymer in 80–90% of solvent, so solvent used largely influences the fiber morphology. Solvent properties such as density, boiling point, volatility, vapor pressure and intermolecular interaction between polymer and solvent have significant effect on the fabrication of nanostructure. In addition, with the decrease in density and boiling point of solvent, fiber dryness increases due to rapid evaporation of solvent [10].

Furthermore, Yang & co-workers [70] examined the influence of different solvents as ethanol, dichloromethane, N,N-dimethylformamide on the polyvinyl pyrrolidone electrospun fiber morphology. They observed smooth texture of fiber with larger average diameter when ethanol was used as solvent whereas, fibers prepared using dichloromethane and N,N-dimethylformamide had beads on a string like shape. Further, in order to obtain nanofibers of small diameter mixed solvents were used.

Moreover, solvent is also required to deliver the polymer molecules to the collector by decreasing the surface tension of the solution; thus, surface tension of the solution largely depends upon the solvent composition. Upon increase in the concentration of acetic acid in chitosan electrospun fibers, surface tension decreases resulting in thinner fibers [21].

15.4.2 PROCESSING PARAMETERS

15.4.2.1 APPLIED VOLTAGE

Voltage is the chief parameter, as fiber formation takes place only after the achievement of required electric field strength across needle tip and collector upon application of high voltage. Applied voltage overcomes the surface tension of solution at the spinneret and increases electrostatic force on the jet which reduces the fiber diameter. Voltage applied largely depends upon feed rate, concentration, and viscosity of the solution. With the increase in the viscosity and feed rate, higher voltage is applied as more electric field strength is required to form a stable Taylor cone and to attenuate the jet [22].

Usually, voltage supply higher than 6 kV is required to overcome surface tension and form stable Taylor cone. However, if voltage applied is beyond its threshold limit, then increase in charge increases the flow rate of solution which results in smaller, unstable Taylor cone and bead formation takes place [10]. Sencadas et al. [51] observed a reduction in chitosan electrospun nanofibers diameter with an increase in applied voltage. Furthermore, several researchers reported that an initial increase in voltage supply reduces fiber diameter but, further increase tend to increase fiber diameter resulting in bead formation [74]. Thus, voltage applied influences the morphology and uniformity of fiber.

15.4.2.2 FLOW RATE

Flow rate indicates the material transfer rate and velocity of polymer coming out of the syringe. Certain volume of polymer is required at the end of the needle to achieve stable Taylor cone and to assist solvent evaporation [57]. Okutan et al. [47] observed that during electrospinning of gelatin, fiber diameter increased, and bead formation occurred when feed rate was increased which could be due to poor solvent evaporation before reaching the collector.

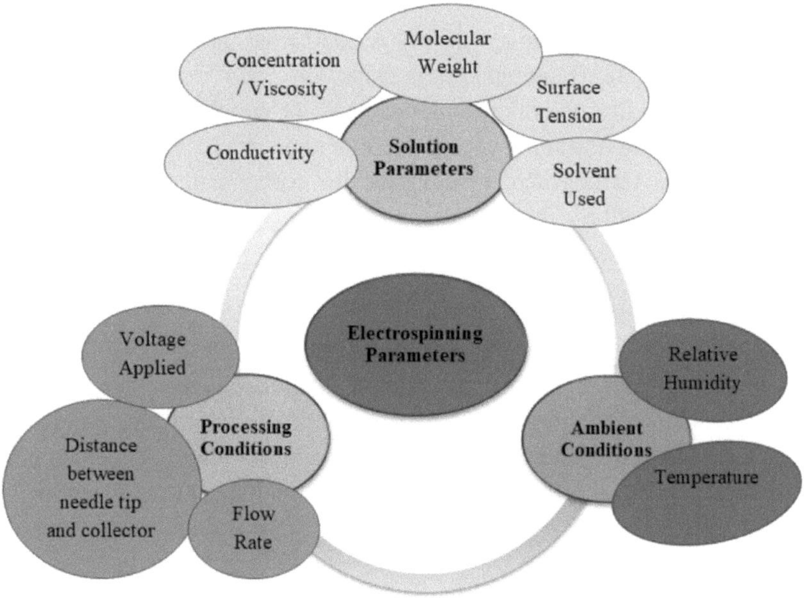

FIGURE 15.3 Various parameters affecting electrospinning.

15.4.2.3 TIP TO COLLECTOR DISTANCE

Tip to collector distance corresponds to the elongation of the spinning jet. This distance between needle and collector must be optimum for adequate drying of fibers before reaching the collector, thereby forming fibers of smaller diameter. On the contrary increase and decrease in the optimum distance results in the formation of beads. For instance, Lee et al. [33] pointed out that due to longer distance, an electrostatic force on the jet reduces and fiber diameter gets increased. Similarly, Sencadas et al. [51] stated that when jet traveling distance was 20 cm, fibers of average diameter of approximately 500 nm were achieved; whereas decrease in distance to 5 cm, average diameter of fibers reduced ~260 nm. However, several researchers found the non-significant effect of change in tip to collector distance on structural morphology and diameter. For example, a decrease in distance produces flatter fibers and small fibers, whereas an increase in distance form spherical beads [21, 30].

15.4.3 AMBIENT CONDITIONS

15.4.3.1 HUMIDITY

Though not much systematic research is conducted on effect of humidity on fiber morphology but Tripatanasuwan et al. [59] discovered that with increase in relative humidity, average diameter of PEO nanofiber decreased which could be due to reduction in evaporation rate of jet because of higher humidity thence charged jet continue to elongate. In contrast, increase in humidity forms beads instead of fiber upon electrospinning of cellulose acetate.

15.4.3.2 TEMPERATURE

Ambient temperature affects the intrinsic parameters of solution such as viscosity, surface tension and conductivity and solvent evaporation rate. Where, conductivity and solvent evaporation rate are directly dependent on the temperature. However, increase in temperature reduces the solution viscosity thereby effects morphology and fiber size. For instance, an increase in temperature from 25 to 60°C during electrospinning of polyamide-6 decreases fiber average diameter resulting in the fabrication of nanofibers [10].

15.5 APPLICATIONS OF ELECTROSPUN FIBERS IN THE FOOD INDUSTRY

In the present scenario, the prime focus of food research or processors is on the bioactive composition of food matrix. It is ascribed to the fact that bioactive compounds are correlated with many health promoting activities, for instance, antioxidant, anticancer as well as anti-inflammatory activities. These compounds include phenolics, flavonoids, bioactive peptides, amino acids, vitamins, minerals, fatty acids, probiotics, etc. [71]. However, the major problems related to bioactive compounds are their instability to the external environment components (light, moisture, oxygen, temperature), poor solubility in water and low gastrointestinal absorption. All such factors may result in the degradation of defined bioactive compounds as well as corresponding functionality.

Various investigations have been executed to overcome the defined problems and encapsulation have been found as a potential tool. In this context, electrospinning has the ability to carry out the nano-encapsulation of bioactive compounds which enhance their thermal stability, photostability, bioavailability as well as resistance to oxidative reactions. Besides, nano-encapsulation can successfully mask the undesired flavor and odors which in turn enhance the overall sensorial attributes of the final product [64]. Furthermore, the electrospun fibers can potentially be used for targeted delivery and controlled release of functional components (Figure 15.4) [41].

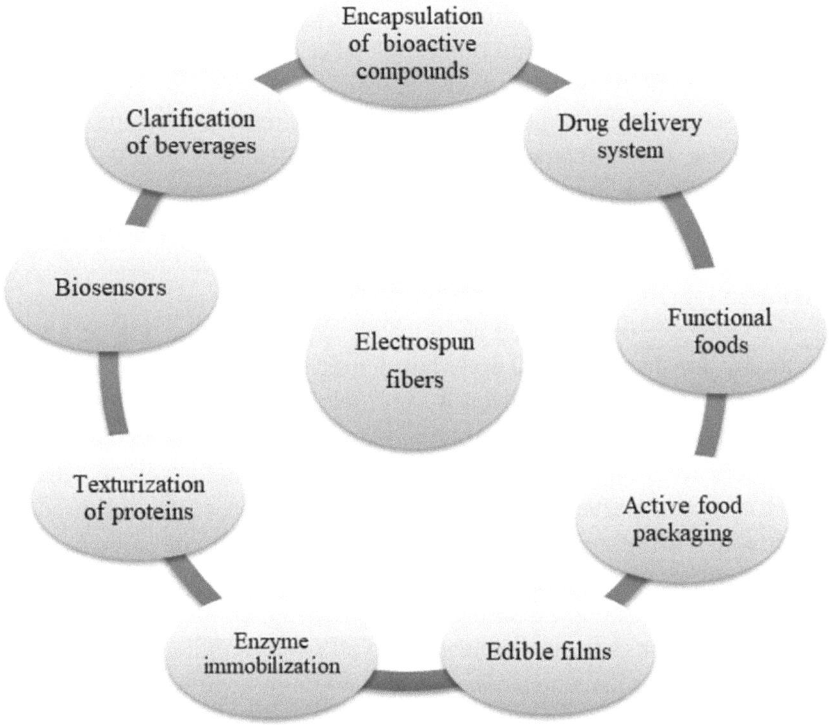

FIGURE 15.4 Food applications of electrospun fibers.

The core functions of any food packaging include containment and to protect the food products from undesired changes caused by internal as well external factors, for instance, light, moisture, microorganisms, enzymes, odors, gases, and physical damage. Over time, the paradigm has been shifted from basic packaging to active/intelligent packaging attribute to the rise in the demand of consumers for safe, healthy, high quality and fresh food products. These active packaging have bioactive compounds incorporated into their structure which protect the food from microbial contamination, oxidation reactions, etc. In such context, electrospun nanofibers-based active food packaging has been developed as an effective tool to serve the purpose [71].

Generally, two approaches have been utilized, first, the electrospinning of naturally active polymers, for instance, chitosan, which has the ability to enhance the shelf life of product through its antimicrobial activity [58]. Second, the incorporation of active agents into polymer solution and the electrospinning process dispense the active compounds throughout the mat [71]. However, the latter approach has certain limitations, i.e., poor insolubility and instability of bioactive compounds to light and temperature conditions. Therefore, certain other molecules, for instance, CD, have been incorporated into the polymer solution, which form inclusion complexes with bioactive compounds and protect them from extreme processing conditions [7]. Besides, the utilization of natural polymers for the development of edible films/packaging can make a significant contribution in the reduction of pollution caused by non-degradable packaging materials. Also, such type of packaging's tends to be non-toxic and maintain the characteristics of food products (Table 15.1).

Enzymes are widely used as biocatalyst in the production of a variety of foods, for instance, bakery, dairy, brewery, and meat products and beverages. They are considered as safe, eco-friendly as well as efficient. However, there are certain problems which need to be considered before their application in developing any food system. These include sensitivity to pH and temperature conditions, high production cost, undesired inactivation and lower recovery rates [71]. Investigations have demonstrated that nanofibers developed through emulsion electrospinning has a great potential in addressing the defined problems [71].

Encapsulation, sustained release as well as immobilization enhance the functionality, stability, and resistance of enzymes to involuntary inactivation. In addition, enzyme immobilization by emulsion electrospinning enhances the efficiency which might be ascribed to the ability of electrospinning to fabricate fine fibers with large surface area which further increase the enzyme loading capacity per unit mass [31, 37]. Also, the immobilization of enzymes makes their recovery easy and efficient.

The population is growing gradually, which forces the food processors to shift from conventional to alternative protein sources. Previous investigation has demonstrated that soy proteins can successfully be texturized to mimic meat proteins using extrusion technology [40]. However, the shear forces and high temperature are the disadvantages of extrusion processing. In such context, electrospinning can serve the purpose by utilizing the alternative

TABLE 15.1 Recent Studies Showing the Application of Electrospun Fibers in Food Science

Bioactive Compound	Polymer Solution	Results	References
Allyl isothiocyanate	PVA + β-CD	• Showed 94% and 100% antibacterial activity against *Escherichia coli* and *Staphylococcus aureus*, respectively.	[6]
α-Tocopherol	Zein + Chitosan	• Improved delivery of α-tocopherol to gastrointestinal tract. • Fibers showed good gastro mucoadhesive property.	[67]
	Gelatin	• Sustained release of vitamins A and E for 60 hours.	[36]
	PCL + β-CD	• PCL + β-CD + α-tocopherol exhibited better UV light stability and antioxidant activity in comparison to PCL + α-tocopherol.	[9]
	Hydroxypropyl-β-CD	• Water solubility of α-tocopherol increased. • Antioxidant activity of vitamin E retained. • Vitamin E displayed stability over the period of three years. • Enhanced UV light stability.	[15]
Bifidobacterium animalis subsp. *lactis BB-12*	PVA	• Maintained cell viability for 40 days at 20°C and for 130 days at 4°C.	[38]
	Pullulan + Whey protein concentrate	• Enhanced cell viability. • Increased survival rate of *Bifidobacterium* during storage at 20°C.	[39]
β-Carotene	Zein prolamin	• Enhanced protection against oxidation reactions. • Encapsulated compound exhibited better light stability.	[19]
Carotenoids extract	Gelatin	• Encapsulation efficiency > 90%. • Encapsulated extract displayed greater retention of lycopene than non-encapsulated. • Retained antioxidant activity during the storage period of 14 days. • Enhanced water solubility of extract.	[25]

TABLE 15.1 *(Continued)*

Bioactive Compound	Polymer Solution	Results	References
	Zein	• High encapsulation efficiency (> 90%). • Greater retention of lycopene in encapsulated extract than non-encapsulated. • Enhanced antioxidant activity. • Improved thermal stability.	[24]
Chamomile essential oil	Gelatin	• Enhanced antioxidant activity. • Potential application as edible packaging.	[55]
Curcumin	Zein	• Possessed significant free radical scavenging activity. • Controlled release of curcumin for seven days.	[13]
		• Showed 83% antibacterial activity against *S. aureus*.	[14]
		• Potential application in detecting the presence of ferric ions in potable water.	[50]
		• Approx. 100% encapsulation efficiency • Encapsulated bioactive agent retained antioxidant activity. • Fibers possessed high antibacterial activity against *E. coli* and *S. aureus*.	[61]
	PVA + β-CD	• Retained the structural integrity if curcumin. • Sustained release over the period of 5 h. • Better thermal stability. • Enhanced stability and solubility.	[54]
	Amaranth protein isolate + Pullulan	• Sustained and controlled release of curcumin. • Encapsulated curcumin exhibited better antioxidant activity than free form.	[12]
	Gelatin + CTAB/ Tween 80	• Sustained release of curcumin. • Exhibited high radical scavenging activity. • Significant antimicrobial activity.	[16]

TABLE 15.1 *(Continued)*

Bioactive Compound	Polymer Solution	Results	References
	Gelatin, zein	• High encapsulation efficiency, i.e., approx. 90% in both gelatin and zein. • Enhanced stability of curcumin. • Controlled and slow release of curcumin in zein fibers. • Gelatin fibers displayed better results in fatty food stimulants. • Zein fibers were suggested for packaging of food products with high moisture level.	[4]
Eugenol	PVA	• Higher antimicrobial activity due to slow release of eugenol. • Significant antibacterial activity against gram-negative bacteria compared to gram-positive.	[32]
Ethyl vanillin	Chitosan + PVA	• Displayed excellent barrier capability to UV. • Exhibited high antimicrobial potential.	[44]
Fish oil	PVA + Whey protein isolate	• High encapsulation efficiency (92.4%) of omega-3. • Enhanced oxidative stability.	[20]
	Zein	• High encapsulation efficiency (96%) of omega-3. • Enhanced oxidative stability during storage.	[42]
		• Better thermal and oxidative stability of oil in coaxial nanofibers than single nanofibers. • Shelf life of oil was 65 days more in coaxial fibers as compared to single nanofibers. • Potential application in development of fish oil based novel functional foods.	[69]
Folic acid	Sodium alginate + pectin + PEO	• After encapsulation, folic acid displayed better retention behavior. • Process can be utilized to address the folic acid deficiency in acidic foods.	[3]

TABLE 15.1 *(Continued)*

Bioactive Compound	Polymer Solution	Results	References
	Amaranth protein isolate + Pullulan	• > 95% encapsulation efficiency. • Enhanced thermal stability of bioactive agent. • Resistant to light degradation when exposed to UV for 2 hours.	[1]
	Zein	• Encapsulation efficiency > 80%. • Fibers possessed high thermal stability at high temperature of 180°C. • Displayed significant photostability during 24 h exposure to UVA irradiation.	[18]
Glucose oxidase	PVA + PCL	• Showed significant antibacterial properties even after electrospinning. • Potential technique for enzyme immobilization.	[5]
Geraniol	PVA + γ-CD	• Minimal loss of geraniol over the period of two years. • Sustained durability. • Enhanced thermal stability.	[28]
Gallic acid	PLA + Hydroxypropyl-β-CD	• No loss in the antioxidant activity of gallic acid. • Slow release of gallic acid.	[8]
	Zein	• Significant antimicrobial activity against gram-positive and gram-negative bacteria. • Potential application as active packaging.	[45]
Nisin	Amaranth protein isolate + Pullulan	• Encapsulation efficiency > 90%. • Better antimicrobial activity against *Salmonella typhimurium*, *Leuconostoc mesenteroides*, and *Listeria monocytogenes*.	[52]
Peppermint essential oil	Gelatin	• Significant antimicrobial activity against *S. aureus* and *E. coli*.	[55]
Proanthocyanidins	Zein	• Approx. 100% encapsulation efficiency. • Proanthocyanidins retained the antioxidant activity.	[60]

TABLE 15.1 *(Continued)*

Bioactive Compound	Polymer Solution	Results	References
		• Enhanced thermal properties. • Sustained releasing behavior.	
Retinyl acetate	PVA + β-CD	• PVA + β-CD better protected the retinyl acetate against oxidation than PVA alone. • Better thermal stability in PVA + β-CD.	[35]
Resveratrol	Zein	• 96.9 and 77.6% encapsulation efficiency at 2 and 10%, respectively, loading of compound. • Enhanced bio accessibility. • Resveratrol + zein nanofibers as edible coating on apple controlled the moisture loss and retained color properties.	[34]
Thymol	Zein + γ-CD	• Zein + γ-CD + thymol showed better antibacterial activity than Zein + thymol. • Zein + γ-CD + thymol maintained the microbiological quality of meat up to five days at 4°C.	[7]
Vanillin	PVA + α-CD/β-CD/γ-CD	• Improved thermal stability. • Prolonged durability. • Better stability and controlled release of vanillin in PVA + γ-CD as compared to PVA + β-CD/γ-CD.	[29]
Yerba mate extract	Zein	• Encapsulated yerba mate extract exhibited better thermal stability. • Enhanced antioxidant activity. • Sustained release of extract. • Potential application in active food packaging.	[49]

protein sources (insect proteins, leaf proteins and plant proteins) to produce fibrous structure which can mimic the texture of meat proteins [46].

There are certain other potential applications, which are less explored or limited to laboratory scale. These include electrospun nanofiber-based biosensors for food and agricultural commodities and filtration membrane for the beverage industry.

15.6 ADVANTAGES OF ELECTROSPINNING

The various functional advantages associated with electrospun fibers have been enlisted in Figure 15.5. Apart from these, there are certain structural advantages including high surface to volume ratio, tailored morphology, desirable porosity and tunable fiber diameters [64]. The combination of functional and structural advantages defines the significance of electrospun fibers in food science and corresponding industries.

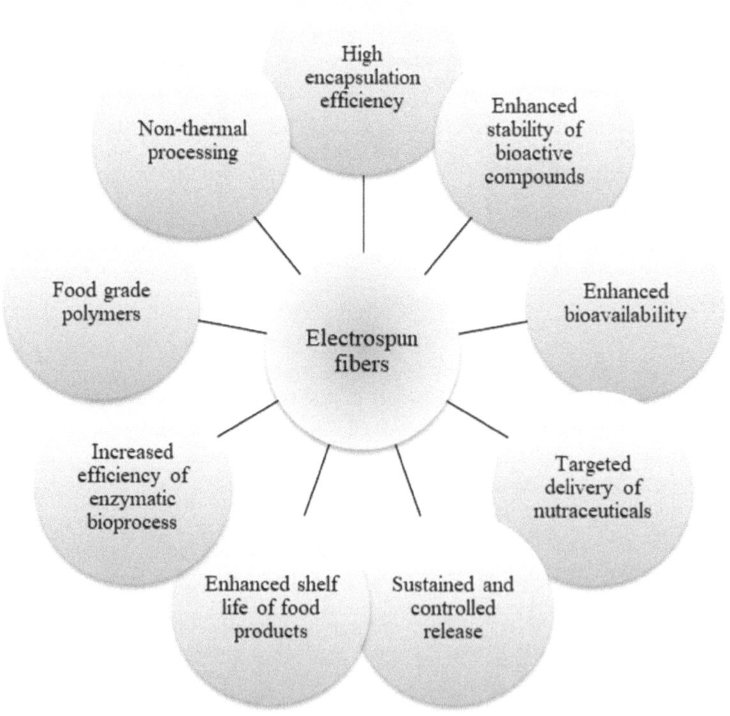

FIGURE 15.5 Advantages of electrospinning.

15.7 LIMITATIONS

In the last two decades, numerous investigations have explored the potential of electrospun fibers in the field of food and agriculture. However, there are certain limitations which have to be addressed for obtaining the desired outcomes. The major limitation is the low throughput which is obstructing the commercial scale application of this technology. Certain modifications in the existing system are needed to be done to overcome this constraint. Secondly, the majority of previous investigations have utilized the synthetic polymers, which is always a concern to the consumers health. Hence, the future investigations must be emphasized on the utilization of biopolymers. Also, the effects of nano-scale food materials on human body and environment are needed to be scientifically analyzed. Furthermore, there are numerous polymers, solvent systems and processing conditions and it is impractical to execute the same experimental conditions for all the bioactive compounds therefore a specific system must be design accordingly. Lastly, scientific investigations are required to evaluate the interactions between electrospun fibers and food materials.

15.8 SUMMARY

Electrospinning is an easy, cost-effective, and adaptable technology to develop submicron to nano-scale fibers. The fibers fabricated by electrospinning process can be tailored as per the desired porosity, diameter, fiber morphology and have the high surface area to volume ratio. Due to structural flexibility and stability, fibers produced have high encapsulation efficiency and high bioavailability with which sustained release and target delivery can be achieved. In the food industry, electrospun fibers are used for the encapsulation of bioactive components, for enzyme immobilization, for development of active packaging material, to develop membrane biosensors and for filtration. In addition to this, electrospun fibers have wide applications in medical science to make dressings for fast wound healing, for drug delivery, in tissue engineering and to make protective clothing. Despite having numerous applications, it is limited to laboratory research area and not extensively used in the industries. This is because electrospinning process depends upon many variables, apart from solution and processing conditions, surrounding environment also affects the morphology of electrospun fiber and also these parameters are inter-linked so it's quite difficult to control all of them at industrial level production. Therefore, more research needs to be done for the scale up of electrospinning process without affecting fiber size and morphology. Lastly, nutraceutical loaded electrospun nanofibers require toxicity testing, *in vivo* bioavailability assessment, analysis of solvent residues entrapped in nanofibers and require investigation on the effect of smart packaging material on physical, chemical, and organoleptic properties of food. Hence, safe consumption of nanofood products still requires detailed analysis and legislation by the government for development.

KEYWORDS

- **bioactive compounds**
- **electrospinning**
- **electrospun fibers**
- **encapsulation**
- **fiber morphology**
- **nanofibers**
- **polymers**

REFERENCES

1. Aceituno-Medina, M., Mendoza, S., Lagaron, J. M., & López-Rubio, A., (2015). Photoprotection of folic acid upon encapsulation in food-grade amaranth (*Amaranthus hypochondriacus* L.) protein isolate–pullulan electrospun fibers. *LWT–Food Science and Technology, 62*(2), 970–975.
2. Aceituno-Medina, M., Mendoza, S., Rodríguez, B. A., Lagaron, J. M., & Lopez-Rubio, A., (2015). Improved antioxidant capacity of quercetin and ferulic acid during *in-vitro* digestion through encapsulation within food-grade electrospun fibers. *Journal of Functional Foods, 12*, 332–341.
3. Alborzi, S., Lim, L. T., & Kakuda, Y., (2013). Encapsulation of folic acid and its stability in sodium alginate-pectin-poly (ethylene oxide) electrospun fibers. *Journal of Microencapsulation, 30*(1), 64–71.
4. Alehosseini, A., Gómez-Mascaraque, L. G., Martínez-Sanz, M., & López-Rubio, A., (2019). Electrospun curcumin-loaded protein nanofiber mats as active/bioactive coatings for food packaging applications. *Food Hydrocolloids, 87*, 758–771.
5. Alonso-González, M., Corral-González, A., Félix, M., Romero, A., & Martin-Alfonso, J. E., (2020). Developing active poly (vinyl alcohol)-based membranes with encapsulated antimicrobial enzymes via electrospinning for food packaging. *International Journal of Biological Macromolecules, 2020*, E-article 100669.

6. Aytac, Z., Dogan, S. Y., Tekinay, T., & Uyar, T., (2014). Release and antibacterial activity of allyl isothiocyanate/β-cyclodextrin complex encapsulated in electrospun nanofibers. *Colloids and Surfaces B: Biointerfaces, 120*, 125–131.
7. Aytac, Z., Ipek, S., Durgun, E., Tekinay, T., & Uyar, T., (2017). Antibacterial electrospun zein nanofibrous web encapsulating thymol/cyclodextrin-inclusion complex for food packaging. *Food Chemistry, 233*, 117–124.
8. Aytac, Z., Kusku, S. I., Durgun, E., & Uyar, T., (2016). Encapsulation of gallic acid/cyclodextrin inclusion complex in electrospun polylactic acid nanofibers: Release behavior and antioxidant activity of gallic acid. *Materials Science and Engineering, 63*, 231–239.
9. Aytac, Z., & Uyar, T., (2016). Antioxidant activity and photostability of alpha-tocopherol/beta-cyclodextrin inclusion complex encapsulated electrospun polycaprolactone nanofibers. *Journal of European Polymer, 79*, 140–149.
10. Bhardwaj, N., & Kundu, S. C., (2010). Electrospinning: A fascinating fiber fabrication technique. *Biotechnology Advances, 28*(3), 325–347.
11. Bhushani, J. A., & Anandharamakrishnan, C., (2014). Electrospinning and electrospraying techniques: Potential food-based applications. *Trends in Food Science and Technology, 38*(1), 21–33.
12. Blanco-Padilla, A., López-Rubio, A., Loarca-Piña, G., Gómez-Mascze, L. G., & Mendoza, S., (2015). Characterization, release and antioxidant activity of curcumin loaded amaranth-pullulan electrospun fibers. *Food Science and Technology, 63*(2), 1137–1144.
13. Brahatheeswaran, D., Mathew, A., Aswathy, R. G., Nagaoka, Y., Venugopal, K., Yoshida, Y., Maekawa, T., & Sakthikumar, D., (2012). Hybrid fluorescent curcumin loaded zein electrospun nanofibrous scaffold for biomedical applications. *Biomedical Materials, 7*(4), E-article 045001.
14. Bui, H. T., Chung, O. L., & Park, J. S., (2014). Fabrication of electrospun antibacterial curcumin-loaded zein nanofibers. *Polymer-Korea, 38*(6), 744–751.
15. Celebioglu, A., & Uyar, T., (2017). Antioxidant vitamin E/cyclodextrin inclusion complex electrospun nanofibers: Enhanced water solubility, prolonged shelf life, and photostability of vitamin E. *Journal of Agricultural and Food Chemistry, 65*(26), 5404–5412.
16. Deng, L., Kang, X., Liu, Y., Feng, F., & Zhang, H., (2017). Effects of surfactants on the formation of gelatin nanofibers for controlled release of curcumin. *Food Chemistry, 231*, 70–77.
17. Dror, Y., Ziv, T., Makarov, V., Wolf, H., Admon, A., & Zussman, E., (2008). Nanofibers made of globular proteins. *Biomacromolecules, 9*(10), 2749–2754.
18. Evangelho Do, J. A., Crizel, R. L., Chaves, F. C., Prietto, L., Pinto, V. Z., Miranda Do, M. Z., Dias, A. R., & Da Rosa, Z. E., (2019). Thermal and irradiation resistance of folic acid encapsulated in zein ultrafine fibers or nanocapsules produced by electrospinning and electrospraying. *Food Research International, 124*, 137–146.
19. Fernandez, A., Torres-Giner, S., & Lagaron, J. M., (2009). Novel route to stabilization of bioactive antioxidants by encapsulation in electrospun fibers of zein prolamine. *Food Hydrocolloids, 23*(5), 1427–1432.
20. García-Moreno, P. J., Stephansen, K., Van, D. K. J., Guadix, A., Guadix, E. M., Chronakis, I. S., & Jacobsen, C., (2016). Encapsulation of fish oil in nanofibers by emulsion electrospinning: Physical characterization and oxidative stability. *Journal of Food Engineering, 183*, 39–49.

21. Geng, X., Kwon, O. H., & Jang, J., (2005). Electrospinning of chitosan dissolved in concentrated acetic acid solution. *Biomaterials, 26*(27), 5427–5432.
22. Ghorani, B., & Tucker, N., (2015). Fundamentals of electrospinning as a novel delivery vehicle for bioactive compounds in food nanotechnology. *Food Hydrocolloids, 51*, 227–2240.
23. Haider, S., Al-Zeghayer, Y., Ahmed, A. F., Haider, A., Mahmood, A., Al-Masry, W., Imran, M., & Aijaz, M., (2013). Highly aligned narrow diameter chitosan electrospun nanofibers. *Journal of Polymer Research, 20*(4), 1–11.
24. Horuz, T. İ., & Belibağlı, K. B., (2019). Nanoencapsulation of carotenoids extracted from tomato peels into zein fibers by electrospinning. *Journal of the Science of Food and Agriculture, 99*(2), 759–766.
25. Horuz, T. İ., & Belibağlı, K. B., (2018). Nanoencapsulation by electrospinning to improve stability and water solubility of carotenoids extracted from tomato peels. *Food Chemistry, 268*, 86–93.
26. Huang, W., Zhan, Y., Shi, X., Chen, J., Deng, H., & Du, Y., (2017). Controllable immobilization of naringinase on electrospun cellulose acetate nanofibers and their application to juice debittering. *International Journal of Biological Macromolecules, 98*, 630–636.
27. Jia, Y. T., Gong, J., Gu, X. H., Kim, H. Y., Dong, J., & Shen, X. Y., (2007). Fabrication and characterization of poly (vinyl alcohol)/chitosan blend nanofibers produced by electrospinning method. *Carbohydrate Polymers, 67*(3), 403–409.
28. Kayaci, F., Sen, H. S., Durgun, E., & Uyar, T., (2014). Functional electrospun polymeric nanofibers incorporating geraniol–cyclodextrin inclusion complexes: High thermal stability and enhanced durability of geraniol. *Food Research International, 62*, 424–431.
29. Kayaci, F., & Uyar, T., (2012). Encapsulation of vanillin/cyclodextrin inclusion complex in electrospun polyvinyl alcohol (PVA) nanowebs: Prolonged shelf-life and high temperature stability of vanillin. *Food Chemistry, 133*(3), 641–649.
30. Ki, C. S., Baek, D. H., Gang, K. D., Lee, K. H., Um, I. C., & Park, Y. H., (2005). Characterization of gelatin nanofiber prepared from gelatin-formic acid solution. *Polymer, 46*, 5094–5102.
31. Kim, J., Grate, J. W., & Wang, P., (2008). Nanobiocatalysis and its potential applications. *Trends in Biotechnology, 26*(11), 639–646.
32. Kriegel, C., Kit, K. M., McClements, D. J., & Weiss, J., (2009). Nanofibers as carrier systems for antimicrobial microemulsions. Part I: Fabrication and characterization. *Langmuir, 25*(2), 1154–1161.
33. Lee, J. S., Choi, K. H., Ghim, H. Do., Kim, S. S., Chun, D. H., & Kim, H. Y., (2004). Role of molecular weight of atactic poly (vinyl alcohol) (PVA) in the structure and properties of PVA nanofabric prepared by electrospinning. *Journal of Applied Polymer Science, 93*, 1638–1646.
34. Leena, M. M., Yoha, K. S., Moses, J. A., & Anandharamakrishnan, C., (2020). Edible coating with resveratrol loaded electrospun zein nanofibers with enhanced bioaccessibility. *Food Bioscience, 2020*, E-article 100669.
35. Lemma, S. M., Scampicchio, M., Mahon, P. J., Sbarski, I., Wang, J., & Kingshott, P., (2015). Controlled release of retinyl acetate from B-cyclodextrin functionalized poly (vinyl alcohol) electrospun nanofibers. *Journal of Agricultural and Food Chemistry, 63*, 3481–3488.

36. Li, H., Wang, M., Williams, G. R., Wu, J., Sun, X., Lv, Y., & Zhu, L. M., (2016). Electrospun gelatin nanofibers loaded with vitamins A and E as antibacterial wound dressing materials. *RSC Advances, 6*, 50267–50277.
37. Lopez-Rubio, A., Gavara, R., & Lagaron, J. M., (2006). Bioactive packaging: Turning foods into healthier foods through biomaterials. *Trends in Food Science and Technology, 17*, 567–575.
38. López-Rubio, A., Sanchez, E., Sanz, Y., & Lagaron, J. M., (2009). Encapsulation of living *bifidobacteria* in ultrathin PVOH electrospun fibers. *Biomacromolecules, 10,* 2823–2829.
39. López-Rubio, A., Sanchez, E., Wilkanowicz, S., Sanz, Y., & Lagaron, J. M., (2012). Electrospinning as a useful technique for the encapsulation of living *Bifidobacteria* in food hydrocolloids. *Food Hydrocolloids, 28,* 159–167.
40. Manski, J. M., Van, D. G. A. J., & Boom, R. M., (2007). Advances in structure formation of anisotropic protein-rich foods through novel processing concepts. *Trends in Food Science and Technology, 18*, 546–557.
41. Mascheroni, E., Fuenmayor, C. A., Cosio, M. S., Di Silvestro, G., Piergiovanni, L., Mannino, S., & Schiraldi, A., (2013). Encapsulation of volatiles in nanofibrous polysaccharide membranes for humidity triggered release. *Carbohydrate Polymer, 98,* 17–25.
42. Moomand, K., & Lim, L. T., (2014). Oxidative stability of encapsulated fish oil in electrospun zein fibers. *Food Research International, 62*, 523–532.
43. Munir, M. M., Suryamas, A. B., Iskandar, F., & Okuyama, K., (2009). Scaling law on particle-to-fiber formation during electrospinning. *Polymer, 50,* 4935–4943.
44. Narasagoudr, S. S., Hegde, V. G., Vanjeri, V. N., Chougale, R. B., & Masti, S. P., (2020). Ethyl vanillin incorporated chitosan/poly (vinyl alcohol) active films for food packaging applications. *Carbohydrate Polymers, 236*, E-article 116049.
45. Neo, Y. P., Swift, S., Ray, S., Gizdavic-Nikolaidis, M., Jin, J., & Perera, C. O., (2013). Evaluation of gallic acid loaded zein sub-micron electrospun fiber mats as novel active packaging materials. *Food Chemistry, 141*, 3192–3200.
46. Nieuwland, M., Geerdink, P., Brier, P., Van, D. E. P., Henket, J. T., Langelaan, M. L., Stroeks, N., et al., (2013). Food-grade electrospinning of proteins. *Innovative Food Science & Emerging Technologies, 20*, 269–275.
47. Okutan, N., Terzi, P., & Altay, F., (2014). Affecting parameters on electrospinning process and characterization of electrospun gelatin nanofibers. *Food Hydrocolloids, 39*, 19–26.
48. Park, J. M., Kim, M., Park, H. S., Jang, A., Min, J., & Kim, Y. H., (2013). Immobilization of lysozyme-CLEA onto electrospun chitosan nanofiber for effective antibacterial applications. *International Journal of Biological Macromolecules, 54*, 37–43.
49. Pinheiro, B. G., Dos Santos, A. T., De Oliveira, J. P., Martins, F. L., Tavares Da, S. F., Martins, G. V., & Da Rosa, Z. E., (2020). Electrospun protein fibers loaded with yerba mate extract for bioactive release in food packaging. *Journal of the Science of Food and Agriculture, 100*, 3341–3350.
50. Saithongdee, A., Praphairaksit, N., & Imyim, A., (2014). Electrospun curcumin-loaded zein membrane for iron (III) ions sensing. *Sensors and Actuators B: Chemical, 202,* 935–940.
51. Sencadas, V., Correia, D. M., Areias, A., Botelho, G., Fonseca, M., & Neves, I. C., (2012). Determination of the parameters affecting electrospun chitosan fiber size distribution and morphology. *Carbohydrate Polymers, 87,* 1295–1301.

52. Soto, K. M., Hernández-Iturriaga, M., Loarca-Piña, G., Luna-Bárcenas, G., & Mendoza, S., (2019). Antimicrobial effect of nisin electrospun amaranth: Pullulan nanofibers in apple juice and fresh cheese. *International Journal of Food Microbiology, 295,* 25–32.
53. Stephansen, K., García-Díaz, M., Jessen, F., Chronakis, I. S., & Nielsen, H. M., (2016). Interactions between surfactants in solution and electrospun protein fibers: Effects on release behavior and fiber properties. *Molecular Pharmaceutics, 13,* 748–755.
54. Sun, X. Z., Williams, G. R., Hou, X. X., & Zhu, L. M., (2013). Electrospun curcumin loaded fibers with potential biomedical applications. *Carbohydrate Polymers, 94*(1), 147–153.
55. Tang, Y., Zhou, Y., Lan, X., Huang, D., Luo, T., Ji, J., Mafang, Z., et al., (2019). Electrospun gelatin nanofibers encapsulated with peppermint and chamomile essential oils as potential edible packaging. *Journal of Agricultural and Food Chemistry, 67*(8), 2227–2234.
56. Tao, Y., Zhang, Z., & Sun, D. W., (2014). Kinetic modeling of ultrasound-assisted extraction of phenolic compounds from grape marc: Influence of acoustic energy density and temperature. *Ultrasonics Sonochemistry, 21*, 1461–1469.
57. Teo, W. E., & Ramakrishna, S., (2006). A review on electrospinning design and nanofiber assemblies. *Nanotechnology, 17*(14), 89–106.
58. Torres-Giner, S., Ocio, M. J., & Lagaron, J. M., (2009). Novel antimicrobial ultrathin structures of zein/chitosan blends obtained by electrospinning. *Carbohydrate Polymers, 77*(2), 261–266.
59. Tripatanasuwan, S., Zhong, Z., & Reneker, D. H., (2007). Effect of evaporation and solidification of the charged jet in electrospinning of poly (ethylene oxide) aqueous solution. *Polymer, 48*(19), 5742–5746.
60. Wang, H., Hao, L., Niu, B., Jiang, S., Cheng, J., & Jiang, S., (2016). Kinetics and antioxidant capacity of proanthocyanidins encapsulated in zein electrospun fibers by cyclic voltammetry. *Journal of Agricultural and Food Chemistry, 64*(15), 3083–3090.
61. Wang, H., Hao, L., Wang, P., Chen, M., Jiang, S., & Jiang, S., (2017). Release kinetics and antibacterial activity of curcumin loaded zein fibers. *Food Hydrocolloids, 63*, 437–446.
62. Wang, S., He, J., & Xu, L., (2008). Non-ionic surfactants for enhancing electrospinability and for the preparation of electrospun nanofibers. *Polymer International,* 1079–1082.
63. Wang, S., Marcone, M. F., Barbut, S., & Lim, L. T., (2013). Electrospun soy protein isolate based fiber fortified with anthocyanin-rich red raspberry (*Rubus strigosus*) extracts. *Food Research International, 52*(2), 467–472.
64. Wen, P., Feng, K., Yang, H., Huang, X., Zong, M. H., Lou, W. Y., Li, N., & Wu, H., (2017). Electrospun core-shell structured nanofilm as a novel colon-specific delivery system for protein. *Carbohydrate Polymers, 169*, 157–166.
65. Wen, P., Zong, M. H., Linhardt, R. J., Feng, K., & Wu, H., (2017). Electrospinning: A novel nano encapsulation approach for bioactive compounds. *Trends in Food Science and Technology, 70*, 56–68.
66. Wongsasulak, S., Patapeejumruswong, M., Weiss, J., Supaphol, P., & Yoovidhya, T., (2010). Electrospinning of food-grade nanofibers from cellulose acetate and egg albumen blends. *Journal of Food Engineering, 98*(3), 370–376.
67. Wongsasulak, S., Pathumban, S., & Yoovidhya, T., (2014). Effect of entrapped α-tocopherol on mucoadhesivity and evaluation of the release, degradation, and swelling characteristics of zeine-chitosan composite electrospun fibers. *Journal of Food Engineering, 120*, 110–117.

68. Yan, J., White, K., Yu, D. G., & Zhao, X. Y., (2014). Sustained-release multiple-component cellulose acetate nanofibers fabricated using a modified coaxial electrospinning process. *Journal of Materials Science, 49*(2), 538–547.
69. Yang, H., Wen, P., Feng, K., Zong, M. H., Lou, W. Y., & Wu, H., (2017). Encapsulation of fish oil in a coaxial electrospun nanofibrous mat and its properties. *RSC Advances, 7*(24), 14939–14946.
70. Yang, Q., Li, Z., Hong, Y., Zhao, Y., Qiu, S., Wang, C., & Wei, Y., (2004). Influence of solvents on the formation of ultrathin uniform poly (vinyl pyrrolidone) nanofibers with electrospinning. *Journal of Polymer Science Part B: Polymer Physics, 42,* 3721–3726.
71. Zhang, C., Feng, F., & Zhang, H., (2018). Emulsion electrospinning: Fundamentals, food applications and prospects. *Trends in Food Science & Technology, 80,* 175–186.
72. Zhang, Y., Ouyang, H., Lim, C. T., Ramakrishna, S., & Huang, Z. M., (2005). Electrospinning of gelatin fibers and gelatin/PCL composite fibrous scaffolds. *Journal of Biomedical Material Research B: Applied Biomaterials, 72,* 156–165.
73. Ziani, K., Henrist, C., Jerome, C., Aqil, A., Mate, J. I., & Cloots, R., (2011). Effect of non-ionic surfactant and acidity on chitosan nanofibers with different molecular weights. *Carbohydrate Polymers, 83*(2), 470–476.
74. Zong, X., Kim, K., Fang, D., Ran, S., Hsiao, B. S., & Chu, B., (2002). Structure and process relationship of electrospun bioabsorbable nanofiber membranes. *Polymer, 43*(16), 4403–4412.

CHAPTER 16

ENCAPSULATION TECHNIQUES AND THEIR APPLICATIONS IN THE FOOD INDUSTRY

GURVEER KAUR

ABSTRACT

Encapsulation is a novel processing method to engulf the food bioactive compounds by using a membrane over it in the food production chain. A number of foodgrade or edible membrane or wall materials (like polysaccharides, proteins, etc.) are accessible from natural origin to use in this process. Research studies have been conducted for selection, optimization, and to improve the efficacy of encapsulation by applying different encapsulation techniques, such as: fluidized bed coating, spray drying, melt injection, etc. This chapter aims to document the need of encapsulation along with the benefits, different wall materials, methods of encapsulating food and encapsulates produced in the food industry. Encapsulation needs to be developed at a larger scale to produce food encapsulates according to the demand and requirement of the consumers to enhance shelf-life and fulfill the nutrition of individual bio-active compound.

16.1 INTRODUCTION

Encapsulation is defined as an operation to engulf or entrap one component (active agent) within another component (coating material). The entrapped component is also known as active agent, encapsulated component or substance, fill, core or payload and the wall material is often called membrane,

Advanced Research Methods in Food Processing Technologies. Junaid Ahmad Malik, Megh R. Goyal, Preeti Birwal, & Ritesh B. Watharkar (Eds.)
© 2024 Apple Academic Press, Inc. Co-published with CRC Press (Taylor & Francis)

external phase, shell, polymeric cover, matrix, or carrier material. Encapsulation process generally create a barrier over the active substance, i.e., bioactive molecules (vitamins, minerals, fatty acids, antioxidants, lycopene, lutein, phytosterols, oils, living cells, i.e., probiotics, etc.) to protect against surrounding environmental factors such as temperature, air composition, etc., and to avoid undesirable chemical interactions and permits the release under control and certain conditions [1, 27].

Encapsulation technologies were developed approximately 60 years ago, for pharmaceutical industry to deliver proper drug and vaccine, time-release, mask unpleasant flavors and enhance and improve stability of formulation. Encapsulation technologies were applied industrially first time to get pressure-sensitive coating almost after two decades for manufacturing carbonless copying paper. Encapsulation technologies were used in biotechnology also to make efficient production processes as efficient and rapid separation can be allowed between metabolites and producer cells by the membrane [28, 33]. Over the last few years, these applications would be applied to the food products also. In recent times, consumers have demanded the incorporation of bioactive compounds in the food, and the bioactive compounds are very sensitive to processing conditions and environmental factors.

Encapsulation is an effective approach that is extensively used to design engineered products in many food industries to develop innovative product, to enhance the preservation of functional compounds like flavor, color, texture, etc., and to include various bioactive compounds with health benefits [47, 60]. Generally, encapsulation refers to the techniques of packaging various ingredients in form of gaseous materials, liquids or solids, liquids in small shells/capsules generally having diameter from few nm to a few mm, that release core/inner substances at controlled rate at specific conditions [28]. The shell can be made of food grade proteins, gums, sugars, lipids, modified, and natural polysaccharides, synthetic polymers, etc. Ingredients, which are encapsulated often include gases to make eating sensations interesting, flavor compounds that are engineered with specific release during mastication, vitamins which need to be protected from moisture and oxygen, enzymes that are designed to immobilize during specific processes, color agents that are designed to release effect of color change at different times [60, 63].

Applications of encapsulation in food industries have been growing at slower rate since the techniques of encapsulation are very specific and much expensive but as the demand of encapsulated food products increased furthermore the production volume increased and this leads to cost-effective production techniques.

Therefore, the aim of this chapter is to provide detailed information related to the advantages of encapsulation techniques, food grade material used for encapsulation with examples, techniques used in encapsulation including applications in the food industry, microencapsulates in food products, measurement of effectiveness of encapsulation and future aspects in the food industry.

16.2 BENEFITS OF ENCAPSULATION

16.2.1 BARRIER ACTION

Encapsulation primarily aims to protect the active components or functional compounds from the degradation. The functional compounds (like polyphenols, minerals, vitamins, etc.) are highly susceptible to the environment (light, heat, moisture, relative humidity, etc.) and sensitive towards processing and gastrointestinal processing conditions [22, 82]. Therefore, a barrier to these functional compounds can help in their protection by reducing (hydrolysis or oxidation) or preventing the degradation (environmentally) [49]. Encapsulation ultimately saves functionality of the functional ingredient by providing time of action and deliver the active agent according to the desired conditions. The functionality of active agent can be fully or partially maintained in this way. The barrier action also helps in separation of the reactive components of the mixture to avoid their reaction with one another. Hence encapsulation techniques can be efficiently and effectively used as novel technique to avoid the undesirable effects on the active or functional components of food products.

16.2.2 CONTROLLED AND SUSTAINED RELEASE OF INGREDIENT

Encapsulation is a unique and novel technique to provide sustained and controlled release of living cells (probiotics) and active agents (e.g., minerals, phytosterols, antioxidants, vitamins, etc.) within desired time period and at a specific place [22, 82]. In recent trends, the primary objective of encapsulation is to release the targeted bioactive component or right stimulus in the right time and place at required rates. Therapeutic proteins, artificial sweeteners and other active agents have been released and achieved at controlled, sustained, and targeted rates by encapsulation for the optimum organoleptic and functional benefits [7, 17, 23, 27, 47].

There are two different types of release, sustained, and delayed. The sustained release meant to be maintained concentration of actives at a constant rate at the targeted place, for example release of the encapsulated sweeteners and flavors at the chewing a chewing gum at a desired rate throughout. The sustained release can be multi-phase release also by releasing the actives at different phases in one process it-self like in bakery processes the first phase release of active ingredient occurs during mixing mixture with liquid and second phase release occurs at the time of heating, but the actives can be same as well as different. The delayed release refers to the mechanism of delayed release of active components at desired site. Examples of the delayed release are encapsulated sodium bi-carbonate is released upon baking, and encapsulated flavors are released from ready-to-eat meals after microwave (MW) heating.

16.2.3 OFF-TASTE AND OFF-FLAVOR MASKING

"Healthy things come in bitter or off-taste packages," is a saying in the food industry and amongst consumers also. To make the healthy and nutritious ingredients full of taste and the healthy food, encapsulation is an innovative tool. The food industry needs to mask unpalatable ingredients to make their consumption easy by using the encapsulation techniques or coating with the desired food grade shells [45]. For example, fish oil and caffeine are encapsulated to mask the off flavor of fish and bitter taste of caffeine.

16.2.4 ENHANCE NUTRITION

Encapsulation is a useful approach in maintaining the nutritional value of the food ingredients by protecting them from the undesirable reactions and by reducing the loss of nutrients [45]. Encapsulation can also enable the addition of proteins, essential amino acids, minerals, vitamins (O_2-sensitive), etc., to various food products and fortification which will ultimately enhance the nutrition of the food product.

16.2.5 IMPROVE SAFETY AND STABILITY

Encapsulation ensures the stability and safety of the bioactive compounds, especially flavonoids and aromatic compounds (reduced flammability).

Encapsulation technologies are used to enhance and preserve flavor stability during storage as well as processing (less evaporation and no degradation of volatile compounds) of the food product. Encapsulated salts, antioxidants, acidulants, etc., are being used to provide microbial safety and enhanced shelf-life by maintaining pH level in meat products. Encapsulation process and structure of shell material also play an important role in the stability of the active agent. A range of matrix encapsulation ultimately used to enhance the shelf-life and make the food product acceptable organoleptically [24, 26, 33, 38].

16.2.6 SUPERIOR HANDLING

Encapsulation is advantageous in handling of the bio-molecules or active agents since they are packed in the shell material therefore solid in state generally. Conversion of liquid or gas phase of active agents to solid phase turned to be useful in easy and superior handling and improved flow properties, especially in case of essential oils and extracts of plants [8, 83].

16.2.7 MODIFICATION OF PHYSICAL CHARACTERISTICS

Properties of the active compounds can be modified using encapsulation techniques in terms of adjusting the physical properties of the compound, i.e., size, color, water/oil soluble, structure/shape of the particle, etc. Textural and visual effects can be created using different encapsulation techniques to make a better appearance and organoleptic properties of the finished product.

16.3 DRAWBACKS OF ENCAPSULATION IN THE FOOD INDUSTRY

- Complexed process and products;
- Highly specific and expensive production;
- Lack of foodgrade wall materials;
- Stability challenges of encapsulates;
- Unnecessary consumer visual and touch notice.

Nevertheless, encapsulation provide many benefits along with the facilitation of tastier, healthy, convenient engineered and innovative food products, which lead to increase in demand of encapsulated product since last few decades.

16.4 SHELL/WALL-MATERIALS FOR ENCAPSULATION IN FOOD APPLICATIONS

A number of compounds of different origins, types, states (gases, liquids, or solids), and properties can be used to coat or entrap the actives. However, a number of certified substances is comparatively limited in number for applications in the food industry than the pharmaceutical industries. The regulations for food additives have to be followed by the food industries or food producers at GRAS (generally recognized as safe) level for each certified wall material and different countries can set their own regulations according to the nutrition demand.

16.4.1 SELECTION CRITERIA FOR WALL MATERIALS

- A biodegradable and food grade material should be used to design the protective wall of actives or encapsulates.
- Behavior of the wall material should be studied under conditions in food formulations before final selection and physicochemical properties of the wall material should also be taken into account.
- Concentration of the wall material should be selected according to desired release and stability.
- Holding capacity of core during storage and processing under several conditions should be within the structure of capsule.
- The material should be natural and able to create a barrier between the core and its surrounding.
- The selected wall material should be cost effective along with the most appropriate material.
- The wall material must be certified for its applications in food and designed to make sure that the safety requirements of government regulation agencies like the European Food Safety Authority or Food and Drug Administration (FDA) are fulfilled [82].
- The wall material should be able to give maximum possible protection to the core against environmental stressors.
- The wall material should be selected based on their functionality to end product with potential restrictions.
- The wall material should have good textural properties even at high concentrations to enhance or maintain workability during application of encapsulation.

16.4.2 TYPES OF WALL MATERIALS

16.4.2.1 CARBOHYDRATE POLYMERS/POLYSACCHARIDES

Carbohydrate polymers are the most commonly availed material for encapsulation in the food industry. The native polysaccharides are generally removed during processing during natural biosynthesis. Hence, they have wide economic importance and their usefulness and applicability expand their application area. Cellulose, starch, and its derivatives (e.g., maltodextrins, dextrin, amylose, etc.) are the first group of wall materials and known as outstanding frontrunner for applications in encapsulation. Polysaccharides are cost effective and they have required physicochemical properties like phase change, melting, solubility, etc., along with the formation of various shapes in a required range of sizes [8]. Volatile compounds from the environment can be absorbed by cyclodextrin (CD) and starch. Flavors and lipids are majorly encapsulated using starch efficiently due to helical structure of amylose [13]. Maltodextrins are recommended as a cheaper substitute (partial) in the application of encapsulation [74].

Ethyl cellulose is used to encapsulate water soluble vitamins and desired thickness of the shell is optimized to reduce the water permeability of vitamins (actives) [8, 73]. Maltodextrin and modified starch are preferred for encapsulation and coating by various encapsulation techniques. Cellulose, carboxymethyl cellulose (CMC), and ethyl methylcellulose (EMC) are preferred as wall material when low humidity storage is required as celluloses are able to absorb water at appropriate relative humidity. Celluloses are colorless, clear, and odorless in aqueous solution, hence they are taken into account for encapsulation to modify texture and visual effects of food.

Plant extracts and exudates, products of gummosis, like gum Arabic (*Acacia senegal*), guar gum (*Cyamopsis tetragonolobos*) gum tragacanth (*Astragalus* spp.), lotus bean gum (*Ceratonia silique*), gum karaya, etc., are widely used as wall materials in food encapsulation. Gum Arabic (GA) has been used to encapsulate oil droplets by creating a strong protective film and it has compatibility with other starches, carbohydrates, protein, and other plant extracts [41]. This protective film is resulted from the stabilization of protein-polysaccharide system, the polypeptides (hydrophobic) anchor arabinogalactan (polysaccharide) onto the oil droplet surface and the aggregation was prevented by the hydrophilic carbohydrates by formation of a thick charged layer [36, 37, 56].

Gum tragacanth and gum Arabic together and individually used as a fixative to the fine spray drying flavor particles, it creates an impenetrable and thin film around the particle and prevent the moisture absorption from the environment and protect it from evaporation and oxidation [51, 52, 55, 62]. A good film formulation has been shown by mesquite gum in encapsulation of food [20]. Soluble soybean polysaccharides are also employed in the food encapsulation for food coating as it can prevent oil oxidation and form a film with colorless, edible, and transparent properties [46].

Alginates and carrageenan are the polysaccharides extracted from marine and commonly used as wall material in the food encapsulation. After reacting with calcium ions, alginates, and hydrocolloids form a stable gel. They are used as wall materials for oleoresins at ambient temperatures. Aroma was retained in crackers during baking when alginate beads were used [14].

Microbiologically or biotechnologically produced polysaccharides comes under the class of animal extracts and many of them are used as wall material in the food encapsulation as they possess biocompatible and biodegradable properties. Microbial polysaccharides such as chitosan, dextran, xanthan, curdlan, and gellan have been extensively used and certified by WHO (World Health Organization) [35, 59, 73, 82]. Chitosan has film forming, gel forming, cationic polymeric and heat shrinking properties, which makes it as a unique wall material to encapsulate essential oils. Essential oil encapsulated with nanoparticles (NPs) of chitosan is a novel process due to its low toxicity and biocompatibility.

16.2.2.2 PROTEINS

Proteins are known to be appropriate wall materials for application of encapsulation apart from polysaccharides, Gluten, gelatin, and milk and whey protein (casein) are the most used proteins for encapsulation. Gluten proteins are less soluble in water, used as the formation of elastic films for coating and it is majorly associated with the bakery industry [75]. Milk proteins (casein and whey) possess excellent surface-active properties. Thermally induced films formation shows excellent gas barrier properties [45]. These films are tasteless, odorless, and translucent to transparent depending on the source of protein used for coating of flavor components [11, 42].

Incorporation of lipids to protein reduces the water vapor permeability. Whey proteins can be yielded by various procedures to use as wall material [70]. Proteins are amphiphilic and flexible in nature and they self-aggregate due to absorption on the emulsion interface furthermore, around oil droplets, they form a homogeneous and continuous membrane through intermolecular interaction [12, 43]. Oil droplets can be coated with protein films, charged layers to provide an electrostatic barrier against coalescence and flocculation. Gelatin in the form of gel or film, used as wall material for encapsulation. Fish gelatin films can be used as wall material in frozen and refrigerated or low temperature food systems to reduce water loss [2]. Stable emulsion with flavor components (volatiles) can be formed by protein. Their application in encapsulation is limited in food due to their ability to react with carbonyls, high cost and low solubility in cold water [73].

16.2.2.3 LIPIDS

Fatty acids, phospholipids, glycerides, and waxes (carnauba wax, bees wax, and candelilla wax) are the common lipid used as wall material in the encapsulation of food and food ingredients [21]. Lipids have advantages to be used as wall material due to their stability and safety, hydrophobic nature and ease of handling. Encapsulation of aroma is done by natural waxes as their melting point is between 60° and 80°C. Natural waxes are the permitted additives and wall materials in the European Union [7, 54].

16.2.2.4 OTHER MATERIALS

There are a number of organic and inorganic polymers/materials have been used as microencapsulation or coatings in the food industry. Food grade polymers (paraffin, polyvinylpyrrolidone (PVP), polyvinyl acetate, polystyrene, etc.) are also employed for encapsulating the food [8, 17, 82]. PVP polymers have been used as excellent coating material in the form of film, where thermal stability is required [7].

16.3 ENCAPSULATION TECHNIQUES IN THE FOOD INDUSTRY

A number of processes are applicable to encapsulate food bio-actives including spray drying, spray chilling, melt injection, extrusion, co-extrusion, coacervation, etc. [64, 65]. Encapsulation can be achieved generally in reservoir and matrix type. The matrix type of encapsulate has dispersed active agent throughout the wall material in the form of small droplets homogeneously. The reservoir type, also known as capsule, core-shell, mono core, etc., has a covered membrane around the active. The active can be released by breaking the shell after applying pressure to it. The active compounds are present in a coring or shell and at the surface in the reservoir and the matrix type encapsulates, respectively.

16.3.1 SPRAY DRYING

Spray drying is known to be common and one of the oldest techniques of encapsulation in the food industries. Spray drying of the core material is principally achieved by dissolving the bioactive compound in water or an aqueous solution of shell to make suspension/emulsion. The suspension is then atomized and sprayed in a hot chamber [5, 6, 26, 27]. The condition of using this technique of encapsulation is that wall or matrix material should be hydrophilic in nature (carbohydrates, lipids, proteins, etc.) or highly soluble in water and it should have good emulsifying and film forming properties [26] and the core material can either be hydrophilic or hydrophobic in nature [81]. A film is formed at the surface of the droplet during the process, water gets evaporated and active molecules size is retarded in few milliseconds to a few seconds, due to less process time it is well suited to heat-liable products like flavors, proteins, living cells, enzymes, etc.

In food industry, for encapsulation a rotary atomizer or a high-pressure nozzle is preferred to atomize the suspension and to maintain the quality of the droplets by minimizing the overheating, co-current air flow is preferred over countercurrent air flow in the hot chamber. The spray-dried powder has diameter of 10 to 100 μm generally [29, 68, 79] and it depends on viscosity of the suspension, velocity of spray, pressure drop across the nozzle, mounting of nozzle (upward or downward), and surface tension of suspension. Smaller particle size of finished product is desired in satisfactory textural and organoleptic point of view.

Granulation or agglomeration have many applications in the food industry and is used when desired size of particles is large (10–150 mm) [5, 6, 58]. Larger particles can be achieved by mounting the nozzle upward side so that it sprays the suspension like fountain to increase the residence time in the dryer. This can be achieved by generating random movements in the equipment and when particles are not fully dried and they remain sticky at the first stage, and this facilitates agglomeration in the second phase [24]. Alternatively, it can be achieved by spraying the binder solution (water) over the particles during tumbling or random movements [5, 6, 44, 58, 80]. The application of this process is more when there is no segregation of encapsulate within food products.

Spray drying method of encapsulation is cost effective and has higher production rate relatively and finished products are stable food powder and flavors [16, 17]. The encapsulated active molecules are released after dilution in water depending upon the porosity and nature of carrier material [10, 16, 29, 33, 67, 68]. The technique has been extensively used in the encapsulation of vitamins, pigments, flavors, oils, enzymes, minerals, living cell and many other bioactive molecules; furthermore, approximately 80–90% encapsulates are prepared by spray drying according to the literature [54, 65].

16.3.2 FLUIDIZED-BED COATING

Fluidized bed coating is a process to coat the powder particle either in batch or continuous process. This method is very similar to spray drying along with agglomeration. Particle or active is suspended in the air at specific terminal velocity and temperature, and then the wall material is sprayed or atomized through a nozzle on the active molecules. The coating material used in fluidized bed coating process must be stable thermally, should have desired viscosity for proper pumping and spraying, and must be able to form a film over powdered active particles. The coating material can be a suspension of starch or cellulose derivatives or gums or proteins and depending on the size of the actives, i.e., generally in solid phase 5–50% of coating is applied [19, 30, 76, 77]. Even molten lipids (fatty acids, vegetable oil, emulsifiers, etc.) can be used coating materials.

The core material should be properly dried and solidify and the coating material should not get solidify before it reaches to the active solid particles, to prevent it heat should be provided to the storage vessel as well as to the atomizing air and nozzle. Coating takes place in uniformity when the shape of active is free from sharp edges, i.e., spherical or circular, narrow, and uniform size distribution, dense, and good flowability. The surface area of the active particle powder determine the requirement of the coating material, very fine; and less-denser particle may create undesirable accumulation on the filter bags (clogging). Pan coating can be used as alternative coating technology [78]. Fluidized bed coating technique can be used to coat second layer on spray-dried products or sensitive core products and it has been used to generate an additional layer of actives for desired and targeted release in the gut [19, 81].

16.3.3 SPRAY COOLING AND SPRAY CHILLING

Spray chilling and spray cooling is an emulsion process which generally produces lipid coated active compounds [29, 38, 80]. The active agents should be dry particles or soluble in lipids or aqueous emulsion. The set of spay cooling and spray chilling is almost same as spray drying but here chamber is maintained at lower temperatures in place of hot chamber and no evaporation takes place here. The coating material in the form of suspension or molten lipid is atomized via spinning disc or nozzle into the chilled chamber. The solidification of lipids on the core material or actives takes place. Lipids used for this encapsulation techniques should possess melting point from 34°C to 42°C and higher for spray chilling and spray cooling technique, respectively [38]. The size of the droplet and melting point depend on the falling height. The size of the finished products depends on the speed and opening of the nozzle or configuration and rotational speed of the disc, core particles, melting temperature and viscosity of molten lipids or coating material. Spray-cooling has a high potential to encapsulate dry particles to conserve flavors, proteins, minerals, and enzymes [81].

16.3.4 VACUUM DRYING AND FREEZE DRYING

Freeze drying and vacuum drying techniques of encapsulation came into account for the temperature-liable bioactive compounds where spray drying cannot be used. The principle of freeze and vacuum drying is same as drying processes, but it is cheaper and faster [6]. Freeze and vacuum drying are used when prolonged release is required, as these processes create a porous barrier between the core material and its surrounding so this porous barrier will not protect the core material effectively but offers a delayed release [18]. The drawbacks of freeze drying that it required long process time and high energy input.

16.3.5 MELT INJECTION AND MELT EXTRUSION

Melt extrusion and injection methods are used to engulf the core materials in a carbohydrate melt. The active agents are mixed with carbohydrates in molten state at higher temperatures (above 100°C) and the molten is pressed or extruded through orifices and then a dehydrating solvent (liquid nitrogen or isopropanol) is used to quenched to form a glass to provide the actives little mobility. Coating material encapsulates the active agents by getting hardened on the actives when comes in contact with liquid nitrogen or isopropanol [63]. Melt injection and melt extrusion processes have few differences only; in melt injection processes vertically positioned screw is used and in melt extrusion horizontally positioned screw is used for pressing or extruding the melt, the surface of extrudates to remove residues of the actives is not washed in case of melt extrusion. Thermoplastic starch is often used melt extrudates in the food products [84]. Commonly used extrudates in the food industry are generally composed of starch and its derivatives, sugars, proteins, lipids or gums, emulsifiers, etc.

16.3.6 EMULSIFICATION

Emulsification is another commonly used technology in encapsulation. Several single and double type of emulsions, i.e., oil/water, water/oil, water/oil/water emulsions are common in the food industries. The size of encapsulate can be controlled by the ratio of oil and water and agitation speed and it majorly depends on geometry of the mixer [34]. Instant food formulations or encapsulates in powder (different shape) form can be prepared by drying the emulsions using freeze or spray drying techniques [63]. Emulsification technique of encapsulation is popular in the food industries for controlled and sustained release of flavors and aromas [1].

16.3.7 EXTRUSION METHODS

Extrusion methods work on the principle of dispersing suspension or polymer aqueous solution and droplets of the active agents are dripped into a gelling bath (sodium alginate as polymer and calcium chloride as the gelling bath). Dripping can be done using a syringe, pipette, vibrating or spraying nozzle, atomizing disc, or jet cutter [82]. Jet cutter extrusion technology is the best and commonly used encapsulation technique in large scale food industries [66]. Electrostatic extrusion technology is extensively used in alginate beads production loaded with yeast and it can produce very small size encapsulates [57].

16.3.8 COACERVATION

Coacervation, a modified encapsulation process, is generally used to produce capsules of different sizes and characteristics. A solution is prepared by mixing the active compounds with the matrix or wall materials and then a complex is formed. The ratio of the actives and wall material or matrix type or by varying ion concentration generally decide the size of the capsule. Gum acacia, polysaccharides, and gelatin are commonly preferred as coating materials [28, 83]. Coacervation is mostly used for encapsulating food active compounds like vitamin A, citrus, flavors. Vegetable oils and water-soluble bioactive compounds than for encapsulating living cells [1, 28]. Coacervation technique of encapsulation is very efficient but has limitations in application because of less availability of desired, approved, and suitable encapsulating material and high operating cost [73].

16.3.9 MOLECULAR INCLUSION

Molecular inclusion techniques are not widely used in encapsulation as these are expensive. CDs are commonly used shell material for encapsulation by this process as they are lipophilic inner pocket in which a core material of correct size can be engulfed reversibly in an aqueous environment [6, 31]. Molecular inclusion protects the actives from degradation by increasing the solubility of lipophilic molecules, but small pocket of the wall material limits the loading capacity of the core material. Liposomes are created by hydrophilic-hydrophobic interaction mechanism and majorly used to preserve the aroma/volatile compounds in the food products [81, 85].

16.4 TYPES OF MICROENCAPSULATES FOR FOOD PRODUCTS

A significant rise in the demand of healthy ad nutritious bioactive compounds and food ingredients has been increased since past few years. Consumers prefer fortified, nutrition enriched or specific bioactive compound containing food product which can be stored for longer period by retaining organoleptic properties. Encapsulation techniques are used to deliver required nutrition to consumers by engulfing food ingredients and fortifying compounds micronutrients, aroma, essential oils, vitamins, living cells, and pigments, etc. [28, 81].

16.4.1 AROMA ENCAPSULATES

Aroma is the most volatile characteristic of food which can be easily deteriorate or release early due to thermodynamics and kinetics according to composition of compound. Molecules of aroma interact with the olfactory receptor which make it an important organoleptic quality attribute [23]. Aroma can be entrapped in the wall material to enhance the stability and functionality of aroma in the food products. Aroma release can be controlled by proper composition and microstructure of food and such engineered food can be designed by the encapsulation tool [64]. Encapsulation is known to be an excellent approach to preserve aroma or flavor compounds by covering with adequate wall material to protect against environmental stressors, chemical reactions, or migration in the food [14, 45].

Encapsulation of flavor or aroma can be achieved using polysaccharides, proteins, waxes, etc., as wall material and spray drying encapsulation process [53]. Natural waxes shown a better retention of aroma and flavor compound for longer time for example ethyl-vanillin is encapsulated by carnauba wax particles using melt injection technique [54]. Lipid particles can be used to encapsulate aroma by spray chilling process [49]. Extrusion and spray drying are common processes to encapsulate aroma at the industrial level when carbohydrates are the carrier materials [82]. The ultimate aim of encapsulating aroma is to enhance stability during processing and control release depending on the physical state of wall materials and characteristics.

16.4.2 VITAMIN ENCAPSULATES

Water- and fat-soluble vitamins are less stable at environment conditions and most of them have an unpleasant taste. Encapsulation of vitamins is done to enhance or improve the stability and to mask the off-flavor for easy consumption. Spray drying, fluidized bed coating and spray chilling are the most commonly used techniques for encapsulating vitamins. Fat soluble vitamins are encapsulated by spray drying of suspension and water-soluble vitamins are encapsulated by fluidized bed coating and spray chilling method. The efficiency of the process depends on the wall material characteristics, temperature in the chamber, vitamins to wall material ratio, etc., and these parameters can be optimized to get stable finished product or stable coated vitamins [8, 83]. Coated vitamins are mainly used in bakery industry to increase nutritional value of the food and used as antioxidant. Encapsulated vitamins are desirable in dry soups, dairy products, baked goods, pet food and cereal bars.

16.4.3 OIL ENCAPSULATES

Essential oils are encapsulated to control the release and delivery of specific compound on desired site. Essential oils are moderately water soluble hence transport taste and odor to water and they carry phenols, aldehydes, ketones, alcohols, esters, terpenes, and wide range of biological properties like antimicrobial properties against food pathogens. Essential oils are generally encapsulated by coacervation techniques to control the release of compounds and to maximize antimicrobial activity. For example, *Thymus serpyllum L.* oil encapsulated into nano-spheres using alginate-inulin beads as carrier material for site specific release of compounds [60].

Fish oil need to be encapsulated primarily to mask the off flavor by restricting the contact of fish oil with environment stressors and metal ions and to convert oil into powder form for easy incorporation in food and superior handling [25]. Fish oil can be entrapped in amorphous compounds to trap off-flavor volatiles and to exclude oxygen. The process of encapsulation of fish oil should avoid inclusion of oxygen or entry of air. The ultimate aim of encapsulating fish oil is to preserve the bio-availability of compounds upon consumption. Spray drying encapsulation process has been majorly used to entrap fish oil but melt injection, extrusion, and coacervation are also being used using sugars, pectin, modified cellulose or starch as the carrier materials [3, 82].

16.4.4 COLOR AND PIGMENT ENCAPSULATES

Coloring agents are most commonly used additives as natural or synthetic substance in the food industry. Coloring compounds are very susceptible to heat and environmental conditions (light, temperature, etc.), and coloring agent tend to migrate into off-color layers, which makes it undesirable for organoleptic selection of food. Encapsulation of color compounds came into account to preserve and enhance the stability of the actives. Coloring compounds are generally encapsulated using carbohydrate wall material [32]. Coloring agents of meat and bakery products are maintained by encapsulating in hydrogenated vegetable oil. Edible coating of fish, meat, snacks, and other products using suspension of pectin and carbohydrates as wall material [73].

Carotenoids are encapsulated using spray drying and freeze drying to avoid degradation (oxidation or isomerization) and maintain nutritional quality by stabilizing the bioavailability of actives [18]. CD and liposomes are used as carrier materials for encapsulating carotenoids, i.e., lycopene, lutein, β-carotene, etc. [73]. Solubility of carotenoids can be improved and protected by using maltodextrin coating [69]. Encapsulation of synthetic pigments is achieved by chitosan emulsion [32].

16.4.5 POLYPHENOL ENCAPSULATES

Polyphenols or other phenolic compounds produce bitter and astringent taste during eating because of their high antioxidant properties. Encapsulation of polyphenols is performed to mast the bitter taste, preserve bioavailability and bioactivity, and maintain stability. Vacuum and spray drying technologies of encapsulation are employed to prepare polyphenol encapsulate which are cost effective [4, 71]. Plant extracts (Amaranthus betacyanin extracts, olive leaf extract, tea leaves extracts, medicinal plant extract, etc.) are encapsulated cost effectively using spray drying and extrusion technology [5, 10, 15, 18, 39].

16.4.6 PROBIOTIC ENCAPSULATES

Probiotics or living microorganisms are used as functional food in form of dietary supplement and are highly susceptible to mechanical and transport conditions, pH variations, and digestive enzymes. Microencapsulation is introduced to probiotics to increase the stability and viability of living cells for targeted delivery and storage purpose [48, 72]. The size and viability of the probiotic strain decide the method of encapsulation. Extrusion and emulsion technologies are used to encapsulate probiotics using polymeric matrix containing gellan, gelatin, carrageenan, chitosan, alginate, etc. [61, 79].

Extrusion technology used alginate with polymer compounds as wall material to produce encapsulated probiotics to avoid degradation by acid [40, 50]. Emulsion technology is used to make small size encapsulates [9]. Probiotic bacteria can be encapsulated by different methods in non-dairy food like biscuits, fruit juices, sausages, mayonnaise, and dry beverages [9]. Fluidized bed coating and spray drying is used to encapsulate probiotics in chocolate [47]. Therefore, encapsulation of probiotics has raised the market due to increasing consumer demands for health-promoting foods.

16.5 SUMMARY

Encapsulation of food products and food ingredients is a safe and convenient approach to develop health promoting foods. It can be documented that encapsulates will play a magnificent role in the food industry in coming years. Development of new and novel approaches to stabilize sensitive food ingredients, to improve nutritional value and bioavailability of compounds is the need of time in various food sectors. Co-encapsulation methodologies can be developed to entrap two or more active compounds to provide a synergistic effect. Future and current research should be focused on reducing the cost of process, easy availability of edible or foodgrade wall material and overcoming the complexities of the encapsulation processes.

KEYWORDS

- **bioactive compounds**
- **carboxymethyl cellulose**
- **coating**
- **encapsulates**
- **ethyl methylcellulose**
- **Food and Drug Administration**
- **gum Arabic**
- **polysaccharides**
- **spray drying**

REFERENCES

1. Augustin, M. A., & Hemar, Y., (2009). Nano-and micro-structured assemblies for encapsulation of food ingredients. *Chemical Society Reviews, 38*, 902–912.
2. Avena-Bustillos, R. J., Olsen, C. W., Olson, D. A., Chiou, B., Yee, E., Bechtel, P. J., & McHugh, T. H., (2006). Water vapor permeability of mammalian and fish gelatin films. *Journal of Food Science, 71*, 202–207.
3. Beindorff, C. M., & Zuidam, N. J., (2010). Microencapsulation of fish oil. In: Zuidam, N. J., & Nedovic, V. A., (eds.), *Encapsulation Technologies for Food Active Ingredients and Food Processing* (pp. 161–187). Dordrecht: Springer.

4. Bell, L. N., (2001). Stability testing of nutraceuticals and functional foods. In: Wildman, R. C., (ed.), *Handbook of Nutraceuticals and Functional Foods* (pp. 501–516). Boca Raton – FL: CRC Press.
5. Belscak-Cvitanovic, A., Stojanovic, R., & Manojlovic, V., (2011). Encapsulation of polyphenolic antioxidants from medicinal plant extracts in alginate-chitosan system enhanced with ascorbic acid by electrostatic extrusion. *Food Research International, 44*(4), 1094–1101.
6. Bilensoy, E., & Hincal, A. A., (2009). Recent advances and future directions in amphiphilic cyclodextrin nanoparticles. *Expert Opinion on Drug Delivery, 6*, 1161–1173.
7. Blecher, L., Lorenz, D. H., Lowd, H. L., Wood, A. S., & Wyman, D. P., (1980). Polyvinylpyrrolidone. In: Davidson, R. L., (ed.), *Handbook of Water-Soluble Gums and Resins* (pp. 17–21). McGraw Hill: New York.
8. Brownlie, K., (2007). Marketing perspective in encapsulation technologies in food applications. In: Lakkis, J. M., (ed.), *Encapsulation and Controlled Release Technologies in Food Systems* (pp. 213–235). Blackwell: Iowa.
9. Burgain, J., Gaiani, C., Linder, M., & Scher, J., (2011). Encapsulation of probiotic living cells: From laboratory to industrial application. *Journal of Food Engineering, 104*(4), 467–483.
10. Cai, Y. Z., & Corke, H., (2000). Production and properties of spray dried *Amaranthus betacyanin* pigments. *Journal of Food Science, 65*(6), 1248–1252.
11. Chen, H., (1995). Functional properties and applications of edible films made of milk proteins. *Journal of Diary Science, 78*, 2563–2583.
12. Chen, L. Y., Remondetto, G. E., & Subirade, M., (2006). Food protein-based materials as nutraceutical delivery systems. *Trends in Food Science and Technology, 17*, 272–283.
13. Conde-Petit, B., Escher, F., & Nuessli, J., (2006). Structural features of starch-flavor complexation in food model systems. *Trends in Food Science and Technology, 17*(5), 227–235.
14. De Roos, K. B., (2003). Effect of texture and microstructure on flavor retention and release. *International Dairy Journal, 13*, 593–605.
15. Deladino, L., Anbinder, P. S., Navarro, A. S., & Martino, M. N., (2008). Encapsulation of natural antioxidants extracted from *Ilex paraguariensis*. *Carbohydrate Polymers, 71*, 126–34.
16. Desai, K. G. H., & Park, H. J., (2005). Recent developments in microencapsulation of food ingredients. *Drying Technology, 23*, 1361–1394.
17. Desai, K. G. H., & Park, H. J., (2006). Effect of manufacturing parameters on the characteristics of vitamin C encapsulated tripolyphosphate-chitosan microspheres prepared by spray-drying. *Journal of Microencapsuation, 23*, 91–103.
18. Desobry, S. S., Netto, F. M., & Labuza, T. P., (1997). Comparison of spray-drying, drum-drying, and freeze drying for β-carotene encapsulation and preservation. *Journal of Food Science, 62*(6), 1158–1162.
19. Dewettinck, K., & Huyghebaert, A., (1999). Fluidized bed coating in food technology. *Trends in Food Science and Technology, 10*, 163–168.
20. Diaz-Sobac, R., Garcia, H., Beristain, C. I., & Vernon-Carter, E. J., (2002). Morphology and water vapor permeability of emulsion films based on mesquite gum. *Journal of Food Processing and Preservation, 26*(2), 129–141.
21. Donhow, I. G., & Fennema, O., (1993). Water vapor and oxygen permeability of wax films. *Journal of the American Oil Chemists' Society, 70*, 867–873.

22. Fang, Z., & Bhandari, B., (2010). Encapsulation of polyphenols – A review. *Trends in Food Science and Technology, 21*(10), 510–523.
23. Firestein, S., (2001). How the olfactory system makes sense of scents. *Nature, 413*, 211–218.
24. Fuchs, M., Turchiuli, C., Bohin, M., Cuvelier, M. E., Ordonnaud, C., Peyrat-Maillard, M. N., & Dumoulin, E., (2006). Encapsulation of oil in powder using spray drying and fluidized bed agglomeration. *Journal of Food Process Engineering, 75*, 27–35.
25. Garg, M. L., Wood, L. G., Singh, H., & Moughan, P. J., (2006). Means of delivering recommended levels of long chain n-3 polyunsaturated fatty acids in human diets. *Journal of Food Science, 71*(5), 66–71.
26. Gharsallaoui, A., Roudaut, G., Chambin, O., Voilley, A., & Saurel, R., (2007). Applications of spray-drying in microencapsulation of food ingredients: An overview. *Food Research International, 40*, 1107–1121.
27. Gharsallaoui, A., Roudaut, G., Beney, L., Chambin, O., Voilley, A., & Saurel, R., (2012). Properties of spray dried food flavors microencapsulated with two-layered membranes: Roles of interfacial interactions and water. *Food Chemistry, 132*, 1713–1720.
28. Gibbs, B. F., Kermasha, S., Alli, I., & Mulligan, C. N., (1999). Encapsulation in the food industry: A review. *International Journal of Food Science and Nutrition, 50*, 213–224.
29. Gouin, S., (2004). Microencapsulation: Industrial appraisal of existing technologies and trends. *Trends in Food Science and Technology, 15*, 330–347.
30. Guignon, B., Duquenoy, A., & Dumoulin, E. D., (2002). Fluid bed encapsulation of particles: Principles and practice. *Drying Technology, 20*(2), 419–447.
31. Harada, A., Takashima, Y., & Yamaguchi, H., (2009). Cyclodextrin-based supramolecular polymers. *Chemical Society Reviews, 38*, 875–882.
32. Higuera-Ciapara, I., Felix-Valenzuela, L., Goycoolea, F. M., & Argu¨elles-Monal, W., (2002). Microencapsulation of astaxanthin in a chitosan matrix. *Carbohydrate Polymers, 56*(1), 41–45.
33. Jafari, S. M., Assadpoor, E., He, Y., & Bhandari, B., (2008). Encapsulation efficiency of food flavors and oils during drying. *Drying Technology, 26*, 816–835.
34. Kailasapathy, K., (2009). Encapsulation technologies for functional foods and nutraceutical product development. *CAB Reviews Perspectives in Agriculture Veterinary Science Nutrition and Natural Resources*, *4*(6), 1–19.
35. Khor, E., & Lim, L. Y., (2003). Implantable applications of chitin and chitosan. *Biomaterials, 24*, 2339–2349.
36. Kim, Y. D., & Morr, C. V., (1996). Microencapsulation properties of gum Arabic and several food proteins: Spray-dried orange oil emulsion particles. *Journal of Agricultural and Food Chemistry, 44*(5), 1314–1320.
37. Kim, Y. D., Morr, C. V., & Schenz, T. W., (1996). Microencapsulation properties of gum Arabic and several food prosteins: Liquid orange oil emulsion particles. *Journal of Agricultural and Food Chemistry, 44*(5), 1308–1313.
38. Kjaergaard, O. G., (2001). Multiple-core encapsulation: Prilling. In: Vilstrup, P., (ed.), *Microencapsulation of Food Ingredients* (pp. 197–214). Surrey – England: Leatherhead Publishing.
39. Kosaraju, S. L., Tran, C., & Lawrence, A., (2006). Liposomal delivery systems for encapsulation of ferrous sulfate: Preparation and characterization. *Journal of Liposome Research, 16*, 347–358.
40. Krasaekoopt, W., Bhandari, B., & Deeth, H., (2003). Evaluation of encapsulation techniques of probiotics for yoghurt. *International Dairy Journal, 13*, 3–13.

41. Krishnan, S., Kshirsagar, A. C., & Singhal, R. S., (2005). The use of gum Arabic and modified starch in the microencapsulation of a food flavoring agent. *Carbohydrate Polymers, 62*(4), 309–315.
42. Lee, S. J., & Rosenberg, M., (2000). Whey protein-based microcapsules prepared by double emulsification and heat gelation. *Food Science and Technology – Lebenson Wiss Technol., 33*(2), 80–88.
43. Lefèvre, T., & Subirade, M., (2003). Formation of intermolecular B-sheet structures: A phenomenon relevant to protein film structure at oil-water interfaces of emulsions. *Journal of Colloid Interface Science, 263*, 59–67.
44. Litster, J. D., (2003). Scaleup of wet granulation processes: Science not art. *Powder Technology, 130*, 34–40.
45. Madene, A., Jacquot, M., Scher, J., & Desobry, S., (2006). Flavor encapsulation and controlled release: A review. *International Journal of Food Science and Technology, 41*, 1–21.
46. Maeda, H., (2000). Soluble soybean polysaccharide. In: Phillips, G. O., & Williams, P. A., (eds.), *Handbook of Hydrocolloids* (pp. 309–320). Cambridge – England: Woodhead Publishing Limited.
47. Maillard, M., & Landuyt, A., (2008). Chocolate: An ideal carrier for probiotics. *AgroFood Industry Hi-Tec., 19*(3), 13–15.
48. Manojlovic, V., Bugarski, B., & Nedovic, V., (2010). Immobilized cells. In: Flickinger, M. F., (ed.), *Encyclopedia of Industrial Biotechnology: Bioprocess, Bioseparation, and Cell Technology* (pp. 1–18). Hoboken – New Jersey: Wiley Publishing.
49. McClements, D., & Lesmes, U., (2009). Structure-function relationships to guide rational design and fabrication of particulate food delivery systems. *Trends in Food Science and Technology, 20*(10), 448–457.
50. McMaster, L. D., & Kokott, S. A., (2005). Micro-encapsulation of bifidobacterium lactis for incorporation into soft foods. *World Journal of Microbiology and Biotechnology, 21*, 723–728.
51. McNamee, B. F., O'Riordan, E. D., & O'Sulivan, M., (2001). Effect of partial replacement of gum Arabic with carbohydrates on its microencapsulation properties. *Journal of Agricultural and Food Chemistry, 49*, 3385–3388.
52. Meer, W., (1980). Gum Arabic. In: Davidson, R. L., (ed.), *Handbook of Water-soluble Gums and Resins* (pp. 8–21). New York–UK: McGraw-Hill.
53. Mellema, M., Van, B. W. A. J., Boer, B., Von, H. J., & Visser, A., (2006). Wax encapsulation of water-soluble compounds in food. *Journal of Microencapsulation, 23*, 729–740.
54. Milanovic, J., Manojlovic, V., Levic, S., Rajic, N., Nedovic, V., & Bugarski, B., (2010). Microencapsulation of flavors in carnauba wax. *Sensors, 10*(1), 901–912.
55. Mohammadifar, M. A., Musavi, S. M., Kiumarsi, A., & Williams, P. A., (2006). Solution properties of targacanthin (water-soluble part of gum *Tragacanth exudate* from *Astragalus gossypinus*). *International Journal of Biological Macromolecules, 38*, 31–39.
56. Nakamura, A., Yoshida, R., Maeda, H., & Corredig, M., (2006). Soy soluble polysaccharide stabilization at oil-water interfaces. *Food Hydrocolloids, 20*(2, 3), 277–283.
57. Nedovic, V. A., Obradovic, B., Leskosek-Cukalovic, I., Trifunovic, O., Pesic, R., & Bugarski, B., (2001). Electrostatic generation of alginate microbeads loaded with brewing yeast. *Process Biochemistry, 37*, 17–22.
58. Ortega-Rivas, E., (2005). Handling and processing of food powders and particulates. In: Onwulata, C., (ed.), *Encapsulated and Powdered Foods* (pp. 75–144). Boca Raton – USA: CRC Press.

59. Park, J. H., Ye, M., & Park, K., (2005). Biodegradable polymers for microencapsulation of drugs. *Molecules, 10*, 146–161.
60. Parris, N., Cooke, P. H., & Hicks, K. B., (2005). Encapsulation of essential oils in zein nano-spherical particles. *Journal of Agricultural and Food Chemistry, 53*, 4788–4792.
61. Petrovic, T., Nedovic, V., Dimitrijevic-Brankovic, S., Bugarski, B., & Lacroix, C., (2007). Protection of probiotic microorganisms by microencapsulation. *Chemical Industry and Chemical Engineering Quarterly, 13*(3), 169–174.
62. Phillips, G. O., Ogasawara, T., & Ushida, K., (2007). The regulatory and scientific approach to defining gum Arabic (*Acacia senegal* and *Acacia seyal*) as a dietary fiber. *Food Hydrocolloids, 22*, 24–35.
63. Porzio, M., (2004). Flavor encapsulation: A convergence of science and art. *Food Technology, 58*(7), 40–47.
64. Porzio, M., (2007). Flavor delivery and product development. *Food Technology, 1*(07), 22–29.
65. Porzio, M., (2007). Spray drying: An in-depth look at the steps in spray drying and the different options available to flavorists. *Perfumer and Flavorist, 32*(11), 34–39.
66. Prusse, U., Bilancetti, L., & Bucko, M., (2008). Comparison of different technologies for alginate beads production. *Chemical Papers, 62*(4), 364–374.
67. Reineccius, G. A., (2001). Multiple-core encapsulation: The spray drying of food ingredients. In: Vilstrup, P., (ed.), *Microencapsulation of Food Ingredients* (pp. 151–185). Surrey–England: Leatherhead Publishing.
68. Reineccius, G. A., (2004). Spray-drying of food flavors. *Drying Technology, 22*(6), 1289–1324.
69. Ribeiro, H. S., Schuchmann, H. P., Engel, R., Walz, E., & Briviba, K., (2010). Encapsulation of carotenoids. In: Zuidam, N. J., & Nedovic, V. A., (eds.), *Encapsulation Technologies for Food Active Ingredients and Food Processing* (pp. 211–253). Dordrecht: Springer.
70. Rosenberg, M., & Lee, S. J., (2004). Water-insoluble, whey protein-based microspheres prepared by an all-aqueous process. *Journal of Food Science, 69*, E50–E58.
71. Schimidt, P. C., (1997). Technological aspects of the development and production of plant extracts. *Die Pharmazeutische Industrie, 59*, 69.
72. Shah, N. P., (2000). Probiotic bacteria: Selective enumeration and survival in dairy foods. *Journal of Dairy Science, 83*, 894–907.
73. Shahidi, F., & Han, X. Q., (1993). Encapsulation of food ingredients. *Critical Reviews in Food Science and Nutrition, 33*(6), 501–547.
74. Shahidi, F. R., & Regg, R. B., (1991). Encapsulation of the pre-formed cooked cured-meat pigment. *Journal of Food Science, 56*, 1500.
75. Singh, H., & MacRitchie, F., (2001). Application of polymer science to properties of gluten. *Journal of Cereal Science, 33*, 231–243.
76. Teunou, E., & Poncelet, D., (2002). Batch and continuous fluid bed coating – review and state of the art. *Journal of Food Engineering, 53*, 325–340.
77. Teunou, E., & Poncelet, D., (2005). Fluid-bed coating. In: Onwulata, C., (ed.), *Encapsulated and Powdered Foods* (pp. 197–212). Boca Raton – USA: CRC Press.
78. Teunou, E., & Poncelet, D., (2005). Dry coating. In: Onwulata, C., (ed.), *Encapsulated and Powdered Foods* (pp. 179–195). Boca Raton – USA: CRC Press.
79. Truelstrup, H. L. T., Allan-Wojtas, P. M., Jin, Y. L., & Paulson, A. T., (2002). Survival of Ca–alginate microencapsulated *Bifidobacterium* Spp. in milk and simulated gastrointestinal condition. *Food Microbiology, 19*, 35–45.

80. Uhlemann, J., Schleifenbaum, B., & Bertram, H. J., (2002). Flavor encapsulation technologies: An overview including recent developments. *Perfumer and Flavorist, 27*, 52–61.
81. Vos, P., Faas, M. M., Spasojevic, M., & Sikkema, J., (2010). Review: Encapsulation for preservation of functionality and targeted delivery of bioactive food components. *International Dairy Journal, 20*(4), 292–302.
82. Wandrey, C., Bartkowiak, A., & Harding, S. E., (2010). Materials for encapsulation. In: Zuidam, N. J., & Nedovic, V. A., (eds.), *Encapsulation Technologies for Food Active Ingredients and Food Processing* (pp. 31–100). Dordrecht: Springer.
83. Wilson, N., & Shah, N. P., (2007). Microencapsulation of vitamins. *ASEAN Food Journal, 14*(1), 1–14.
84. Yilmaz, G., (2003). *Thermoplastic Starch Matrices for Encapsulation and Controlled Release of Volatile Compounds* (p. 219). PhD Thesis; Utrecht University, Utrecht, The Netherlands.
85. Zuidam, N. J., Van, W. E., De Vrueh, R., & Crommelin, D. J. A., (2003). Stability, storage and sterilization of liposomes. In: Torchilin, V. P., & Weissig, V., (eds.), *Liposomes* (pp. 149–165). Cambridge – Oxford: Oxford University Press.

CHAPTER 17

IMPACT OF FREEZING AND THAWING PROCESSES ON TEXTURAL PROPERTIES OF FOOD PRODUCTS

LAVANYA DEVARAJ, NIKITHA MODUPALLI, C. K. SUNIL, R. MAHENDRAN, and N. VENKATACHALAPATHY

ABSTRACT

Texture and structure are important parameters in judging the final end quality of food products, or plant-/animal-/dairy-based or nondairy products. Since these products are categorized as highly perishable to semi-perishable due to the presence of high moisture content that promotes microbial growth thus leading to spoilage. In this context, freezing and thawing helps to retain the texture and structure of food products by converting the excessive moisture into ice crystals, thus making free water unavailable for the growth of microbes. In general, the freezing refers to the liberation of heat, which is mainly used to remove the field heat from the food produce, such as fruits and vegetables (FAV) after harvest, meat, and meat products after post- and pre-slaughter. On the other hand, the mechanism of thawing is the reverse of the freezing process. Both unit operations occur simultaneously in the food processing industries to maintain, to preserve the texture and structure of end-product.

17.1 INTRODUCTION

Freezing and thawing have been used for thousands of years in food preservation to enhance the shelf-life, to nurture the nutritive and organoleptic

Advanced Research Methods in Food Processing Technologies. Junaid Ahmad Malik, Megh R. Goyal, Preeti Birwal, & Ritesh B. Watharkar (Eds.)
© 2024 Apple Academic Press, Inc. Co-published with CRC Press (Taylor & Francis)

properties of food products. During ancient days, freeze-preservation of food commodities was achieved by means of natural ice, snow, cold nights, and cold waves. The freezing mechanism reduces the activity of microorganisms and enzymes to a considerable extent [2, 82]; and it involves the formation of ice crystals through nucleation stage followed by further growth of ice crystals [26]. On the other hand, thawing is entirely an opposite process of the freezing operation; and it involves the phase change from ice to water by means of providing heat to the product. Thawing process usually occurs at the gliding temperature (the temperature difference between the saturated vapor and saturated liquid at constant pressure) of the food products.

Low temperature food preservation methods can be broadly distinguished into chilling (0–8°C) and freezing (<–18°C). Both the processes are distinguished based on the temperature differences in inactivating microbial and enzymatic load. Food freezing is one of the most satisfactory methods over any other conventional method, which requires the shortest processing time in terms of energy use, cost, and maintaining the quality of FAV (fruits and vegetables). Therefore, freezing cost can be kept as low as any other method of food preservation. In this regard, modern freezing techniques have been implemented in the area of food preservation, such as air blast freezing, immersion freezing (IF), cryogenic freezing, etc., among which ultrasound-assisted freezing is the best option in food industries because of its wide array of advantages over other methods [27]. There are various methods to achieve the freezing processes especially practiced at industrial levels (Table 17.1).

Thawing process is opposite to that of the freezing process, which consumes a lot of energy. In food commodities (such as: FAV, meat products), freezing removes the heat from the produce, whereas in the thawing process, heat is provided to the product. Thawing is much more important in the case of the fish industry, as fish meat is usually more susceptible to outer factors regarding deterioration and damage of the raw material than other meat products. The principle involved in thawing is the conversion of ice to liquid or water form and this phase change requires a lot of energy. There are various conventional and modern thawing methods that are adopted for food commodities and the temperature of thawing varies from one source of food to another [41]. A brief summary of conventional and novel methods of thawing practiced in food industries are discussed in Table 17.2.

TABLE 17.1 Description of Conventional and Novel Methods of Freezing

Type of Freezer	Description	Types of Food	References
Air-blast freezer	• Batch or continuous flow machines depending on excessive 'dead times' in production. • Most used freezing system. • It is based on the forces, convection of cold air into rooms of various dimensions by means of a fan. • Air speed: 5 m/s. • Larger industries: 10–15 m/s. • Reduces freezing times in continuous process. • Batch and continuous types with belt type freezers. • Temperature: –35 to –45°C; with freezing time of 12–48 hours.	• Regular and irregular shaped foods. • e.g., Meat, poultry, fish, FAV, other products like – cheese, butter, dough, bread, and baked products, pre-cooked ready meals.	[27]
Anti-freeze protein and ice nucleation protein	• Added directly to food and interact with ice. • Functionally distinct and opposite classes of proteins. • Antifreeze protein lowers the freezing temperature and retard the crystallization on frozen storage. • Ice nucleation protein raises the temperature of ice nucleation and reduces the degree of super cooling. • Limitation: High cost.	Inhibit crystallization of ice in dairy products (such as ice cream and de-icing agents) to preserve the smooth and creamy texture of ice creams, chilled, and frozen meat, beef, fish, etc.	[54]
Cryogenic freezer	• Alternate to immersion freezing. • Common cryogens are liquid N_2 and liquid CO_2. • Low boiling points: –196°C (LN_2) and –79°C (LCO_2). • Cryogens are colorless, odorless, chemically inert, and non-toxic.	Baked goods, fish, shellfish, meat, and poultry products.	[35]

TABLE 17.1 *(Continued)*

Type of Freezer	Description	Types of Food	References
	• Safe for direct contact with food. • Either sprayed or immersed. **Advantages:** • High freezing stabilities. • Ease of operation. • Low equipment cost. **Disadvantages:** • High cost of cryogens.		
De-hydro freezing	• Food is dehydrated to a desirable moisture and then frozen. • Promising way to preserve FAV. • Lowers cost of packaging, distribution, and storage. • Maintain product quality comparable to conventional methods. • Osmotic agents: glucose, fructose, lactose, maltodextrins, and corn syrup. **Limitations:** • Care should be taken when choosing an aqueous solution of high osmotic pressure, which leads to product composition modification with a negative impact on sensory qualities.	Fruits like kiwi, litchi, apples, melons, and vegetables (e.g., carrot, peas, beans, potatoes, etc.).	[54]
High pressure freezing	• Change in the physical state of food can be achieved by using external manipulation of temperature and pressure. • Pressure and temperature: 200–900 MPa at –20°C.	Vegetables, such as carrot, cabbage, peas, beans, mushroom, and some fruits.	[54]

TABLE 17.1 *(Continued)*

Type of Freezer	Description	Types of Food	References
	• High pressure freezing at 200, 340 and 400 MPa would be effective in increasing both the texture and historical structure of vegetables. • Uniform and small sized ice crystals both at the surface and at the central zones. **Limitations:** • Freezing should adopt to low pressure levels to minimize pressure effect on proteins.		
Immersion freezing (IF)	• Based on direct contact of food products with a refrigerant fluid: Nitrogen gas (–50 to –196°C), liquefied CO_2 (–50 to –70°C). • Rapid freezing can be achieved. • Products are immersed directly or sprayed with a cold liquid, such as brine, and glycol. • Products are packed to prevent cross contamination. • Irregular shape products can be handled.	Fruits and vegetables, meat, and meat products; pre-packed in packaging materials are suitable for IF.	[63]
Plate freezer	• Also called contact freezer. • Food products are sandwiched between 2 cold metal plates. • Cooling by conduction. • Limited to foods of thickness about 8 cm. • Freezers are either horizontal or vertical type. **Disadvantages:** • High capital investment. • Not suitable for packed foods, as it delays freezing time.	Meat patties, fish filets and chopped leafy vegetables.	[46]

TABLE 17.2 Summary of Conventional and Novel Methods of Thawing Process

Method	Description	Product	References
	Conventional Methods of Thawing		
Ambient temperature thawing	• Also called counter thawing. • Rapid thawing technique. • Monitoring the temperature regularly to keep food below 4°C. • Requires protection against flies, pests, and domestic pests. • Cook foods immediately to avoid surface spoilage by microorganisms.	Meat and poultry products, frozen peas, beans, apples, bananas, and other frozen sea foods.	[54]
Cold water thawing	• Use cold water to achieve fast thawing of frozen foods. • Frozen foods are placed in leak proof bags and are submerged in cold tap water. • Change the tap water every 30 minutes until frozen food is thawed. • Time varies as per the product size, e.g., 1–2 pounds package of beef or poultry takes 1–2 hours to thaw.	Meat and poultry products, beef, and sea foods	[1]
Microwave thawing	• Another fast method to thaw frozen foods. • Microwave thawing leads to cooking of frozen foods, where penetrated electromagnetic waves get converted into heat. • To avoid microbial spoilage, foods need to be cooked soon after MW thawing.	Lean beef cuts, shrimp, fish, and fish products	[1]
Refrigerator thawing	• Uses refrigerator temperature of 40°F or below. • Takes 24 hours for complete thawing. • Different sizes of meat and poultry packages will require different amount of time to achieve thawing, e.g., One pound of hamburger will take less time than large cuts of meat. • To avoid drip loss during thawing, place the frozen meat and poultry product on a plate or cookie sheet.	Meat and meat products like hamburger, beef; fresh and processed vegetables	[1]

TABLE 17.2 *(Continued)*

Method	Description	Product	References
	Novel Methods of Thawing		
Acoustic thawing	• Utilizes acoustic energy to thaw frozen food stuffs. • Thawing takes place under relaxation frequency. • Poor penetration, localized heating and high-power requirement hinders the development of this technology.	Fish and fish products, meat, poultry, frozen peas, broccoli, asparagus.	[54]
High pressure thawing	• Thawing rate depends only on the conduction of heat to preserve food quality and to reduce the necessary thawing time. • Frozen meat requires only one-third of time necessary at atmospheric pressure. • Improves frozen product texture. • No negative impact on color, penetration force or cooking loss of thawed beef. • Pressure level, treatment time can affect thawing rate and product quality. • Can thaw large amount of product. **Limitations:** • High cost.	Frozen milk products, tofu, meat, and poultry products.	[54]
Ohmic thawing	• Also called electro heating. • Heat is generated instantly inside the food when electric current passes through the food with high electrical resistance. • Aseptic processing and pasteurization of food can be achieved. • More efficient compared to MW heating. • No limitation of penetration depth. • Certain advantages live high heating rate, high energy conversion efficiency, volumetric heating, etc.	Frozen tuna, shrimp, beef, and eggs; Frozen chunks of beef (60 Hz to 60 kHz)	[86]

This chapter focuses on the impact of freezing and thawing unit operations on the textural and structural properties of food products.

17.2 MEASUREMENT METHODS FOR TEXTURAL AND STRUCTURAL PROPERTIES OF FOODS

Texture comprises of touch, feel, perception, appearance, taste, and structure, whereas on the other side, the structure of food component is influenced by its texture; and both occur simultaneously when food products are subjected to various processing parameters. Different food products exhibit different textural and structural properties, due to the nutritional compositions, physiological properties, moisture distribution, and other mechanical and structural properties. When food commodities are subjected to freezing and thawing processes, great changes in the structural and textural properties take place, which are highly influenced by the temperature of exposure, retention time, nature of the product, etc. [7]. Hence, it is most important to measure these properties by means of either subjective or objective methods. Subjective method requires an experienced person to notice these changes, which can be accomplished by means of visual observation and sensing the products to notice the differences. Whereas objective method involves the utilization of an instrument called texture analyzer to observe the texture and structural changes [23].

Depending on the nature of food products to be analyzed, the instrumental methods are categorized as physical methods (which include various devices, such as: shear, compression, penetration, biting, grinding, tensile, and fragmentation devices), structural and chemical methods. Based on the shearing force, shear devices (such as: Lee-Kramer shear press and Warner Bratzler devices) were designed. In relation to this, Wheeler et al. [91] examined the effect of shear force on steaks thawed at different temperatures and they found that steaks thawed at 12°C were measured significantly tenderer than those thawed at –2°C before cooking. In recent days, consumers are more conscious about the product and are willing to pay more based on the freshness and other esthetic values of the product [10]. Therefore, it is necessary to take care of the overall texture and other functional changes in food products when subjected to chilling/freezing/thawing process, such as: drip loss, shrinkage, loss of turbidity, lack of tenderness, juiciness in case of FAV, meat, and sea foods. The methods for the assessment of tenderness using textural analyzer are shown in Table 17.3.

TABLE 17.3 Methods for Assessment of Tenderness/Texture in Meat, Poultry, and Fish Products

Methods	References
Physical Methods	
Biting Devices:	
• Volodkevich;	[89]
• MIRINZ-mechanical;	[31]
• MIRINZ-pneumatic;	[31]
• MIT denature tenderometer;	[70]
• KT biting device;	[48]
• General foods texturometer;	[32]
• Instron RVI;	[79]
• Nip tenderometer.	[71]
Chemical Methods:	
• Trace elements;	[88]
• Collagen characteristics;	[75]
• 30,000-Da component.	[57]
Compression Methods:	
• Swift tenderness;	[64]
• Orifice device;	[45]
• Sinusoidal compressor.	[73]
Grinding Methods:	
• Mincer;	[61]
• Fragmentation method;	[61]
• Myofibrillar fragmentation index.	[61]
Penetration Devices:	
• Christal texture meter;	[61]
• Armor tenderometer;	[38]
• Slice tenderometer;	[51]
• Instron compression.	[13]
Sensory Ratings:	
• Chew count;	[40]
• Panel scores.	[39]
Shear Devices:	
• Shear jaw;	[78]
• Lee-Kramer shear process;	[84]

TABLE 17.3 *(Continued)*

Methods	References
• Warner-Bratzler shear;	[49]
• Rotating knife.	[9]
Structural Methods:	
• Fiber diameter;	[43]
• Fiber size;	[42]
• Shortening.	[55]
Tensile Methods:	
• Wang fiber extensibility;	[90]
• Smith tensiometer;	[14]
• L and W yield meter;	[55]
• Connective tissue-tearing device.	[68]

17.3 EFFECTS OF FREEZING AND THAWING ON STRUCTURE OF FOODS

17.3.1 MEAT AND SEA FOODS

The structure of meat and sea products contains large amount of water held within and in between the muscle-fiber tissues, which lead to the creation of compartments in tissues thus making it a very complicated process when subjected to freezing and thawing processes. Meat and other seafood products are rich sources of proteins, vitamins, minerals, other essential fatty acids, and amino acids. They are highly perishable in nature that leads to quick spoilage if proper storage and preservation methods are not implemented [7]. Freezing and thawing in this context are considered as the most promising technologies because of their advantages in retaining the nutrients and organoleptic properties followed by microbial growth inhibition, less or limited modification of sensory attributes without giving up the original freshness [25, 86].

Improper freezing and thawing leads to thaw loss/drip loss and loss of protein components have been observed in beef cuts and sheep muscle. Although there are several freezing methods have been adopted, the standard freezing temperature among them was found to be –18°C. However, large ice crystal formation is one of the major drawbacks in conventional freezing that leads to serious quality issues (such as: lipid oxidation, protein, moisture, texture, tenderness, and color), which disrupts meat muscle homeostasis. In recent days, modern meat processing industries have adopted ultrasound temperature freezing at –40°C that can help in the formation of small ice crystals, thus minimal deterioration of texture of the sample [28].

Diamante & Tran [28] examined the effect of freezing method and thawing temperature on size and shape of meat in connection to drip loss and protein content of thawed beef brisket exudates. Beef samples were cut into different shapes and subjected to standard freezing (–15°C) followed by air blast freezing (–36°C) and deep freezing (–18°C). The frozen samples were thawed at 25°C for 5 hours after 24 hours of freezing. They noted that drip loss of beef briskets was highly influenced by the size and shape of the meat. The drip loss in these samples was more when thawed at 45°C than at 25°C. The freeing method and thawing temperature, size, and shape had no much effect on protein content of thawed exudates from beef briskets. According to Leygonie et al. [53], freezing and thawing not only modify the other components of meat, but also modifies the moisture distribution in the meat tissue, which brings about the quality changes.

Hong et al. [44] showed that no significant difference was noticed in bighead carps frozen by ultrasound freezing at –40°C, freezing at –40°C for 12 hours and –18°C for storage of 3 months.

Jelena et al. [7] concluded that the freezing rate and time showed a significant effect on water activity and humidity of frozen chicken meat samples. The research also stated that the product under slow freezing conditions showed higher toughness due to formation of larger ice crystals. This causes the formation of larger pores in the matrix and greater histological damage to the muscle tissues. Another study on super-chilling chicken breast meat showed that there was drip loss after 28 days of storage, which caused shrinkage of muscle tissues, causing an increase in toughness and hardness [47].

Similarly to retain the quality of sea foods after catchment, freezing, and thawing techniques were most promising for preserving the quality and to prolong the shelf-life of sea products. The quality changes in sea foods (like shrimp, mollusks, and other shellfish) stored at freezing conditions were mainly due to lipid oxidation, denaturation of proteins, sublimation, and recrystallization of ice, which brings about abnormal flavor changes, off-odor development, rancidity, dehydration, loss in weight, loss of juiciness, drip loss and muscle toughening. These changes lead to microbial attack, which causes autolysis and finally the entire product becomes unfavorable for consumption [56].

The quality changes in tiger shrimp subjected to air blast freezing at –28 ± 2°C with an air velocity of 4–8 m/s and cryogenic freezing at –70 to –100°C followed by microwave (MW) thawing, refrigeration temperature and freeze thaw cycles were studied by Boonsumrej et al. [11] to evaluate considerable changes in terms of thiobarbituric acid (TBA), salt soluble protein (SSP), freezing losses, thawing losses and cutting force (CF). A study on chicken filets undergoing freeze-thaw cycles at –20°C for freezing and 4°C for thawing reported that the textural values including hardness were significantly reduced with increase in a number of cycles [21]. The repeated freeze-thaw cycles caused repeated formation and melting of ice crystals, leading to damage in myofibrils of muscle tissues and caused drip loss. Many researchers suggested that the structure of meat and other seafood products are highly influenced by freezing and thawing time and temperature; the faster the freezing process, the lower will be the drip loss. Hence, it is necessary to adopt advanced freezing and thawing techniques in meat and seafood processing industries to improve the structure and other physicochemical properties of processed meat products.

17.3.2 FRUITS AND VEGETABLES (FAV)

The FAV are classified under perishable groups due to the presence of 85–95% of moisture that attracts fruit spoilage due to microbial growth, if proper processing and storage methods are not followed. There are certain disadvantages of freezing, such as: damage to the cell structure due to ice crystal development followed by structural losses, loss of rigidity, shrinkage, freeze burn, loss of turgor pressure and loss in the firmness of FAV. Vegetables cells are comparatively thicker than the fruit cells to resist the damage caused by the growth of ice crystals inside the cells.

There are two processes, which mainly occur during freezing that cause damage to cellular structure: (i) firstly formation of ice crystals outside the cells especially in the intercellular spaces, where the solute concentrations are lowest that leads to puncturing of cell membrane, loss of turgor and; and (ii) secondly, cell damage caused by growing ice crystals, because fruits tend to have thin cell-wall, which leads to loss of firmness and are easily damaged.

Freezing process also brings about chemical changes, such as: breakdown of cells, development of off-flavor, colors, denaturation of protein structure due to alteration in tertiary structure, conversion of ascorbic acid (AA) into 2,3-diketogulonic acid, which is inactive in state. As an evidence of improper freezing, FAV can experience “freeze burn,” which completely alters the structural and textural properties.

FAV also suffers from shrinkage, loss of juiciness, spongy texture in case of blanched potatoes, loss of flavor, drip loss due to puncture in the cell-wall and increased syneresis [20, 29]. Similarly, in case of thawing, falsified textural changes occur in fruits as a result of fast freezing and thus leading to drip loss and collapse in the structure due to lack of water holding capacity in the cells.

Innovation in the area of freezing and thawing mainly aims to preserve the cell viability in tissues of FAV that can be accomplished by effective adoption of cryoprotectants. Gomez-Galindo et al. [34] studied the effects of cryoprotectants in preserving the FAV cell viability after subjecting to freezing and thawing. Their results revealed that spinach samples treated with vacuum impregnation (VI) alone or in combination with pulsed electric field (PEF) along with cryoprotectant solution (like trehalose (40% w/w)) proved drastic improvement in the treated spinach tissues. A study on freezing and thawing of apple tissues showed that by freezing at a temperature of –20°C, the texture of apple tissues exhibited vacuole rupture, modification, and collapse of cell-walls, greater intercellular pockets due to ice crystals and tissue rupture observed by imaging, leading to irreversible damage to the tissue strength and texture [20]. The study has also detected enzyme assisted cell-wall dissolution in apple tissues during the thawing process, resulting in softening of structure as revealed in the puncture test.

Another study on free-thaw cycles of papaya cubes (freezing –20 for 72 h and thawing 4°C for 4 h) showed weakening of cell integrity, thus losing the firmness of the fruit and increased drip loss volume [67]. Dehydro-freezing and thawing of apple tissues revealed that the textural properties of apple were significantly superior than that of normal freeze-thawed apple slices [36]. The reason for this can be prior removal of moisture content of the sample, thus avoiding the formation of intracellular gaps during the thawing cycles.

Van Buggenhout et al. [16] worked on freezing and thawing of cut strawberries using high-pressure shift freezing, and they concluded that HPSF exerted good effect in maintaining the texture of freeze-thawed strawberries frozen below –18°C.

17.3.3 DAIRY AND NON-DAIRY PRODUCTS

Dairy and non-dairy products are classified under highly perishable commodities as these contain about 87% of moisture. The main aim of freezing followed by thawing is to preserve the freshness of the produce and to reduce the spoilage caused by microorganisms. Further, milk can be processed and converted into various consumable commodities (e.g., cheese, paneer, khoa, skimmed milk, peda, and other dairy-based sweets) in combination with suitable ingredients along with proper freezing and storage. Nowadays, consumers prefer more non-dairy products, which are lactose-free by replacing milk with peanuts, soy milk, etc. Since, these products are helpful to lactose intolerance people, whose intestinal microflora lacks lactase enzyme to digest lactose.

Various attempts have been done by scientists in the area of freezing and thawing to nurture the physicochemical properties of dairy products. In this context, an experiment was conducted to analyze the physicochemical properties of goat curd cheese by freezing at 4–6°C with a storage interval of 3 months after freezing. It was concluded that minor variations were noticed in physicochemical parameters and not much significant differences were noticed [69]. Similar studies on the physicochemical properties of Ewe's milk subjected to freezing, refrigeration (5°C) and frozen storage (–4°C) showed an increase in the acidity with a significant decrease in pH that affected the overall milk stability. Alvarenga & Sousa [4] concluded that damage during freezing in cheese can be minimized through proper storage temperature of –20°C at the end of cheese ripening stage.

Non-dairy products are more in demand due to the lack of lactose in these and can be utilized in the development of healthy no-dairy beverages. For example, soy milk obtained from soybeans contains greater protein content. Soy milk can be used as a healthful beverage due to its rich nutritional profile and it is used as a base material in preparation of tofu also called "soy curd."

Protein and lipid content of soymilk subjected to freezing and its impact on freeze-thaw fraction was studied by Morita et al. [62], who found that the upper and lower layers of soy milk after freezing process had significantly equal proportions of protein and lipid ratio (w/w) irrespective of freezing temperature and they suggested that freeze-thaw fraction of soymilk were not affected with the freezing temperature.

The microstructure and textural parameters of tofu subjected to freeze-thaw treatment at –16°C followed by thawing at 20°C for 20 min were studied. From the structure analysis of tofu using Gibson Ashby model, it was shown that more water reduction in freeze-thawed tofu than that in the fresh sample [94]. From these studies, it can be concluded that freezing followed by thawing has a significant impact in maintaining the structural features of both dairy and non-dairy commodities.

17.3.4 STARCH AND DOUGH

Starch when subjected to freezing and thawing process leads to considerable changes in the microstructure and physicochemical properties. In the past few decades, there is a vast demand for starch-based frozen meals, especially in Southeast Asian countries. The structure of the starch-based commodities is mainly due to the interaction between the starch and other polysaccharides. When stored at low temperature conditions, starch gelatinization occurs that leads to change in original structure and phase separation process as a result of formation of ice crystals.

Water is present as a dispersing medium that controls the properties like sensory, processing, and storage. Hence, freezing should be done in such a way to control the moisture loss during further storage. The morphological features of starch granules in different food sources (such as: cereals, pulses, roots, or tubers) differ. The structural changes in freeze-thawed starchy materials are mainly due to retrogradation of amylose and amylopectin followed by syneresis after thawing. Due to structural changes, the product becomes soggy and unacceptable. Therefore, to improve the functional properties of starch and to achieve improved texture and rheological properties, a small number of hydrocolloids are added. The hydrocolloids help to improve the structure and texture by making the product more viscous by lowering the syneresis [77]. In rice gels subjected to freeze thaw cycle, reduction in spongy structure and syneresis (%) can be achieved by the addition of konjac glucomannan [19].

These days, dough prepared from wheat, rye, etc., are utilized in the preparation of bread, cake, biscuits, pastas, noodles, and other ready-to-eat fast foods. Irreversible transformation takes place during the conversion of starch to dough. When dough is subjected to the freezing process, there are serious physicochemical changes. Starch contains about 75% of solid particles in a larger fraction, which is responsible for the absorption of moisture during dough formation in bread-making. The dough quality utilized in bread preparation largely depends on freezing and thawing. In the research studies, it was noted that bread dough frozen at 30–40°C retrogradation of starch results in the increased elasticity in sweet frozen dough [60]. Szczesniak [83] studied the changes in physicochemical properties of potato starch granules during deep-freezing. It was concluded that pressure exerted on potato starch by freezing water can bring about phase change by making the granule more crumble and compressed.

17.4 EFFECTS OF FREEZING AND THAWING ON TEXTURE OF FOODS

17.4.1 MEAT AND SEA FOODS

The texture of meat and sea foods is a highly influential parameter in the meat processing industries. In the case of meat products, slaughtering leads to many changes either in terms of texture, functional or structural changes that make it more perishable when exposed to environmental conditions. These parameters can be retained through proper processing procedures like low temperature storage to retain the freshness of these foods. The textural changes (like loss of surface glossiness, tenderness in meat products, loss of firmness, chewiness) lead to serious changes in the overall appearance of the produce. There are various advanced technologies in the area of freezing like IF; cryogenic freezing to retain the freshness of meat and sea foods; however, cryogenic freezing is not widely practiced due to high cost of cryogenic fluids. The quality of salmon subjected to IF using dichlorodifluoromethane was studied by Bucholz & Pigott [15], who concluded that the overall quality of fish can be retained using liquid refrigerants and can be accomplished in fatty fishes to prevent rancidity due to the presence of unsaturated fatty acids without giving up the prior texture.

Chilled-aged meat and 2-months old frozen meats were studied for the difference in shearing forces during the texture measurement. The experiment revealed that shear force was lower in the case of 2 months old frozen meat compared to that of the aged, chilled meat. This might be due to the deterioration of muscular cells in 2-months old frozen meat due to the formation of large ice crystals in the muscle fibers of the fish [76]. The Salmon fish from Atlantic, Northern, and Western Norway were processed by smoking followed by storage at frozen temperature and thawing. The studies showed that expansion in extracellular spaces with muscle shrinkage was noticed in salmon subjected to smoking, freezing, and thawing process than that of fresh fish muscle. The results concluded that the texture of smoked salmon was unaffected in combination with freezing and thawing process [80]. Even after slaughtering, the meat texture could remain unaffected for more than three days, when properly frozen and thawed. Further aging of meat proceeds after the thawing process.

Crouse et al. [24] conducted an experiment on the effect of shear force on aged-meat samples before and after the freezing process. They noted that aged-meat samples after freezing showed lower shear force than that of the meat samples before freezing. This change in shearing force might be due to the loss of calcium dependent protease inhibitor; and also, the post-mortem proteolysis can be increased in meat before aging than in after aging followed by freezing process.

17.4.2 FRUITS AND VEGETABLES

Freshness of FAV is considered as one of the important parameters for consumer acceptance. Since these commodities are highly perishable proper freezing conditions are necessary to retain texture, and in most of cases, tissue softening occurs. The structure of most FAV consists of cell-wall along with the components of middle lamella like pectin, cellulose, and hemicellulose. The moisture percentage in most of the FAV is about 95% along with soluble solutes to maintain the structure of the edible commodities. If these commodities are subjected to freezing, then the rupture in the cell wall due to ice crystal formation expands the tissues leading to textural changes. During thawing process after freezing, the tissue of FAV becomes soft, which is reliable in case of fruits that are consumed as raw; whereas in case of vegetables, softened tissues are acceptable because the tissues become still softer during cooking process.

Recent studies on IF of green beans indicated that using liquid nitrogen as a refrigerant had a positive effect in maintaining the texture of beans even after cooking [22]. However, decrease in freezing rate showed no damage on texture of frozen beans. Marti & Aguilera [59] evaluated the three freezing methods for strawberries and wild blueberries: static freezing at –23±2°C, plate freezing at –50°C, and IF in liquid nitrogen. They reported that cell structural damage in strawberries was mainly due to slow freezing, which can be avoided using liquid nitrogen to obtain a better texture and lower drip loss. Similar studies on IF over conventional air-blast freezing were conducted to study the textural behavior of fresh beans harvested at the green stage. According to the observations, IF showed 4 times faster freezing rate of –30C than air-freezing rate with improvement in textural behavior of beans. The IF was found to be better and most preferred method by sensory panelists over other conventional freezing types to maintain the overall quality in case of FAV.

Tissue softening is considered as one of the major issues in fruit and vegetables processing industries. Charoenrein & Owcharoen [18] studied the microstructure, texture, and pectin substances in the ripe mangoes subjected to repeated freezing and thawing process. They observed tissue degradation during slow freezing process compared to that of medium and fast freezing, which brings about shrinkage in the cell wall by forming ice crystals. The quality of vacuum packed, seed-free pulp from durian subjected to freezing at –20°C followed by ice water thawing at 0°C and hot water blanching at 90°C was investigated to study the quality parameters based on physicochemical attributes. The studies revealed slight variations in terms of smell and increase in moisture content was noted in hot water thawed samples after freezing. In another case, not much variation in pH, color, total soluble solids (TSS), and organic components was observed in durian samples with respect to ice water thawing and freezing method [85].

17.4.3 DAIRY AND NON-DAIRY PRODUCTS

In case of dairy products, texture resembles the acceptability either in case of paneer, cheese, khoa, channa/chakka, ice creams, etc.; and these parameters will vary from one source to another source. Dairy commodities like milk and its products are highly perishable due to high moisture and fat content. If proper preservation methods are not adopted, then it will lead to rancidity of fat in case of butter, ghee that can be identified with the help of sensory tests. While discussing about non-dairy products (such as: soy milk, peanut milk, tofu, miso, etc.) are also rich source of protein, fat, and other nutrients. Textural deterioration in non-dairy products is mainly due to the denaturation of protein during handling and due to improper packaging, processing, and storage attributes.

Arora & Gupta [6] evaluated paneer manufactured with different levels of fat concentration (4, 5 and 6%) during low temperature freezing to study the physicochemical behavior of the samples. During this process, slighter reduction in moisture with an appreciable rise in NPN was observed in paneer samples with various fat levels at –13°C than in –32°C and also resulted in substantial increase in penetration depth implying the softer texture of frozen paneer after thawing. Similarly, the major issue in cheese manufacturing units was found to be textural changes and partial loss of whey protein after thawing. This problem can be overcome by adding salt directly to the frozen cheese followed by thawing during further ripening process of cheese.

The flavor and texture of Cottage and Manchego cheese were affected with the prolonged exposure to freezing conditions than those samples frozen for a short period under frozen storage. Similar observations were identified by Fontecha et al. [30] on the texture and microstructure of Ewe's milk cheese frozen at –80°C and stored at –20C and they found that proper freezing method does not cause any changes in the texture and microstructure of cheese; but slow-frozen cheese samples showed marked changes during 4 months of storage at frozen conditions.

The investigation conducted by Alonso et al. [3] on frozen Cabrales cheese at –20°C for 4–8 months showed no adverse changes in physico-chemical and organoleptic properties. The level of proteolysis and lipolysis was found to be lower in case of Cabrales cheese at frozen storage. The flavor textural changes in butter stored at refrigerated and frozen conditions were investigated by Krause et al. [50], who observed no off-flavor development in frozen butter, whereas stale off-flavor was identified in butter stored at refrigeration temperature.

Tofu is one of the non-dairy products obtained by soybeans used as a major ingredient, and it is popular in most of the East Asian countries. Fuchigami et al. [33] investigated the implication of high pressure freezing at 20°C for 90 minutes followed by high pressure thawing at 100–600 MPa to study the structural response of Kinu-tofu (called soybean curd). Their findings showed that better retainment of texture in Kinu-tofu can be achieved when subjected to high pressure thawing at of 200–500 MPa, because freezing and thawing pressure disrupts the growth of ice crystals. Further studies on the structural and textural properties of freeze-thawed tofu were conducted by Xu et al. [94], who noticed that the texture of tofu became harder, springier, and cohesive at –16°C prior to thawing at 20°C for 20 minutes as per consumer's expectation.

17.4.4 STARCH AND DOUGH

Starch obtained from various plant sources serves as a base ingredient in dough formation, which is used in bakery for other new product development. The texture and structural composition of starch differs from one source to another source. For example, in rice flour has less binding capacity compared to wheat flour, which exhibits higher elasticity than rice flour due to the gluten content.

When starch gels are subjected to freezing and thawing process, formation of ice crystals makes the starch to be concentrated in a non-ice phase. During the particular freeze-thaw cycles followed by centrifugation process, the liquid get is separated from the gel. This separation is mainly by starch retrogradation, which can be measured by measuring the enthalpy of the freeze-thaw cycle; and the structure of starch gel can be assessed using differential scanning calorimeter (DSC).

White et al. [92] studied the freeze-thaw stability of waxy and regular maize starch by heating at 120°C after 10 repeated freeze-thaw cycles. Due to exposure to high temperature, gelatinization was noted in these samples and the percentage recovery from the initial gelatinization enthalpy for waxy maize starch was 58% followed by 59% in the regular maize starch. They concluded that gelatinization percentage in starch can be minimized through proper freezing and thawing.

Maquenne [58] observed an increase in the rate of retrogradation in potato starch due to decrease in freezing temperature. Also, Woodruff & MacMasters [93] observed the changes in the microstructure of any starch paste due to retrogradation effect of decreased temperature. The lower percentage of syneresis was found in cryogenic frozen starch samples of tapioca, and sago at –18°C helped to retain its textural stability [87]. Most of the researchers concluded that stability of texture and structure can be obtained in starch (like potato, tapioca, sago, etc.) through proper freezing, thawing, and frozen storage conditions.

Also, the texture of the dough is an important parameter to obtain the acceptable product quality and also to optimize process and equipment costs. Proper freezing and thawing temperature helps to maintain the shelf-life of dough. The quality of frozen β-glucan rich rice-barley flour incorporated batter and dough were utilized in the baking of bread and cake. The outcome of the observation indicated that there were no remarkable changes in textural properties in terms of hardness, stickiness in the dough and spreadability in batter during 4 months of frozen storage. Hence, dough and batter used in bakery industries can be stored under frozen conditions to get good quality of final product [37].

Over exposure to freeze-thaw conditions leads to undesirable fermentation followed by increased proofing time, and less rise in the loaf volume leads to the development of coarser texture in bread baked from frozen dough than that of the fresh dough. Also, formation of ice crystals due to prolonged freezing weakens the protein network structure of dough and is responsible for retention of gas during fermentation [12].

The changes in specific volume and bread height in frozen dough might be due to the release of liquid in the frozen dough after thawing process [74]. In contrast, the inelastic properties in cookie batter might be due to the minimum development of gluten network in batter products. To avoid this, soft wheat flour is preferred because of formation of weak gluten network and low hydration properties [8]. It was evident that the quality of batter and dough can be retained even after 6 months by storing at –18°C without any deterioration changes in product density and surface characteristics. The final quality of the product is influenced by the amount of ingredients present in the formed dough [65]. Weakening in the structure of frozen stored dough was found due to the reasons, such as:

- Release of reducing substances in yeast added dough;
- Decrease in water binding capacity due to formation of ice crystals which redistributes the water in dough; and
- Reduction in cross linking.

Hamed [37] indicated that the frozen dough stability can be improved by addition of food gums (such as: guar gum, carrageenan gum, locust gum bean gum, etc.) as stabilizers to the dough or batter in the preparation of pies, ice cream and other dough/batter-based products.

17.5 CORRELATION BETWEEN THE STRUCTURE AND TEXTURE OF FOOD PRODUCTS

The texture and structure of food products will differ based on its nature. Structure of a food product is defined as the sensory manifestation of a texture, which is modified accordingly when subjected to various processing conditions, such as: heating, cooling, freezing, etc. Retention of original food texture and structure is an irreversible process. Based on the subjective and objective method (also called instrumental method), one can judge the behavior of the mechanical, geometric, and superficial behavior of any food product. For this purpose, texturometers are used to sense the texture of any food product with the help of a particular probe to identify the gumminess, juiciness, springiness, cohesiveness, and adhesiveness. Based on the depth of penetration, distance, and applied forces to the texture of different food products will differ. Instrumental methods have shown excellent correlation among the texture and structure of food components with accurate results. Adequate correlation among sensory and texture of food in food industries using instrumental method depends on:

- Based on consumer's acceptance, the degree of acceptance level in newly developed product must be judged;
- Instruments should be developed based on the proper knowledge prior to subjective method;
- Proper selection of an instrument to measure food quality;
- The developed methods must be improved and optimized based on sensory evaluation from experienced panelists.

In case of snack bars, bakery, fruit, and vegetables products, texture is one of the most dominating attributes in judging the final quality [5]. The factors affecting the structural and textural properties of various foods are depicted in Figure 17.1.

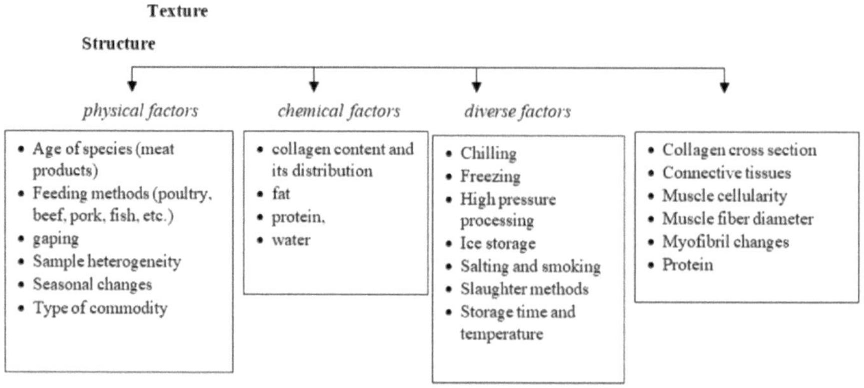

FIGURE 17.1 Factors affecting the structural and textural properties of foods.

The post-mortem textural changes in sea bream subjected to diverse storage temperatures were examined by Suárez et al. [81], who observed the induced firmness of fish muscle when stored at 1°C compared to that at 4°C and they also noticed that the rate of degradation of muscle collagen was more at 4°C than that at 1°C. These changes might be due to the cross-linkage of connective tissues, which are major contributors to the loss of firmness. In support of this post-mortem, change in sea bream during ice storage for seven days showed greater changes in hardness, gumminess, and chewiness than the brittleness [17].

The textural properties of different fruit varieties (such as: sour cherries, apples, and peaches) were studied during ripening, freezing, and thawing. The texture profile of these fruits was analyzed using a penetrometer, and organoleptic methods were found to be useful in classifying the fruits of different degree of maturity and variety [52].

Similarly, the texture and shelf-life of frozen salmon were analyzed by Remington [72], who found a decrease in the pore size of surface ice crystals as a result of an increase in freezing rate. For a better understanding of the texture and structure of any food products prior to any processing parameters, the instrumental method was found to be promising to obtain the expected results. For this purpose, 11 samples of extruded products were analyzed for texture based on the instrumental method using different probes. The textural properties of snacks were described in terms of hardness, fracturability, crispiness, adhesiveness, and chewiness, etc. Through the correlation of texture and structural properties, instrumental analysis helps to choose which probe is best in performing sensory test to differentiate the product sensory attributes [66].

17.6 SUMMARY

Today, consumers are more conscious about the quality of food products, which should be acceptable in appearance, texture, structure, color, etc. These quality parameters among the food commodities can be judged based on the texture, structure, and other functional ingredients that were added during new-product formulation. Hence, it is important to maintain the overall quality through proper preservation and storage methods, because all the food products are not available throughout the year for consumption. Therefore, it is necessary to store the perishable and seasonal commodities at low temperature conditions. At low temperature, most of the free water converts into ice crystals, making them not available for microbial activity. Freezing and thawing must be done simultaneously, which are opposite processes to retain the functional properties of food commodities. Some instances, over-exposure of food commodities to elevated freezing conditions leads to deterioration and damage. After processing of food products, it is necessary to measure its textural attributes. In this regard, advanced texture measuring devices (e.g., texture analyzer), which give accurate results for textural and structural properties by the utilization of suitable probes that have been developed to replace the manual method.

KEYWORDS

- **dairy products**
- **dough**
- **food structure**
- **food texture**
- **freezing**
- **fruits and vegetables**
- **non-dairy products**
- **seafood**
- **starch**
- **thawing**

REFERENCES

1. Akhtar, S., Khan, M. I., & Faiz, F., (2013). Effect of thawing on frozen meat quality: A comprehensive review. *Pakistan Journal of Food Sciences, 23*(4), 198–211.
2. Al-Jasser, M. S., (2012). Effect of cooling and freezing temperatures on microbial and chemical properties of chicken meat during storage. *Journal of Food Agriculture and Environment, 10*(1), 113–116.
3. Alonso, L., Juarez, M., Ramos, M., & Martin-Alvarez, P. J., (1987). Effects of changes during ripening and frozen storage on the physicochemical and sensory characteristics of cabrales cheeses. *International Journal of Food Science and Technology, 22*(5), 525–534.
4. Alvarenga, N., & Sousa, I., (2011). Effect of freezing on the rheological, chemical and color properties of serpa cheese. *Journal of Dairy Research, 78*(1), 80–87.
5. Anton, A. A., & Luciano, F. B., (2007). Instrumental texture evaluation of extruded snack foods: A review (evaluación instrumental de textura en alimentos extruidos: Una revisión). *CYTA – Journal of Food, 5*(4), 245–251.
6. Arora, V. K., & Gupta, S. K., (1980). Effect of low-temperature storage on paneer. *Indian Journal of Dairy Science, 33*(3), 374–380.
7. Babić, J., Cantalejo, M. J., & Arroqui, C., (2009). The effects of freeze-drying process parameters on broiler chicken breast meat. *LWT–Food Science and Technology, 42*(8), 1325–1334.
8. Barak, S., Mudgil, D., & Khatkar, B. S., (2013). Effect of composition of gluten proteins and dough rheological properties on the cookie-making quality. *British Food Journal, 115*(4), 564–574.
9. Bjorksten, J., Anderson, P., Ka, B., & Kapsalis, J., (1967). A portable rotating knife tenderometer. *Food Technology, 21*(1), 84–90.
10. Boleman, S. J., Boleman, S. L., Savell, J. W., West, & Johnson, D. D., (1997). Consumer evaluation of beef of known tenderness levels. *Journal of Animal Science, 75*, 1521–1524.
11. Boonsumrej, S., Chaiwanichsiri, S., Tantratian, S., Suzuki, T., & Takai, R., (2007). Effects of freezing and thawing on the quality changes of tiger shrimp (*Penaeus monodon*) frozen air-blast and cryogenic freezing. *Journal of Food Engineering, 80*(1), 292–299.
12. Bot, A., & De Bruijne, D. W., (2003). Osmotic properties of gluten. *Cereal Chemistry, 80*(4), 404–408.
13. Bouton, P. E., & Harris, P. V., (1972). A comparison of some objective methods used to assess meat tenderness. *Journal of Food Science, 37*, 218–221.
14. Bramblett, V. D., Hostetler, R. L., Vail, G. E., & Draudt, H. N., (1959). Qualities of beef as affected by cooking at very low temperatures for long periods of time. *Food Technology, 13*, 707–711.
15. Bucholz, S. B., & Pigott, G. M., (1972). Immersion freezing of fish in dichlorodifluoromethane. *Journal of Food Science, 37*(3), 416–419.
16. Van, B. S., Messagie, I., Maes, V., Duvetter, T., Van, L. A., & Hendrickx, M., (2006). Minimizing texture loss of frozen strawberries: Effect of infusion with pectin methylesterase and calcium combined with different freezing conditions and effect of subsequent storage/thawing conditions. *European Food Research and Technology, 223*(3), 395–404.
17. Caballero, M. J., Betancor, M., & Escrig, J. C., (2009). Postmortem changes produced in the muscle of sea bream (*Sparus Aurata*) during ice storage. *Aquaculture, 291*(3, 4), 210–216.

18. Charoenrein, S., & Owcharoen, K., (2016). Effect of freezing rates and freeze-thaw cycles on the texture, microstructure and pectic substances of mango. *International Food Research Journal, 23*(2).
19. Charoenrein, S., Tatirat, O., Rengsutthi, K., & Thongngam, M., (2011). Effect of konjac glucomannan on syneresis, textural properties and the microstructure of frozen rice starch gels. *Carbohydrate Polymers, 83*(1), 291–296.
20. Chassagne-Berces, S., Poirier, C., & Devaux, M. F., (2009). Changes in texture, cellular structure and cell wall composition in apple tissue as a result of freezing. *Food Research International, 42*(7), 788–797.
21. Chen, T. H., Zhu, Y. P., & Han, M. Y., (2017). Classification of chicken muscle with different freeze-thaw cycles using impedance and physicochemical properties. *Journal of Food Engineering, 196*, 94–100.
22. Chourot, J., Lauwers, J., Massoji, N., & Lucas, T., (2001). Behavior of green beans during the immersion chilling and freezing. *International Journal of Food Science and Technology, 36*(2), 179–187.
23. Chrystall, B., (1994). Meat texture measurement. In: *Quality Attributes and Their Measurement in Meat, Poultry and Fish Products* (pp. 316–336). Cham: Springer.
24. Crouse, J. D., & Koohmaraie, M., (1990). Effect of freezing of beef on subsequent postmortem aging and shear force. *Journal of Food Science, 55*(2), 573–574.
25. Dawson, P. L., Chaves, B. D., Northcutt, J. K., & Han, I. Y., (2013). Quality and shelf life of fresh chicken breasts subjected to crust freezing with and without skin. *Journal of Food Quality, 36*(5), 361–368.
26. Degner, B. M., Chung, C., Schlegel, V., Hutkins, R., & McClements, D. J., (2014). Factors influencing the freeze-thaw stability of emulsion-based foods. *Comprehensive Reviews in Food Science and Food Safety, 13*(2), 98–113.
27. Dempsey, P., & Bansal, P., (2010). Air blast freezers and their significance to food freezing: A review. In: *Proceedings of ENCIT 2010, 13th Brazilian Congress of Thermal Sciences and Engineering* (pp. 5–10).
28. Diamante, L. M., & Tran, N. T. M., (2016). Effects of meat shape and size, freezing method and thawing temperature on the drip loss of beef brisket and the protein content of its thaw exudates. *Journal of Food Chemistry and Nanotechnology, 2*(1), 14–20.
29. Ferrero, C., Martino, M. N., & Zaritzky, N. E., (1993). Stability of frozen starch pastes: Effect of freezing, storage and xanthan gum addition. *Journal of Food Processing and Preservation, 17*(3), 191–211.
30. Fontecha, J., Kaláb, M., Medina, J. A., Peláez, C., & Juárez, M., (1996). Effects of freezing and frozen storage on the microstructure and texture of ewe's milk cheese. *Zeitschrift für Lebensmittel-Untersuchung und Forschung (J. Food Study & Research), 203*(3), 245–251.
31. Frazerhurst, L. F., & MacFarlane, P., (1983). *A Device for Measuring the Tenderness of Meat.* NZ Patent, 1983, E-article 190945.
32. Friedman, H. H., Whitney, J. E., & Szczesniak, A. S., (1963). The texturometer—A new instrument for objective texture measurement. *Journal of Food Science, 28*(4), 390–396.
33. Fuchigami, M., Teramoto, A., & Ogawa, N., (1998). Structural and textural quality of kinu-tofu frozen-then-thawed at high-pressure. *Journal of Food Science, 63*(6), 1054–1057.
34. Gomez, G. F., Dejmek, P., Phoon, P. Y., Velickova, E., & Tylewicz, U. (2011). Preservation of cell viability in fruit and vegetables tissues after freezing and thawing. In: *11th International Congress on Engineering and Food.* Althens, Greece.

35. Goswami, T. K., (2010). Role of cryogenics in food processing and preservation. *International Journal of Food Engineering, 6*(1). doi: 10.2202/1556-3758.1771.
36. Ben, H. S. L., Bellagha, S., & Allaf, K., (2016). Dehydrofreezing of apple fruits: Freezing profiles, freezing characteristics, and texture variation. *Food and Bioprocess Technology, 9*(2), 252–261.
37. Hamed, A., Ragaee, S., Marcone, M., & Abdel-Aal, E. M., (2015). Quality of bread and cookie baked from frozen dough and batter containing βeta-glucan-rich barley flour fraction. *Journal of Food Quality, 38*(5), 316–327.
38. Hansen, L. E. O. J., (1972). Development of armor tenderometer for tenderness evaluation of beef carcasses. *Journal of Texture Studies, 3*(2), 146–164.
39. Harries, J. M., Rhodes, D. N., & Chrystall, B. B., (1972). Meat texture, part I: Subjective assessment of the texture of cooked beef. *Journal of Texture Studies, 3*(1), 101–114.
40. Harrington, G., & Pearson, A. M., (1962). Chew count as a measure of tenderness of pork loins with various degrees of marbling. *Journal of Food Science, 27*(1), 106–110.
41. Haugland, A., (2002). *Industrial Thawing of Fish: To Improve Quality, Yield and Capacity*. Doctoral dissertation. Norwegian University of Science and Technology,
42. Herring, H. K., Cassens, R. G., Suess, G. G., Brungardt, V. H., & Briskey, E. J., (1967). Tenderness and associated characteristics of stretched and contracted bovine muscles. *Journal of Food Science, 32*(3), 317–323.
43. Hiner, E. L., Hankins, O. G., Sloane, H. S., Fellers, C. R., & Anderson, E. E., (1953). Fiber diameter in relation to tenderness of beef muscle. *A Journal of Food Science, 18*(1–6), 364–376.
44. Hong, H., Luo, Y., Zhou, Z., Bao, Y., Lu, H., & Shen, H., (2013). Effects of different freezing treatments on the biogenic amine and quality changes of bighead carp (*Aristichthys nobilis*) heads during ice storage. *Food Chemistry, 138*(2, 3), 1476–1482.
45. Howe, P. E., (1927). The relation of cooking to the study of the quality and palatability of meat. *Journal of Home Economics, 19*(8), 8–15.
46. Johnston, W. A., (1994). *Freezing and Refrigerated Storage in Fisheries* (p. 340). Rome: Food and Agriculture Organization (FAO).
47. Kaewthong, P., Pomponio, L., Carrascal, J. R., Knøchel, S., Wattanachant, S., & Karlsson, A. H., (2019). Changes in the quality of chicken breast meat due to super-chilling and temperature fluctuations during storage. *The Journal of Poultry Science*, E-article 180106.
48. Kelly, R. F., Taylor, J. C., & Graham, P. P., (1960). Preliminary comparisons of new tenderness measuring device with objective and subjective evaluations of beef. *Journal of Animal Science, 1960*, 645–650.
49. Kragten, S. A., & Gil, M., (2015). Instrumental tenderness–shear force. In: *A Handbook of Reference Methods for Meat Quality Assessment* (p. 45).
50. Krause, A. J., Miracle, R. E., Sanders, T. H., Dean, L. L., & Drake, M. A., (2008). The effect of refrigerated and frozen storage on butter flavor and texture. *Journal of Dairy Science, 91*(2), 455–465.
51. Kulwich, R., Alsmeyer, R. H., & Decker, R. W., (1963). Use of slice-tenderness evaluation device with pork. *Food Technology, 17*(2), 201–209.
52. Lásztity, R., Sebők, A., & Major, J., (1992). Textural properties of FAV and their changes during freezing and storage at low temperatures. *Polytechnic News on Chemical Engineering, 36*(4), 225–238.
53. Leygonie, C., Britz, T. J., & Hoffman, L. C., (2012). Impact of freezing and thawing on the quality of meat. *Meat Science, 91*(2), 93–98.

54. Li, B., & Sun, D. W., (2002). Novel methods for rapid freezing and thawing of foods–A review. *Journal of Food Engineering, 54*(3), 175–182.
55. Locker, R. H., & Wild, D. J. C., (1982). Machine for measuring yield point in raw meat. *Journal of Texture Studies, 13*(1), 71–82.
56. Londahl, G., (1997). Technological aspects of freezing and glazing shrimp. *InfoFish International, 1997*, 49–56.
57. Macbride, M. A., & Parrish, Jr. F. C., (1977). The 30,000-dalton component of tender bovine longissimus muscle. *Journal of Food Science, 42*(6), 1627–1629.
58. Maquenne, L., (1903). *Sur La Retrogradation De L'empois D'amidon* (The south retrogradation of *Lempois damidon*). *Comptes Rendus Chimie, 137*(797), 16–23.
59. Marti, J., & Aguilera, J. M., (1991). Effects of freezing rate on the mechanical characteristics and microstructure of blueberries and wild blackberries. *Revista de Agroquimica y Tecnologia de Alimentos (J. of Agrochemicals & Food Technology, Spain), 31*(4), 493–504.
60. Meziani, S., Jasniewski, J., & Gaiani, C., (2011). Effects of freezing treatments on viscoelastic and structural behavior of frozen sweet dough. *Journal of Food Engineering, 107*(3, 4), 358–365.
61. Miller, A. J., Strange, E. D., & Whiting, R. C., (1989). Improved tenderness of restructured beef steaks by a microbial collagenase derived from vibrio B-30. *Journal of Food Science, 54*(4), 855–857.
62. Morita, K., Shimoyamada, M., & Nakajima, M., (2015). Effects of freezing temperature on the freeze-thaw fractionation of soymilk. *Food Science and Technology Research, 21*(6), 863–867.
63. Nagare, B., Marcolli, C., Welti, A., Stetzer, O., & Lohmann, U., (2016). Comparing contact and immersion freezing from continuous flow diffusion chambers. *Atmospheric Chemistry and Physics, 16*(14), 8899–8914.
64. Palmer, W. E., (1962). *Tenderness Testing Device* (pp. 639–364). Canadian Patent.
65. Pareyt, B., & Delcour, J. A., (2008). The role of wheat flour constituents, sugar, and fat in low moisture cereal-based products: A review on sugar-snap cookies. *Critical Reviews in Food Science and Nutrition, 48*(9), 824–839.
66. Paula, A. M., & Conti-Silva, A. C., (2014). Texture profile and correlation between sensory and instrumental analyses on extruded snacks. *Journal of Food Engineering, 121*, 9–14.
67. Phothiset, S., & Charoenrein, S., (2014). Effects of freezing and thawing on texture, microstructure and cell wall composition changes in papaya tissues. *Journal of the Science of Food and Agriculture, 94*(2), 189–196.
68. Pool, M. F., (1967). Objective measurement of connective tissue tenacity of poultry meat. *Journal of Food Science, 32*(5), 550–553.
69. Pricop, E. M., & Popa, M. E., (2012). Effect of different refrigeration treatments on the physicochemical characteristics of raw goat milk curd-style cheese. *Journal of Agroalimentary Processes and Technologies, 18*(2), 175–180.
70. Proctor, B. E., Davison, S., & Brody, A. L., (1956). Recording strain gage denture tenderometer for foods, part 2: Studies on the masticatory force and motion, and the force-penetration relationship. *Food Technology, 10*(7), 327–331.
71. Purchas, R. W., (1973). Some aspects of raw meat tenderness: A study of some factors affecting its change with cooking and a new means of measurement. *Journal of Food Science, 38*(4), 556–559.

72. Remington, M. C. A., (2017). *The Effect of Freezing and Refrigeration on Food Quality* (p. 132). Masters Dissertation. Clemson University.
73. Salé, P., Noel, Y., Lasteyras, A., & Oleon, C., (1984). Sinusoidal compression system to study rheological properties of foods in transient state. *Journal of Texture Studies, 15*(2), 103–114.
74. Seguchi, M., Nikaidoo, S., & Morimoto, N., (2003). Centrifuged liquid and breadmaking properties of frozen-and-thawed bread dough. *Cereal Chemistry, 80*(3), 264–268.
75. Seideman, S. C., (1986). Methods of expressing collagen characteristics and their relationship to meat tenderness and muscle fiber types. *Journal of Food Science, 51*(2), 273–276.
76. Shanks, B. C., Wulf, D. M., & Maddock, R. J., (2002). The effect of freezing on Warner-Bratzler shear force values of beef longissimus steaks across several postmortem aging periods. *Journal of Animal Science, 80*(8), 2122–2125.
77. Shi, X., & BeMiller, J. N., (2002). Effects of food gums on viscosities of starch suspensions during pasting. *Carbohydrate Polymers, 50*(1), 7–18.
78. Shockey, C. F., McKee, L. G., & Hamm, W. S., (1944). Instrument for measuring changes in texture of dehydrated fish. *Industrial and Engineering Chemistry Analytical Edition, 16*(10), 638–645.
79. Shorthose, W. R., & Harris, P. V., (1990). Effect of animal age on the tenderness of selected beef muscles. *Journal of Food Science, 55*(1), 1–8.
80. Sigurgisladottir, S., Ingvarsdottir, H., Torrissen, O. J., Cardinal, M., & Hafsteinsson, H., (2000). Effects of freezing/thawing on the microstructure and the texture of smoked Atlantic salmon (*Salmo Salar*). *Food Research International, 33*(10), 857–865.
81. Suárez, M. D., Martínez, T. F., Saez, M. I., Alferez, B., & García-Gallego, M., (2011). Changes in muscle properties during postmortem storage of farmed sea bream (*Sparus aurata*). *Journal of Food Process Engineering, 34*(3), 922–946.
82. Sunil, C. K., Chidanand, A. R., & Pare, D. V. A., (2016). Application of emerging technologies for freezing and thawing of foods. In: *Food Process Engineering* (pp. 81–112) Burlington–ON: Apple Academic Press.
83. Szczesniak, A. S., (2002). Texture is a sensory property. *Food Quality and Preference, 13*(4), 215–225.
84. Szczesniak, A. S., & Torgeson, K. W., (1965). Methods of meat texture measurement viewed from the background of factors affecting tenderness. *Advances in Food Research, 14*, 33–165.
85. Tagubase, J. L., Ueno, S., Yoshie, Y., & Araki, T., (2016). Effect of freezing and thawing on the quality of durian (*Durio zibethinus* Murray) pulp. *Transactions of the Japan Society of Refrigerating and Air Conditioning Engineers, 2016*, 16–17.
86. James, C., & James, J. S., (2010). Freezing/thawing. In: Toldrá, F., (ed.), *Handbook of Meat Processing* (pp. 105–124). New York: John Wiley & Sons.
87. Varavinit, S., Anuntavuttikul, S., & Shobsngob, S., (2000). Influence of freezing and thawing techniques on stability of sago and tapioca starch pastes. *Starch-Stärke, 52*(6, 7), 214–217.
88. Vavak, L. D., Satterlee, L. D., & Anderson, P. C., (1976). The relationship of cardiac shear and trace element content to beef muscle tenderness. *Journal of Food Science, 41*(4), 729–731.
89. Volodkevich, N. N., (1938). Apparatus for measurements of chewing resistance or tenderness of foodstuffs. *Journal of Food Science, 3*(1, 2), 221–225.

90. Wang, H., Doty, D. M., Beard, F. J., Pierce, J. C., & Hankins, O. G., (1956). Extensibility of single beef muscle fibers. *Journal of Animal Science, 15*(1), 97–108.
91. Wheeler, T. L., Shackelford, S. D., & Koohmaraie, M., (1996). Sampling, cooking, and coring effects on Warner-Bratzler shear force values in beef. *Journal of Animal Science, 74*(7), 1553–1562.
92. White, P. J., Abbas, I. R., & Johnson, L. A., (1989). Freeze-thaw stability and refrigerated-storage retrogradation of starches. *Starch-Stärke, 41*(5), 176–180.
93. Woodruff, S., & McMaster, M. M., (1938). Gelatinization and retrogradation changes in corn and wheat starches shown by photomicrographs. *Bulletin (University of Illinois (Urbana-Champaign campus): Agricultural Experiment Station), 445*, 1–16.
94. Xu, Y., Tao, Y., & Shivkumar, S., (2016). Effect of freeze-thaw treatment on the structure and texture of soft and firm tofu. *Journal of Food Engineering, 190*, 116–122.

CHAPTER 18

3D PRINTING OF FOODS: CONCEPTS, APPLICATIONS, AND PROSPECTS

CINU VARGHESE, M. S. ROOPESH, PREM PRAKASH SRIVASTAV, and ANJALI SUDHAKAR

ABSTRACT

3D printing is carried out in different sectors like medical, clothing, architecture, packaging, education, biological tissue, commercial manufacturing; however, the application and implementation of 3D printing in the food industry is an ongoing process. 3D food printers are the application of 3D printers in which different forms of food materials can be extruded. This chapter focuses on the concept, principles, and different techniques of 3D food printing (3DFP) and its application in the food processing industry. The innovation in the 3DFP sector is growing up and needs further research related to the process and printing parameters.

18.1 INTRODUCTION

Technology always leveraged food industry in every phase, including production, processing, safety, nutrition, and delivery. 3D printing, often known as additive manufacturing (AM) opened to a new era of the food sector as 3D food printing (3DFP). 3D printing is a process of building the product, adding a thin layer of material upon a thin layer using a 3D object digital design (usually a computer-aided design (CAD) file) [26]. Materials commonly used for traditional 3D printing are polymers, rubber, concrete, plaster, and even wood. The technology was first implemented for rapid prototyping, followed by tooling [15]. The advantage and adaptation of

Advanced Research Methods in Food Processing Technologies. Junaid Ahmad Malik, Megh R. Goyal, Preeti Birwal, & Ritesh B. Watharkar (Eds.)
© 2024 Apple Academic Press, Inc. Co-published with CRC Press (Taylor & Francis)

application areas of AM continued to increase, and now it has evolved to be applied in the food industry. 3DFP is the most challenging and complex adaptation of AM. The ability to extrude the material to a defined 3D space without compromising the quality, texture, taste, and morphology makes 3DFP more challenging and at the same time more interesting.

3D food printers work on the same principle of traditional 3D printers. The syringe containing food material deposits food material layer by layer in an additive fashion [23]. Chuck Hull invented 3DFP, and the first part of the 3D food object was developed in 1983 [12]. The first 3D printer was developed with a method called stereolithography. The main objective for the setting of the 3D food printer was for the military field. Recently, this emerging technology was experimented in different sectors in the food industry, mainly layered food manufacturing. In 2006, Cornell University introduced Fab@home (a multi-material 3D food printer) to print different material, such as cheese, chocolate, and cookie dough [11]. The first commercial 3D chocolate printer Choc Edge came in the limelight. *In vitro* meat was first printed using a bioprinter by Modern meadow in 2014 [15]. In the same year, Natural Machines introduced Foodini with an addition that allows the user to develop the designs remotely [12].

The most evolved 3D food printers are equipped with predesigned models, and they are user-friendly. Potential benefits, including customized geometry, shortened time, reduced process cost, and customized nutritious product have marked 3DFP as the third revolution of the food processing industry [24, 25]. A variety of extruded products is possible with the help of 3D food printers, namely customized confectionaries, pancakes, pizzas, chocolate, sugar cubes, etc. 3DFP technologies introduced customized shape, novel texture and even complex structures that have not been imagined before [8, 15]. 3DFP has widely influenced different sector, including military field, space, and formulation of food for special groups such as elders, undernourished children, and those with nutritional intolerance [12]. In this context, 3DFP is a new hope for producing creative and attractive food shapes, which can lead to a different change in the monotony of traditional food. According to the demand, the raw materials can be blended, to fulfill the nutritional requirement of the target group and printed with 3DFP for customized nutrition [20].

This chapter aims to highlight an overview of the concepts, applications, and current and future aspects of 3DFP. It also illustrates the different technologies used in printing and the factors affecting the quality of the 3D printed product.

18.2 TECHNIQUES OF 3D FOOD PRINTING (3DFP)

Different types of 3D printing techniques have been adopted in the food sector to reach quality and precision within food processing design. These techniques are categorized based on the type of food material used such as semi-solid, paste powder and cell culture [20]. Nowadays, the four most commonly used 3D printing technologies are extrusion-based printing, selective laser sintering (SLS), binder jetting and inkjet. Extrusion-based 3DFP is the most commonly used technique in the food processing industry. According to the state of the printing material, extrusion-based printing has been classified into room temperature extrusion, fused deposition manufacturing and gel-forming extrusion [12].

18.2.1 EXTRUSION-BASED PRINTING

Previously this technique was used to manufacture plastic products and was named fused deposition modeling (FDM), invented by Crump in 1989. After introducing FDM, the same extrusion process also applied to the AM [4]. During extrusion-based 3DFP, the material extrudes continuously through an extrusion nozzle and deposits to adhere to the previous layer on cooling (Figure 18.1). 3DFP with extrusion-based method has been used in the printing of chocolates and soft materials such as meat paste, cookie dough, mashed vegetables such as potato, carrot, etc., and cheese. Depending on the extrusion temperature and the nature of the printing material used, extrusion-based printing can be further classified into soft material extrusion, Hot-melt extrusion, and Hydrogel forming extrusion.

18.2.1.1 SOFT MATERIAL EXTRUSION

Soft-material extrusion works on the principle of extrusion and deposition. Materials that are able to self-support are extruded in this process. There will be no phase change, and the stability of the 3D structure correlates with the rheological properties of the printing material. The flowability of the material through the nozzle and stability of the structure after printing is determined by the viscosity of the material. The deposition of soft material on a complex 3D structure is difficult and limited because it is prone to distortion and instability.

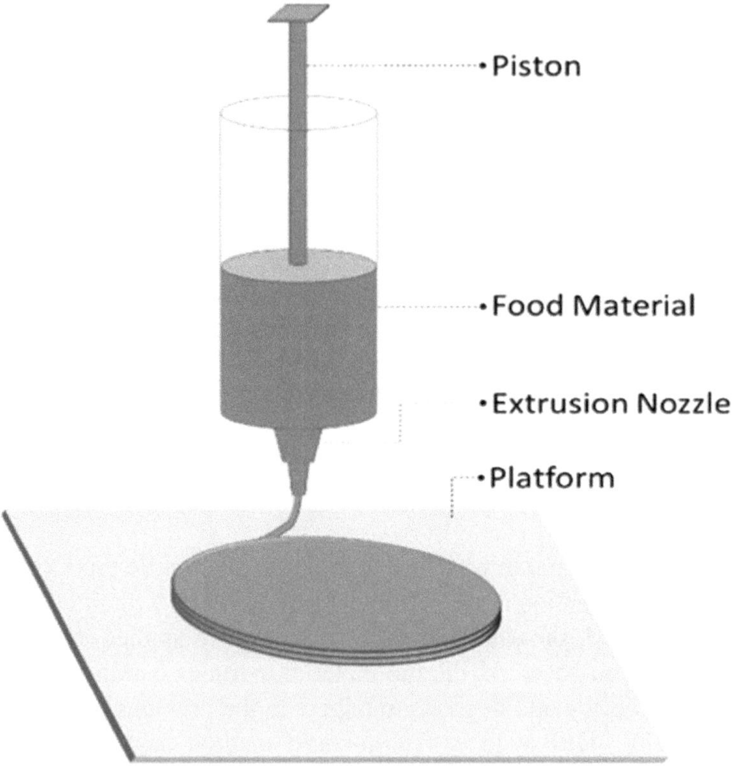

FIGURE 18.1 Schematic diagram of extrusion-based printing.

Source: Adapted from Ref. [8]

To fabricate a complex structure, it is necessary to provide a supporting structure to retain the stability of the 3D design during extrusion-based printing. The supporting structure has to be removed manually after printing. Thus, the printing of additional structure leads to high material cost and is time-consuming, followed by slow down printing [14]. The material commonly used in soft material extrusion is dough, meat puree, processed cheese.

18.2.1.2 HOT MELT EXTRUSION

Hot melt extrusion (HME) is commonly used in the printing of chocolates. The working temperature of extrusion ranges from 28 to 40°C. The main challenge in the printing of chocolates is the complex crystallization of the major ingresdient, the cocoa butter used in the chocolate fabrication. Following are the factors that directly depend on the quality and accuracy of chocolate printing [10]:

- Diameter of the nozzle;
- Height difference between the extrusion nozzle and platform;
- Extrusion axis movement.

18.2.1.3 HYDROGEL-FORMING EXTRUSION

In hydrogel-forming extrusion, with the aid of the syringe, vibrating nozzle and, the jet cutter, the hydrocolloid solutions are dispensed into a gel setting bath. The critical parameters for hydrogel-forming extrusion are the rheological parameters and gel-forming capacity. Firstly, the viscoelastic properties of the hydrocolloid solution should be in such a way that it has the capacity of self-support prior to the deposition of the next layer. The binding mechanism of the hydrogel-forming extrusion is mainly classified into three categories: (i) chemical cross-linking; (ii) ionotropic cross-linking; and (iii) complex coacervate formation. This technique is commonly used to print fruits for aged people having swallowing issues. Gelatin and xanthan gum can also be printed using this technique [3].

18.2.2 SELECTIVE SINTERING

Selective sintering is the second most commonly used technique after the extrusion process. This can be divided based on the type of sintering used; the first one is SLS, and the second is selective hot air sintering and melting (SHASAM). These processes work on the principle of binding the powder using the sintering source to form a solid layer. The 3D object develops in a layer-by-layer manner within the powder bed and is later retrieved from under the powder bed. After depositing the first layer, a roller moves on the previous layer and deposit the new layer of powder (Figure 18.2). The remarkable benefit that makes selective sintering different from other techniques is the ability to build complex food products within a short span.

CandyFab 400 made the first attempt to print a 3D structure using sugar. CandyFab worked with SHASAM technique by using a pointed heat source that passed through a sugar bed and created a caramelized sculpture due to the melting of the sugar particles [6].

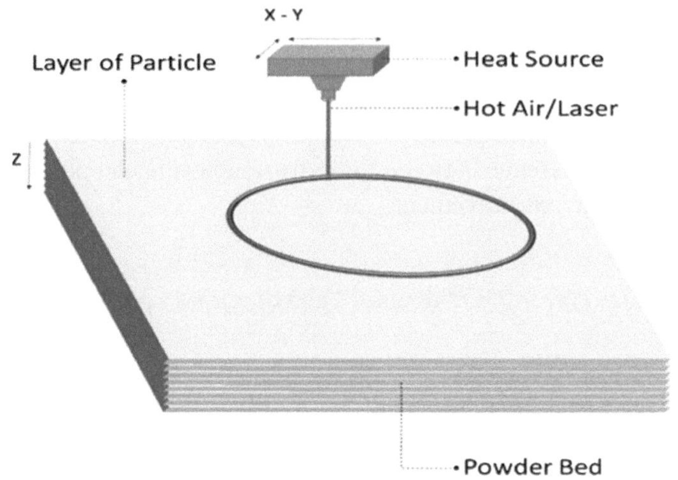

FIGURE 18.2 Schematic diagram of selective laser/hot air sintering.

Source: Adapted from Ref. [8]

18.2.3 BINDER JETTING

In this AM mechanism, the powder material is evenly spread over the fabrication platform, and the two consecutive powder layers are bonded using a liquid binder. The binding mechanism is accelerated by the adhesive force and chemical reaction between the powder and binder (Figure 18.3).

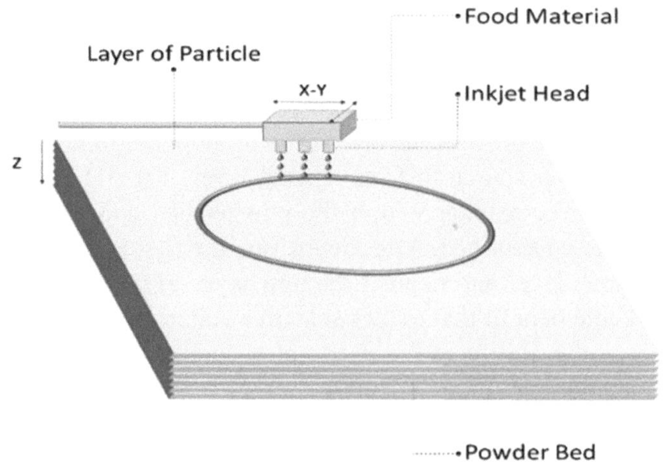

FIGURE 18.3 Schematic diagram of binder jetting.

Source: Adapted from Ref. [8]

The binder drops on demand, and adhesion starts due to the cross-linking of the particles between the materials. To minimize the disturbance caused by the binder, water mist is used to evenly stabilize the powder layer. One of the main advantages of binder jetting is the faster creation of 3D structures and low material cost. On the other side, poor surface finishing and high machine cost decrease its demand.

3D system's Chefjet printers applied Z-Corp inkjet mechanism for confectionery products including sugar, fondant, sweet, and flavored candy with sculptural structures [22]. SugarLab used sugar and different flavors cinders to create a customized cake for weddings and other special events [27].

18.2.4 INKJET PRINTING

Inkjet printing works with an array of food materials and dispenses drops on demand through inkjet nozzle head (Figure 18.4). The shapes form a digital image, and this technique is used for surface fill and cavity deposition. Inkjet printing mainly handles low viscous food. It is commonly used on flat surfaces or for surface finishing rather than on the complex shape. The temperature during printing is a significant factor, as it can alter the rheological properties of the dough according to the formulation [30].

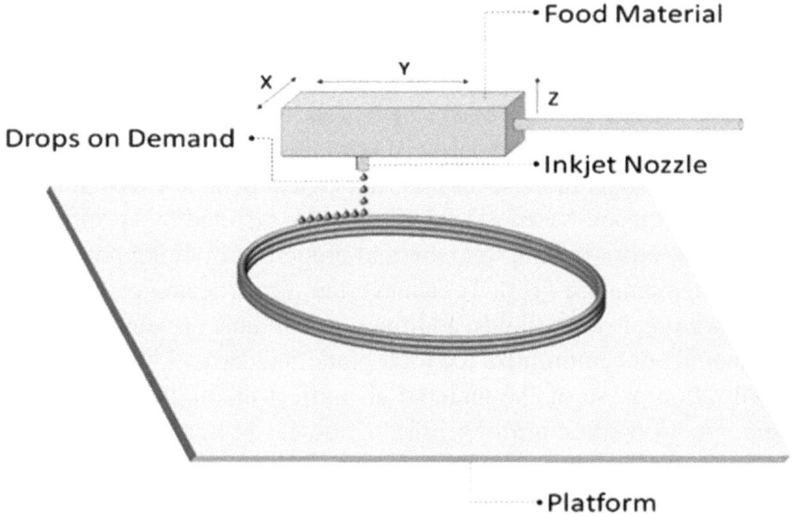

FIGURE 18.4 Schematic diagram of inkjet printing.

Source: Adapted from Ref. [8]

The 3D edible food products, which require pre-patterning at multiple layers during the process can be printed using this technique. Food jet printer printed cupcake, biscuit, and pizza bases using pneumatic membrane jets to deposit the drops in 2012.

Nowadays, these four different 3DFP techniques have been modified for commercial machine designs. The food safety concerns in the food industry limited the usage of the electron beam, laser, and unsafe additives in food printing. Although printing techniques are available, yet there is still a long way to adapt them for commercial usage further.

18.3 FACTORS AFFECTING 3D FOOD PRINTING (3DFP)

The critical challenge faced by 3DFP is to align the food safety issues of the food-grade materials with the printing process. The printing accuracy and precision critically influence the production of an appealing object. The main factors for the quality of printing are:

- Extrusion mechanisms including syringe-based extrusion, air-pressure extrusion, screw-based extrusion;
- Material properties;
- Processing factors;
- Post-printing process.

The material property-related factors are: (i) the capability of the material that has to be extruded is critical criteria before 3D printing; (ii) self-supporting strength of the printed layers of the entire 3D structure determines the applicability of the material in 3D printing; and (iii) printability of the material depends on the viscous and rheological parameters of the material. Moreover, extrusion-based 3D printing can be affected by specific gelation mechanisms (cross-linking) and thermal properties (melting point and glass transition temperature (T_g)). To achieve the ideal rheological property and smooth printing hydrocolloids, additives, and soluble proteins can be added, but it should not compromise the food-grade standards. Although the T_g and crystallization phase of the material also affect the ability of the structure to support itself after printing [28]. These factors can be minimized by modifying the rheological, physicochemical, structural, and mechanical properties of the food material [30].

The effect of printing parameters on the printability of the material varies with the ingredients. In hot melt extrusion, the processing parameters such as nozzle height, size, moving speed, fill density also play a vital role in the printability of the final product. According to Hao et al. when a lower nozzle height was applied than the normal height, the volume of the extruded product exceeded the 3D space between the platform and nozzle. Conversely, an increase in the nozzle height led to incomplete and rough surface finishing of the product [10]. Temperature also has a critical role in the case of hot-melt extrusion to achieve a decent finishing in the final product, especially in chocolate 3D printing. The nozzle diameter has directly impacted the resolution of the final product.

As a thumb rule, the smallest nozzle diameter to allow the extrusion of the material has to be considered, as it gives the most acceptable resolution of the printed product [1]. Moreover, the larger nozzle diameter increases the extrusion rate and printing speed and thus resulted in a rough surface fill product [22]. The other processing factors such as laser type, its power and energy density in SLS techniques have to be fine-tuned to get a quality of fabricated construct. Higher laser density leads to the formation of strong parts with more mechanical stability due to longer reaction time while, the lower density results in a porous and brittle structure [1].

Most of the food consumed in day-to-day life is going through different post-treatment like baking, drying, and frying. The ideal 3D food printed products should be resistant to all post-processing methods. In order to apply at a broader scale, 3D food products should be compatible with traditional post-treatments. Otherwise, they will get distorted after post-treatments. The stability of the printed product is possible in two ways, either with recipe modification or adding additives.

Lipton et al. reported the impact of transglutaminase on the stability of printed lean beef paste. The addition of 0.5% of transglutaminase resulted in the formation of a protein matrix with time. The printed lean beef subjected to deep frying and turkey meat to sous-vide cooking resulted in the successful product after post-treatment [13]. In another research, the ingredient composition of the cookie was modified to get more finished cookies. The study showed that with an increase in butter content, the printability was good, but the stability of the product decreased, Similarly, the increase in yolk concentration increased the shape stability of the product [24].

Pulatsu et al. [20] highlighted the effects of sucrose, butter, and milk level on the extrudability and 3D-printability of the cookie dough and concluded that the formulation with low sugar level yielded structurally stable cookies after baking [20]. The fracturability and the self-supporting capacity of the final product after the post-treatment have to be explored in the future to expand and sustain 3DFP in the food processing sector.

In general, the printability, stability, and quality of the printed product can be gained by modifying and controlling the material properties (such as: rheological, physicochemical, and mechanical properties). In addition, processing parameters, mainly, nozzle diameter, nozzle height, printing speed, and fill density, have to be optimized. The fracturability and the instability of the layers will may increase after post-processing such as baking, steaming, and frying, and thus it is required to either additives in a safe manner or modify the recipes accordingly.

18.4 APPLICATIONS OF 3D PRINTING IN FOOD PROCESSING INDUSTRY

Food printers introduce the artistic touch with fine dining, mass extension and customization of food to the food industry. They add high value and more creativeness that till now was considered impossible. This technology is in the infant stage, and its sole potential is yet to be exploited. The barriers could be overcome when there will be more freedom to be creative with flavors, shapes, and colors for household users. The investigation regarding the potential use of 3DFP and customer satisfaction has to be correlated with the progress in the technology. Following are essential advantages of 3DFP in the food processing sector:

1. **Customized Food Design:** The participation of customers in food designing is complicated but 3DFP permits the customer's aspects in creativity in terms of color, shape, and flavor. The introduction of 3D printing for home users will further generate more design solutions in the field of 3DFP [9].
2. **Food on the Go and Food as Pharma:** 3DFP can be applied in the production of food for astronauts and military personnel and also in hospitals for the patient and in veterinary food for pets [21].
3. **Personalized Nutrition:** The possibility to print nutrient-dense food is an attractive feature of 3DFP. Food could be personalized according to consumer's health, requirements, appetite, and lifestyle [5]. The inclusion of nutrients according to the need of the consumer is possible with 3DFP. Personalized food can control the composition, micro, and macronutrients, shape, size, sensory factors, calories, and the textural properties of the printed product.
4. **Simplify the Production and Supply Chain:** 3DFP reduces many of the food manufacturing processes. An industrialist can

concentrate on food ingredients production rather than focusing on food production [7].

5. **Simplifying the Customized Food Supply Chain:** By utilization of fewer sources, products can reach consumers within a short time [29].
6. **Solution for Food Shortage with a Reduction in Food Waste:** By utilizing the unpreserved vegetables and fruits and also the byproducts such as meat cut off, that normally go to waste, 3DFP technology can effectively reproduce them. The byproducts can be mixed with other materials and made into puree to get printed and baked, resulting in a crunchy and durable food material [18].
7. **Treating Malnutrition by Customizing Food:** One of the major concerns with tackling malnutrition is providing adequate nutrients which can satisfy the day-to-day requirements of the consumers. The advantage of customization of food can easily help the target groups (infant, adult, elderly people, pregnant women and athletes) by controlling the composition with the addition of micro- and macro-nutrients and also controlling the minerals [2]. 3DFP paved the way for new geometrics and thus has led to more acceptability and interest within the children.
8. **Zero-Waste Production Technology:** The contribution to waste from 3DFP is zero. Moreover, it promotes the addition of algae and insects in common food, as they are more nutritious and not easy to consume directly [21].

18.5 CHALLENGES IN 3D PRINTING FOR FOOD PROCESSING SECTOR

3DFP is more complex as we get to know more about the technology. There are a lot of factors that have to be reconsidered and optimized. The material properties, such as rheological properties vary according to ingredients as it is a major factor in the printability of the material. Proper use of mechanical properties and innovations and investigations in the digital recipe has to be done to progress the 3DFP technology. Different food formulations have to be experimented and applied with different pressure and temperature combinations.

The processing parameters, nozzle size, and nozzle height have critical importance in 3DFP. The definition and resolution of this promising technology are controlled by the nozzle size. The nozzle size has to be compromised with the flowability of the material. Sometimes even if the nozzle size is perfect, there may be chances of blockage in the syringe due to the fresh material containing fat, and in some cases, water may form a void at the bottom of the syringe, making it difficult in printing [5]. Modification of traditional recipes and molding them via 3DFP is a new to 3DFP technology. So, while choosing the recipe, the ingredient should have the stability to fit in the 3D space and self-support the entire structure. In addition, consumer acceptability is a significant factor of concern for 3D printed food products. The awareness regarding the technology is low, and thus the investigations related to the introduction of this technology for domestic users are yet to be done [16].

18.6 FUTURE OF 3D FOOD PRINTING (3DFP)

3DFP has been found to benefit food science and food technology as it matures with the food sector. The introduction of 3DFP for home users will take this technology to the next level of creativity. Global food companies took a lot in research on 3DFP, and few companies have already developed 3D food printers for home. Natural machines developed Foodini which can print food from fresh ingredients. With the development of an interactive open web-based user interface, 3D food printers can become a part of our life, like a home appliance to prepare favorite food on demand and can collaborate with health workers for suggesting a healthy diet [19, 31]. 3DFP assures more freedom in composition, texture, and nutrition in the creation of food [31]. Ranging from molecular to microscopic and then macroscopic levels, 3DFP can be explored and enable the design of novel texture. The 3D printer will be a digital food machine concept for the future. In the next few years, 3DFP is expected to develop into a promising technology by providing a new dimension of enjoying food [17].

18.7 SUMMARY

The 3DFP sector research is growing day by day and poise a massive impact in the future. However, the optimization of the product and process parameters has not yet been conducted. Potential safety issues and consumer acceptance have been considered in widescreen, but the research in these areas is limited. 3DFP is a promising technology for the upcoming era and also for the food industry. The major challenges in the process of productivity and product innovation, and functionality have to be addressed.

KEYWORDS

- **3D food printing**
- **additive manufacturing**
- **computer-aided design**
- **fused deposition modeling**
- **inkjet printing**
- **selective laser sintering**
- **solid freeform fabrication**

REFERENCES

1. Amorim, F. L., Lohrengel, A., Neubert, V., Higa, C. F., & Czelusniak, T., (2014). Selective laser sintering of mo-cuni composite to be used as EDM electrode. *Rapid Prototyping Journal, 20*(1), 59–68.
2. Azam, R. S. M., Zhang, M., Bhandari, B., & Yang, C., (2018). Effect of different gums on features of 3-D printed object based on vitamin-d enriched orange concentrate. *Food Biophysics, 13,* 250–262.
3. Cohen, D. L., Lipton, J. I., Cutler, M., Coulter, D., Vesco, A., & Lipson, H., (2009). *Hydrocolloid Printing: A Novel Platform for Customized Food Production* (pp. 1–5). Unpublished Paper Presentation at 20th Annual International Solid Freeform Fabrication (SFF) Symposium (Austin – Texas).
4. Crump, S. S., (1992). *Apparatus and Methods for Creating 3-Dimensional Objects* (p. 17). Google Patents.
5. Derossi, A., Caporizzi, R., Ricci, I., & Severini, C., (2019). Critical variables in 3-D food printing. *Fundamentals of 3-D Food Printing and Applications, 3,* 41–91.
6. Fina, F., Goyanes, A., Gaisford, S., & Basit, A. W., (2017). Selective laser sintering (SLS): 3-D printing of medicines. *International Journal of Pharmaceutics, 529*(1, 2), 285–293.
7. Gholamipour-Shirazi, A., Kamlow, M. A., Norton, I. T., & Mills, T., (2020). How to formulate for structure and texture via medium of additive manufacturing: A review. *Journal of Foods, 9*(4), 1–20.
8. Godoi, F. C., Prakash, S., & Bhandari, B. R., (2016). The 3-D printing technologies applied for food design: Status and prospects. *Journal of Food Engineering, 179,* 44–54.
9. Hao, L., Mellor, S., Seaman, O., Henderson, J., Sewell, N., & Sloan, M., (2010). Material characterization and process development for chocolate additive layer manufacturing. *Innovative Developments in Design and Manufacturing – Advanced Research in Virtual and Physical Prototyping, 5*(2), 57–64.
10. Hao, L., Seaman, O., Mellor, S., Henderson, J., Sewell, N., & Sloan, M., (2010). Extrusion behavior of chocolate for additive layer manufacturing. *Innovative*

Developments in Design and Manufacturing–Advanced Research in Virtual and Rapid Prototyping, 5(12), 53–73.
11. Hop, L., & Kurman, M., (2013). *Fabricated New World of 3-D Printing* (1st edn., p. 302). New York: John Wiley & Sons Inc.
12. Le-Bail, A., Maniglia, B. C., & Le-Bail, P., (2020). Recent advances and future perspective in additive manufacturing of foods based on 3-D printing. *Current Opinion in Food Science, 35*, 54–64.
13. Lipton, J., Arnold, D., Nigl, F., Lopez, N., & Cohen, D., (2010). *Multi-Material Food Printing with Complex Internal Structure Suitable for Conventional Post-Processing* (pp. 1–6). Unpublished Paper Presentation at 21st Annual International Solid Free-form Fabrication Symposium–An Additive Manufacturing Conference, SFF 2010 (Austin–Texas).
14. Lipton, J. I., Cutler, M., Nigl, F., Cohen, D., & Lipson, H., (2015). Additive manufacturing for the food industry. *Trends in Food Science and Technology, 43*(1), 114–123.
15. Liu, Z., Zhang, M., Bhandari, B., & Wang, Y., (2017). The 3-D printing: Printing precision and application in food sector. *Trends in Food Science and Technology, 69,* 83–94.
16. Lupton, D., (2017). Download to delicious: Promissory themes and sociotechnical imaginaries in coverage of 3-D printed food in online news sources. *Futures, 93,* 44–53.
17. Noort, M., Van, B., K., & Renzetti, S., (2017). 3-D-printed cereal foods. *Cereal Foods World, 62*(6), 272–277.
18. Peppel, C. V. D., (2015). *3-D Food Printing* (p. 143). MSc Thesis, Wageningen University, Netherlands.
19. Periard, D., Schaal, N., Schaal, M., Malone, E., & Lipson, H., (2007). Printing of foods. In: *18th Solid Freeform Fabrication Symposium, SFF-2007* (Vol. 8, pp. 12–18). Austin–TX.
20. Pulatsu, E., Su, J. W., Lin, J., & Lin, M., (2020). Factors affecting 3-D printing and post-processing capacity of cookie dough. *Innovative Food Science and Emerging Technologies, 61*, 102–314.
21. Severini, C., Derossi, A., & Azzollini, D., (2016). Variables affecting the printability of foods: Preliminary tests on cereal-based products. *Innovative Food Science and Emerging Technologies, 64*(13), 112–119.
22. Shirazi, S. F. S., Gharehkhani, S., Mehrali, M., Yarmand, H., Metselaar, H. S. C., Adib, K. N., & Osman, N. A. A., (2015). A review on powder-based additive manufacturing for tissue engineering: Selective laser sintering and inkjet 3-D printing. *Science and Technology of Advanced Materials, 32*(5), 123–142.
23. Sun, J., Zhou, W., Huang, D., Fuh, J. Y. H., & Hong, G. S., (2015). An overview of 3-D printing technologies for food fabrication. *Food and Bioprocess Technology, 8*(8), 1605–1615.
24. Varghese, C., Wolodko, J., Chen, L., Doschak, M., Srivastav, P. P., & Mohandas, R., (2020). Influence of selected product and process parameters on microstructure, rheological, and textural properties of 3-D printed cookies. *Foods, 9*(7), 907.
25. Von, H. K. W., Von, H. E. M., Williams, D. X., & Gale, R. R., (2014). Making an edible component comprises depositing successive layers of food material according to digital data that describes the edible component, and applying edible binders to regions of the successive layers of the food material. *3-D Systems Inc. (Thde-C), 12*, 113–130.
26. Yang, F., Zhang, M., & Bhandari, B., (2017). Recent development in 3-D food printing. *Critical Reviews in Food Science and Nutrition, 57*(14), 3145–3153.

27. Yang, J., Wu, L. W., & Liu, J., (2001). *U.S., Patent No. 6,280,785* (pp. 1–15). Washington, DC: U.S. Patent and Trademark Office.
28. Zhang, M., Vora, A., Han, W., Wojtecki, R. J., Maune, H., Le, A. B. A., Thompson, L. E., et al., (2015). Dual-responsive hydrogels for direct-write 3-D printing. *Macromolecules, 65*(2), 123–156.
29. Zhang, L., Lou, Y., & Schutyser, M. A. I., (2018). 3-D printing of cereal-based food structures containing probiotics. *Food Structure, 17*(4), 23–35.
30. Ziaee, M., & Crane, N. B., (2019). Binder jetting: A review of process, materials, and methods. *Additive Manufacturing, 62*(2), 132–145.
31. Zoran, A., & Coelho, M., (2011). Cornucopia: The concept of digital gastronomy. *Leonardo, 5*, 13–19.

CHAPTER 19

POTENTIAL OF ENZYME-LINKED IMMUNOSORBENT ASSAY (ELISA) FOR FOOD AUTHENTICATION

SADAF NAZIR, SYED AMIR ASHRAF, INSHA ZAHOOR,
Z. R. AZAZ AHMAD AZAD, and FARHANA MEHRAJ ALLAI

ABSTRACT

Enzyme-linked immunosorbent assay (ELISA) is the preferred method for qualitative and quantitative detection of food contaminants, due to its high sensitivity to provide the desired portability, fast response, sensitivity, and specificity. The basic principle of ELISA involves molecular recognition between antigen and antibody, which is coupled to an enzyme assay. Antibody recognizes and binds to the selective target molecule. And the conjugated enzyme assay generates fluorescence colorless substrate, which is later detected as a signal and quantified. The entire process involves immobilization of an antigen to a surface, which is complexed with the antibody linked to an enzyme. Direct/indirect, sandwich, and competitive are different variants of ELISA to detect food contaminants. ELISA can be used to analyze the presence of infectious agents (such as *E. coli, Salmonella*, and *Staphylococcus*). The lowest level of detection of contaminants can be as low as one ppm for ELISA. The only demerit for ELISA is the sample preparation step, which extends its total analysis time. To validate the claim of ELISA as an alternative analysis tool for food safety, comparative studies need to be conducted.

Advanced Research Methods in Food Processing Technologies. Junaid Ahmad Malik, Megh R. Goyal, Preeti Birwal, & Ritesh B. Watharkar (Eds.)
© 2024 Apple Academic Press, Inc. Co-published with CRC Press (Taylor & Francis)

19.1 INTRODUCTION

In recent years, a serious concern emerged for food borne illnesses and food allergies due to the presence of chemical components, process components, and microbial toxins in food. The presence of such components can lead to various intolerances in the human body. And the intolerances are toxin and allergen specific. Therefore, these unwarranted components in food must be monitored to confirm the safety of food. Various conventional techniques used for food authentication include [1]: high performance liquid chromatography (HPLC), mass spectrometry (MS), reverse transcription-polymerase chain reaction (RT-PCR), and microbiological techniques (Figure 19.1). Such techniques are accurate and sensitive with a bottleneck of being selective and time consuming. Therefore, there is a need for a rapid, simple, and efficient approach to screen and to analyze food for contaminants.

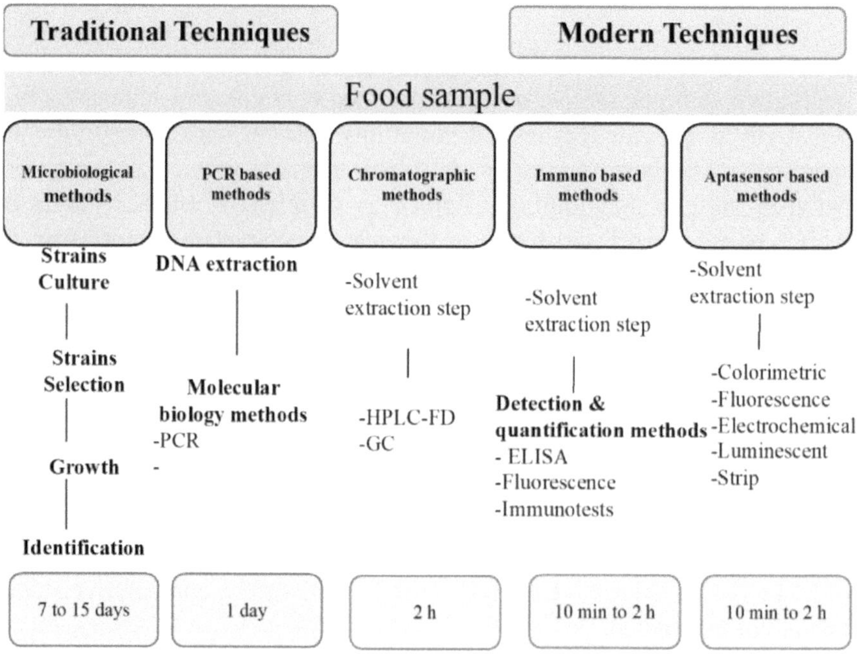

FIGURE 19.1 Traditional and modern techniques for food authentication.

The immunoanalytical techniques have emerged within the field of food safety. Immunoanalytical techniques are rapid, sensitive, convenient to implement, and have a high throughput. These techniques are based on biochemical reactions and employ binding of globular proteins (antibodies). Immunoassays involve the use of antibodies, which recognize the chemical structure of a target molecule. Initially immunoassays involved radioisotope markers which posed many health risks [3]. Later, enzyme-linked immunosorbent assay (ELISA) was described by Engvall & Perlmann [3]. It is the most widely preferred immunoassay due to its rapid and robust screening. ELISA is more selective and suitable for detecting and quantifying all types of analytes from low molecular weight (< 1,000 Da) to complex substances in the food [22].

On comparison, with other traditional techniques, ELISA several merits, such as: small sample volume requirement, quantitative analysis, simplicity of operation in sample processing, and fewer sample extraction steps. Antibodies used for ELISA are immunoglobulins, which are obtained from any vertebrate. Both monoclonal and polyclonal antibodies are used for identification of food components. However, polyclonal antibodies have wide suitability and are tolerant to a number of changes. The analyte (contaminant) binding of antibody in ELISA may be competitive or non-competitive. Competitive ELISA employs competition between analyte and an analogous molecule, for the binding sites of the antibody, which is added in limited quantity. Alternatively, non-competitive ELISA involves the measurement of bound sites [7].

The commercial applications of ELISA include the detection of food adulterants, pesticide residues, species adulteration of meat and milk products. It is also an essential tool to identify quality of fish and sea foods.

This chapter focuses on principal, mechanism, different variants, action, and reactions and validation process of ELISA.

19.2 PRINCIPLE OF ELISA: MOLECULAR RECOGNITION AND BIOSENSING

The principle of ELISA is based on molecular recognition and biosensing. It involves an interaction between different antigen and antibody at specific sites through noncovalent bonding (halogen bonding, hydrogen bonding, metal coordination, van der Waals forces, hydrophobic forces, π-π interactions, and electrostatic or electromagnetic effects). The antigen and antibody should exhibit a molecular complementarity (Figure 19.2). Biosensing is based on molecular recognition. A specific molecular recognition probe targets an analyte of interest which is converted into a measurable signal. With the advent of research in molecular recognition probes and biosensing, the technique of ELISA has evolved as an alternative biosensor [24].

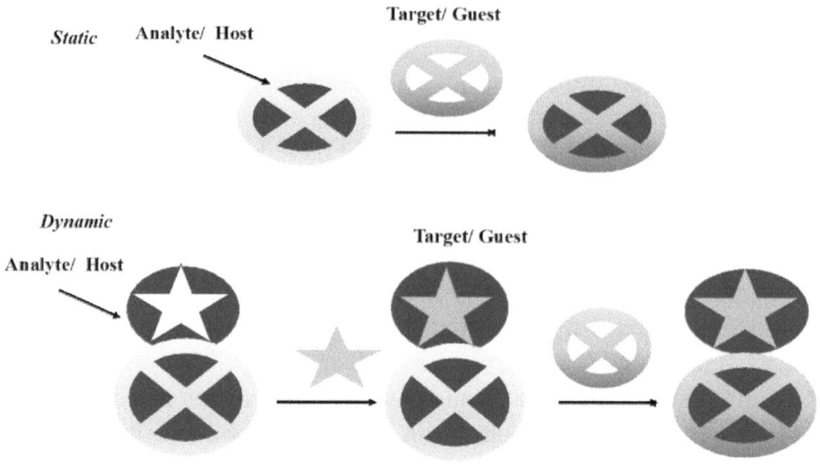

FIGURE 19.2 Interaction between target and analyte.

The principle of ELISA is just like other immunoassays. The ELISA employs binding of antibodies to target antigen that leads to the detection and quantification of antigen. Antibodies used in ELISA are immunoglobulins (IgG). IgG are easy to extract, purify, and conjugate with enzymes. The antibodies are target specific and have high affinity for target antigens. For the maximum sensitivity and precision, the substrate must be coated with antibodies of high affinity. ELISA substrates are colorimetric, and their selection depends on the enzyme used in labeling at the preceding step. The enzyme label is the signal converter and forms bioconjugates. The detection signal in the assay is achieved by using a chromogenic reagent.

19.3 MECHANISM OF ELISA

Various variations of ELISA have been used for authenticating food samples. All consist of a basic mechanism. A reaction between antibody and antigen occurs as described in the following steps in subsections (Figure 19.3).

19.3.1 COATING/CAPTURE

An ELISA kit consists of a specialized 96-well polystyrene plates, a multi-channel precision pipette with disposable plastic tips. A plate reader (luminometer) to detect the target substance. The target substance (sample) and standards are prepared in standard diluents, i.e., 2% Bovine serum albumin (BSA) dissolved in wash buffer. The process requires removal of excess liquid to prevent the dilution of solution added at each step [3]. The process is initiated by coating a layer of target (antigen or antibody). Polystyrene binds to a wide variety of proteins and further binding depends on the concentration of coating solution. Coating formulation is prepared by placing capture antibody in coating buffer, i.e., 0.2 M sodium bicarbonate or sodium carbonate (pH 9.4). Alkaline conditions initiate an binding of antigen to the surface of microplate wells. The antigens/antibody is coated on sorbent substrate by physical adsorption effect. The binding takes place by passive adsorption or covalent linkage. Later, specific recognition causes antigen/antibody immobilization on polystyrene microplate wells. Coating depends on the diffusion coefficient of attaching molecule, concentration of antigen being absorbed, temperature, time of adsorption and ratio of surface area coated to volume of coating solution.

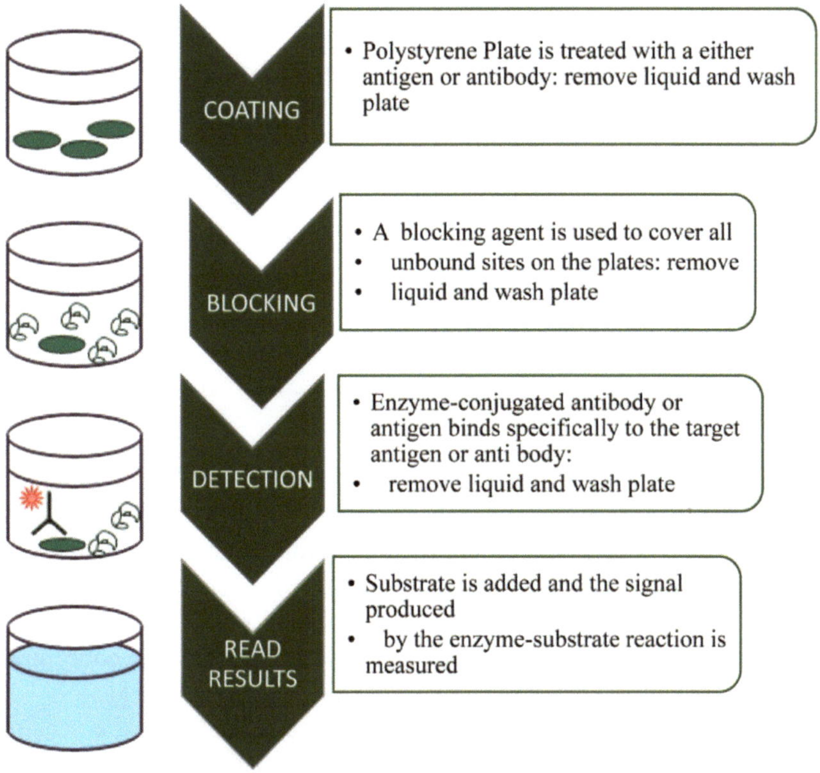

FIGURE 19.3 Mechanism of ELISA.

19.3.2 PLATE BLOCKING

The addition of a molecule such as protein to envelop the unsaturated surface-binding sites of the microplate wells is called plate blocking. The surface of the sorbent surface is blocked to avoid any non-specific adsorption and enzyme labeled antibody added.

A blocking agent (2% w/v BSA in wash buffer) is coated onto the unbound sites. A series of washes is carried out after blocking. And incubation carried out with enzyme-conjugated antibody. And unbound antibody is removed by a series of washes. Wash buffer is used for the series of washes. Upon substrate addition, desired color intensity is produced. The reaction is stopped by adding a small amount of stop solution (2 M sulfuric acid).

19.3.3 PROBING/DETECTION

Specific antibodies for antigens are incubated for the binding process to take place. Later, the enzyme-labeled antibody is incubated with antigen to form bioconjugates. The bioconjugates catalyze the chromogenic reaction. Washes are followed at each particular step to remove the loosely bound material and excess liquid. The presence of excess proteins may lead to false positives, so the substrate must be washed with washing buffer after coating and incubation, respectively.

A calorimetric signal is produced, and the plate is read. The detection of signal is a quite complex step in the overall ELISA process. The process gets more complicated if numerous layers of antibodies are cast to amplify a signal.

19.3.4 SIGNAL MEASUREMENT

A signal is initiated by the tag on the specific antibody and color variations are induced by chromogenic reagents. The color intensity is quantitively measured by a microplate reader. The signal for standard is measured and a standard sigmoidal curve (from 0 to 1,000 pg/ml) is plotted between optical density (or fluorescence) and concentration, This range may vary with the amount of standard protein and antigen to be quantified. The concentration of unknown samples is then estimated from the standard plot [13].

Novel techniques that can be incorporated to enhance the sensitivity of ELISA include the use of fluorescent signals, electrochemical signals, chemiluminescent signals, and plasmonic signals [28].

19.4 VARIANTS OF ELISA

ELISA is carried out differently for different food samples with just a few modifications in the basic procedure. The assays of ELISA involve direct, indirect, sandwich or competitive antigen-antibody binding (Figure 19.4). The direct ELISA is achieved by adsorption to the assay plate. Indirect ELISA is accomplished by adhering a capture antibody to the plate Detection is direct (via enzyme-labeled primary antibody) or indirect (via enzyme-labeled secondary antibody) (Table 19.1).

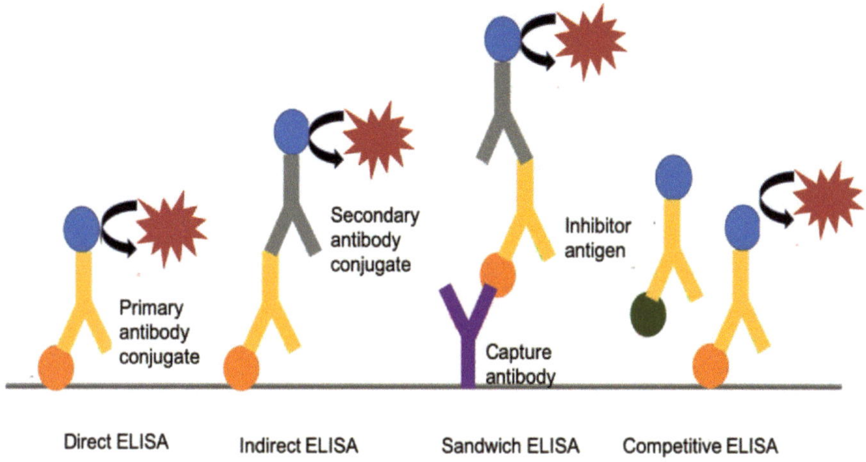

FIGURE 19.4 Variants of ELISA.

TABLE 19.1 Steps in Various ELISA Types

Mechanism	Indirect	Direct	Sandwich	Competitive
Antibody coating	Absent	Absent	Present	Absent
Antigen coating	Present	Present	Absent	Present
Blocking	Present	Present	Present	Present
Antigen incubation	Absent	Absent	Present	Present
Primary antibody incubation	Present	Present	Present	Present
Secondary antibody incubation	Present	Absent	Present	Present
Substrate preparation	Present	Present	Present	Present
Detection	Present	Present	Present	Present
Analysis	Present	Present	Present	Present

Sandwich ELISA utilizes a primary antibody to capture antigens which is immobilized to the substrate. Another enzyme-labeled antibody is used to produce a signal after identifying the primary antibody Competitive ELISA is involved when there is a deficiency of binding positions for antibodies. Recently, the introduction of nanoparticles (NPs) has significantly improved the performance of ELISA. Nanofibers and nanobeads have been used to improve adsorbent substrate. Nanocomposites (CuS, CeO_2, graphene oxide) are used as enzyme labels. In addition, gold and silver (Ag) particles are incorporated as chromogenic reagents [27].

19.5 APPLICATIONS OF ELISA FOR FOOD AUTHENTICATION

The applications of ELISA involve authentication and analysis of food. ELISA is used to detect various allergens, antinutritional factors, microorganisms, microbial toxins, and pesticide residues. It is also used to authenticate the quality of meat, fish, milk, and irradiated food products.

19.5.1 DETECTION OF ALLERGENS

Allergens are a group of toxic compounds present in various food products which can lead to various health issues. ELISA is a successful tool for the estimation of allergens, such as gliadin in wheat flour, lupine allergens, egg allergens, orange allergens, apple allergens, peanut allergens, glycoalkaloids, and saponins.

Coeliac patients present sensitivity towards gluten, on consumption of wheat flour products. Gliadin component of gluten is majorly responsible for this type of allergy. Consequently, ELISA has proved successful in detecting the adulteration of gluten free products. Recently, single antibody fragments have been isolated and assessed for use in indirect phage ELISA to detect gluten in foods [4]. In addition, lupine allergens from various legumes have been analyzed [12]. Immunoreactivity of peanut allergen (Ara-h-1) has been estimated using ELISA [26]. Egg allergens from processed foods (baked cookies) have been evaluated as they trigger allergenicity in certain sensitive individuals [20]. An optimized ELISA was developed to quantify major allergens in orange (Cit-s-2) and apple (Mal-d1) [11, 14]. ELISA has also shown positive results in detection of antinutritional factors such as glycoalkaloids in fresh tubers and saponins in yams [25].

19.5.2 DETECTION OF MICROBES AND MICROBIAL TOXINS

ELISA is able to detect both viable and non-viable microorganisms and is more sensitive for fungal toxins. Microorganisms such as *Erwinia carotovora, Psuedomonas syringae* have been determined by immunoassays for potato tubers and peas, respectively [25]. Mycotoxins are the fungal toxins present in food and can pose dangerous effects to health. ELISA can be used to quantify toxins in various food products at low concentrations of parts per million (ppm) [2]. Enterotoxins from various bacteria such as *Staphylococcus aureus* and *Salmonella* are hazardous and can be estimated using ELISA. Also, botulin a toxin from *Clostridium botulinum* has been detected and estimated using ELISA [16].

Marine dinoflagellates produce toxins, which pose a threat to humans by contamination from several marine food sources. Various antibodies that act against neurotoxins, such as brevetoxin can be used in an ELISA to detect the toxins at low levels [6]. Studies report detection of Listeria monocytogenes at an acceptable sensitivity of 3.7×10^6 CFU/mL by using an AuNP-based ICG strip assay (nano-ELISA) The detection limits of 0.04 tp 1.05 μg/kg has been reported [7]. Recently, aflatoxigenic variability among *Aspergillus* species has been determined using ELISA [19].

19.5.3 AUTHENTICATION OF MILK AND MILK PRODUCTS

The technique is widely used for the authentication of milk and its products. Nowadays species adulteration of milk either bovine or camel is adulterated with goat milk. And ELISA can successfully detect low percentages of adulterant [8]. Various growth promoters have been used to promote growth in milk producing animals have been successfully investigated in milk. Clenbuterol is widely used as in both humans and animals and is sometimes misused as a growth promoter for food-producing animals, which decreases fat accumulation and increases muscular mass The maximum residue limits set by the Food and Agriculture Organization (FAO) are 0.05 μg/l in milk. The accumulation of clenbuterol (an illegal growth promoter) in milk was screened using a micro-fluidic fiber-based ELISA [17]. A sandwich-based model with an antibody similar to IgE was developed to detect an allergen, bovine β-lactoglobulin (BLG) in cow's milk [5].

19.5.4 AUTHENTICATION OF MEAT AND MEAT PRODUCTS

Meat is also prone to species adulteration by beef or pork. ELISA can detect the low percentages of adulterant and authenticate labeled products present in the market [15]. Amantadine is a banned drug used to treat influenza in poultry. Residues of amantadine in poultry were investigated using colorimetric ELISA. Such ELISA was based on hybridization reaction and aggregation of non-cross-linking gold NPs [30]. Mammalian meat adulteration can also be detected by ELISA. Recently, monoclonal antibody-based sandwich ELISA was used to authenticate mammalian meat [9]. IgG polyclonal antibody-based ELISA is also used to authenticate the adulteration of meat using pork [18]. Indirect ELISA (polyclonal antibody) has applications in differentiating between porcine and bovine gelatin in mixtures [23]. IgG polyclonal antibody-based ELISA can detect and distinguish adulteration at 0.1%. The use of swine specific DNA as target and two complementary DNA sequences as probes shows as a limit of detection (LOD) of 0.1 μm. Such ELISA can be used for the identification of pork [10, 27].

19.5.5 AUTHENTICATION OF FISH AND FISH PRODUCTS

Lower -value fish is often used as an adulterant in fish products made from high-value fish. Monoclonal and polyclonal antibodies are used to detect such type of adulteration. Reports suggest detection of okadaic acid, saxitoxin, and brevotoxin with the help of ELISA [24]. It is reported that ELISA can successfully authenticate the gene of *Katsuwonus pelamis* and *Thunnus* species in food via PCR-coupled ELISA [1]. A monoclonal antibody-based ELISA can analyze the presence of okadaic acid in fish with a LOD of 10 ng/ml. A competitive ELISA for the detection of brevetoxin-2 is reported [1, 25].

19.5.6 AUTHENTICATION OF IRRADIATED FOODS

Irradiation is done on various food products to minimize the number of spoilage microorganisms present in food. But higher doses of irradiation can be hazardous. So recently guidelines have appeared that radiated food must be properly labeled. ELISA is an efficient tool to detect the dose of irradiation applied to food [21].

The irradiation of food is based on radiolysis of lipids to form 2-dodecylcyclobutanone (2-DCB), and 2-tetradecylcyclobutanone (2-TCB). Various methods are used to authenticate irradiated foods, but none of the methods is ELISA based. However, an ELISA or nano-ELISA based on 2-DCB, and monoclonal antibodies can be used to detect irradiated food. A biosensor (immuno-rotary) based on 2-DCB and ε-subunit monoclonal antibodies which links through biotin-avidin-biotin linker is reported [1].

19.5.7 AUTHENTICATION OF GENETICALLY MODIFIED FOODS (GMF)

ELISA is also used to identify a particular genetically modified food (GMF). The GMF may be raw material, semi-processed foods, and processed ingredients. However, ELISA has low recognition ability than PCR and is less authentic for testing GMFs [16]. Recently, a silicon coated magnetic NPs probe based on carbon nanospheres was developed. The developed ELISA was used for Cry1Ac protein detection.

Also, graphene oxide/gold nano-urchins have been developed for authentication of transgenic crops. In addition, a label-free electroluminescence immunosensor was used for the analysis of Cry1Ab level in GMF. The sensor showed a linear range of 0.010 to 1.0 ng/mL and LOD of 3.0 pg/mL for Cry1Ab [1].

19.5.8 DETECTION OF PESTICIDE RESIDUES

Assessment of crops for pesticide residue is a major concern and various chromatographic techniques have been used. Recently immunoassay like ELISA has proved a time saving technique for its determination [25]. ELISA has recently detected acetamiprid, a neonicotinoid pesticide in tea leaves, triazophos, paraquat, diflubenzuron, and triadimefon, parathion in food [1, 25]. ELISA detection of various pesticides such as acetamiprid, azoxystrobin, chlorothalonil, fenitrothion, and imidacloprid has also been validated [29].

19.6 ADVANTAGES OF USING ELISA FOR FOOD SAFETY

- Adequately detects contaminants for analysis from real samples;
- It has high selectivity range;
- Portable devices and on-site testing;

- Reproducibility (error: 5–10%);
- Storage stability-satisfactory;
- Time required for each analysis is 15–45 minutes.

19.7 FUTURE TRENDS

To develop multi-analyte sensors and arrays that rely on immobilizing the sensor surface and bind a particular target with various affinities: Amplification of obtained signals can be improved to attain high sensitivity in detection.

19.8 SUMMARY

ELISA is a sensitive, cost-efficient, and a portable analysis tool; but involves the single of the kit for detection of contaminants. It is nowadays most common method for food authentication and traceability. It is used to detect the species adulteration of milk, meat, and their products. It is able to detect mycotoxins in trace amounts. In addition, various allergens such as gliadin and lysozyme can be detected using ELISA. The technique is of much importance in avoiding economic losses in fraudulent substitutions.

KEYWORDS

- **antibody**
- **antigen**
- **authentication**
- **biosensors**
- **contaminants**
- **immunoassays**
- **molecular recognition**

REFERENCES

1. Cespedes, A., Garcıa, T., Carrera, E., Gonzalez, I., Fernandez, A., & Asensio, L., (1999). Indirect ELISA for the identification of sole (*Solea solea*), European plaice (*Pleuronectes

platessa), flounder (*Platichthys flesus*), and Greenland halibut (*Reinhardtius hippoglossoides*). *Journal of Food Protection, 62,* 1178–1182.

2. Dewey, F. M., (1988). Development of immunological diagnostic assays for fungal plant pathogens. In: *Proc. Brighton Crop Protection Conference Pests and Disease* (Vol. 2, pp. 776–781).
3. Engvall, E., & Perlmann, P., (1971). Enzyme linked immunosorbent assay (ELISA) quantitative assay of immunoglobulin G. *Immunochemistry, 8*, 871–875.
4. García-García, A., Madrid, R., González, I., García, T., & Martín, R., (2020). A novel approach to produce phage single domain antibody fragments for the detection of gluten in foods. *Food Chemistry, 321*, E-article 126685.
5. He, S., Li, X., Gao, J., Tong, P., & Chen, H., (2017). Development of sandwich ELISA for testing bovine β-lactoglobulin allergenic residues by specific polyclonal antibody against human IgE binding epitopes. *Food Chemistry, 227*, 33–40.
6. Hokama, Y., Abad, M. A., & Kimura, L. H., (1983). A rapid enzyme immunoassay for the detection of ciguatoxin in contaminated fish tissues. *Toxicon, 21*, 81–88.
7. Immer, U., & Lacorn, M., (2015). Enzyme-linked immunosorbent assay (ELISA) for detecting allergens in food. In: Flanagan, S., (ed.), *Handbook of Food Allergen Detection and Control* (pp. 199–217). Cambridge: Woodhead Publishing.
8. Ivens, K. O., Baumert, J. L., & Taylor, S. L., (2016). Commercial milk enzyme-linked immunosorbent assay (ELISA) kit reactivities to purified milk proteins and milk-derived ingredients. *Journal of Food Science, 81*, T1871–T1878.
9. Jiang, X., Rao, Q., Mittl, K., & Hsieh, Y. H. P., (2020). Monoclonal antibody-based sandwich ELISA for the detection of mammalian meats. *Food Control, 110*, E-article 107045.
10. Kich, J. D., Schwarz, P., Eduardo, S. L., Coldebella, A., Piffer, I. A., Vizzoto, R., & Cardoso, M. R. I., (2007). Development and application of an enzyme-linked immunosorbent assay to detect antibodies against prevalent *Salmonella serovars* in swine in southern Brazil. *Journal of Veterinary Diagnostic Investigation, 19*, 510–517.
11. Kiyota, K., Kawatsu, K., Sakata, J., Yoshimitsu, M., Akutsu, K., Satsuki-Murakami, T., Ki, M., et al., (2017). Development of monoclonal antibody-based ELISA for the quantification of orange allergen cit-s-2 in fresh and processed oranges. *Food Chemistry, 232,* 43–48.
12. Koeberl, M., Sharp, M. F., Tian, R., Buddhadasa, S., Clarke, D., & Roberts, J., (2018). Lupine allergen detecting capability and cross-reactivity of related legumes by ELISA. *Food Chemistry, 256*, 105–112.
13. Kohler, C., & Milstein, C., (1975). Continuous cultures of fused cells secreting antibody of predefined specificity. *Nature, 256,* 495–497.
14. Kschonsek, T. J., Dietz, A., Wiegand, C., Hipler, U., & Böhm, V., (2019). Allergenicity of apple allergen *Mal-d*-1 as effected by polyphenols and polyphenol oxidase due to enzymatic browning. *LWT – Food Science and Technology, 113*, E-article 108289.
15. Liu, L. H., Chen, F. C., Dorsey, J. L., & Hsieh, Y. H. P., (2006). Sensitive monoclonal antibody-based sandwich ELISA for the detection of porcine skeletal muscle in meat and feed products. *Journal of Food Science, 71*, M1–M6.
16. Luthy, J., (1999). Detection strategies for food authenticity and genetically modified foods. *Food Control, 10*, 359–361.
17. Ma, L., Nilghaz, A., Choi, J. R., Liu, X., & Lu, X., (2018). Rapid detection of clenbuterol in milk using microfluidic paper-based ELISA. *Food Chemistry, 246,* 437–441.

18. Mandli, J., El Fatimi, I., Seddaoui, N., & Amine, A., (2018). Enzyme immunoassay (ELISA) for a sensitive detection of pork adulteration in meat. *Food Chemistry, 255*, 380–389.
19. Micah, L. K., Toroitich, F. J., & Obonyo, M. A., (2020). Development of an ELISA-based method for testing aflatoxigenicity and aflatoxigenic variability among *Aspergillus* species in culture. *Scientific African, 7*, E-article 00266.
20. Nguyen, A. V., Williams, K. M., Ferguson, M., Lee, D., Sharma, G. M., Do, A. B., & Khuda, S. E., (2019). Enhanced quantitation of egg allergen in foods using incurred standards and antibodies against processed egg in a model ELISA. *Analytica Chimica Acta, 1081*, 157–167.
21. Olson, D. G., (1998). Irradiation of food. *Food Technology, 52*, 56–61.
22. Poms, R. E., Klein, C. L., & Anklam, E., (2004). Methods for allergen analysis in food: A review. *Food Additives and Contaminants: Part A, 21*(1):1–3.
23. Rohman, A., Windarsih, A., Erwanto, Y., & Zakaria, Z., (2020). Review on analytical methods for analysis of porcine gelatin in food and pharmaceutical products for halal authentication. *Trends in Food Science and Technology, 101*, 122–132.
24. Rotariu, L., Lagarde, F., Jaffrezic-Renault, N., & Bala, C., (2016). Electrochemical biosensors for fast detection of food contaminants: Trends and perspective. *Trends in Analytical Chemistry, 79*, 80–87.
25. Samarajeewa, U., Wei, C. I., Huang, T. S., & Marshall, M. R., (2009). Application of immunoassay in the food industry. *Critical Reviews in Food Science and Technology, 29*, 403–434.
26. Tian, Y., Liu, C., Zhang, K., Tao, S., & Xue, W., (2020). Glycosylation between recombinant peanut protein Ara-h-1 and glucosamine-T could decrease the allergenicity due to the protein aggregation. *LWT – Food Science and Technology, 127*, E-article 109374.
27. Wu, L., Li, G., Xu, X., Zhu, L., Huang, R., & Chen, X., (2019). Application of nano-ELISA in food analysis: Recent advances and challenges. *Trends in Analytical Chemistry, 113*, 140–156.
28. Xiong, Y., Leng, Y., Li, X., Huang, X., & Xiong, Y., (2020). Emerging strategies to enhance the sensitivity of competitive ELISA for detection of chemical contaminants in food samples. *Trends in Analytical Chemistry, 126*, E-article 115861.
29. Yamasaki, T., Inoue, T., Hirakawa, Y., Miyake, S., Ueno, E., & Saito, I., (2015). Validation of ELISA kits for pesticide residue analysis in vegetables and fruits. *Food Hygiene and Safety Science (Shokuhin Eiseigaku Zasshi), 56*, 240–246.
30. Zhu, F. F., Peng, J., Huang, Z., Hu, L. M., Zhang, G. G., Liu, D. F., Xing, K. Y., et al., (2018). Specific colorimetric ELISA method based on DNA hybridization reaction and non-crosslinking gold nanoparticles aggregation for the detection of amantadine. *Food Chemistry, 257*, 382–387.

CHAPTER 20

APPLICATIONS OF NANOTECHNOLOGY IN FOOD PROCESSING

SOUMYA RATHORE and ANAND KUMAR PANDEY

ABSTRACT

Nanotechnology (NT) poses a huge impact on foods by evolving the conventional technologies involved in food science and food processing. NT has overturned the food sector over the past few decades and has contributed several benefits like increased shelf-life, prevention from contamination and spoilage, improves quality of food products, etc. The emerging applications of NT in the area of food science and technology are the use of nanoparticles (NPs) in various operations like processing, packaging, development of functional foods, enhancement in food quality, improvement in food safety, detection in food-borne pathogens and shelf-life enhancement of food products. Another recent innovation of NT in food science is food nano-sensing. Nano-sensing provides a better evaluation of food safety and quality parameters, resulting in the production of safe food and thus boost human health and well-being. Rapid development of NT is expected to transform many areas of food science and industry with increasing investment and market share by providing effective new methods to improve safety, textural, and sensory properties, and the nutritional value of food products.

20.1 INTRODUCTION

Nanotechnology (NT) can be elaborated as the production and designing technology of structures and their systems, which have materials of shape and

Advanced Research Methods in Food Processing Technologies. Junaid Ahmad Malik, Megh R. Goyal, Preeti Birwal, & Ritesh B. Watharkar (Eds.)
© 2024 Apple Academic Press, Inc. Co-published with CRC Press (Taylor & Francis)

size measured at nanoscale (i.e., 10^{-9} of a meter), where unique phenomena take place and thus enable applications, which cannot be made feasible by macro-materials [2]. The optical, magnetic, electronic, and quantum properties of materials vary widely at this scale and provide exclusive characteristics which are totally different from that of the bulk materials. The application of this technology covers the vast areas of different branches of science and technology, the latest one is the food sector. NT in food is an emerging technology which was unexplored in past decades, but it has gained recognition by the research community in the 21st century to lead the food industries as compared to pharmaceuticals and drug deliveries technologies [29].

NT has potential in food science diligently from food processing to preservation, packaging to food safety, curbing food wastage to employ nano-delivery systems, the contributions of NT in the food sector are remarkably great. The novel applications of NT in food science are shown in Table 20.1. The prime focus of NT in food processing area is to produce the food that is thermally stable, have improved solubility, quality, and bioavailability in order to provide a better and healthy diet, to consumers, containing beneficial functional elements of food. It also focuses on the food preservation to increase the shelf-life of foods by detecting the contaminants at early stages, upgrading the storage systems, and incorporating the antimicrobial agents in foods. NT also provides a vast scope in food packaging area like use of nanocomposites, nano-plastics and other bioactive materials that has good barrier properties and high biodegradability [7, 15].

Although numerous researchers have reported the vital role of NT, still its significance and use in the food industry is limited due to toxicity and food safety related issues which demands further research for the development of safe and eco-friendly NT-based innovations.

This chapter discusses a variety of nanomaterials and nanostructures along with their wide applications in the different areas of the food industry. Nanoemulsions, nanocapsulation, and metallic nanoparticles (NPs) are the most investigated counterparts and have been illustrated below. The toxic effects of these nanoforms which limits the applications of NT propose further research and have been briefly described.

20.2 NANOMATERIALS AND NANOSTRUCTURES IN FOOD

NT includes the investigation, analysis, control, and applications of materials at the nanoscale. These nanomaterials range from 1–100 nm in size in order to develop novel and innovative techniques. The properties of increased

TABLE 20.1 Recent Applications of Nanotechnology in Food Sectors

Product Category	Example of Product	Nanomaterial	Functions
Food contact material	Nano-silver baby mug	Nanoparticles of silver	Increment in anti-bacterial properties of nanoparticles of silver.
Food packaging	Adhesive for containers of burger of McDonalds.	Nanospheres of 50–150 nm	The surface area of nanospheres is around 400 times compared to that of starch particles which are natural. Also, they require low content of water as adhesive and thus takes low energy and less time to get dried.
Food packaging	Durethan KU 2–2601 plastic wrapping, Bayer.	Silica nanoparticles in nanocom-posites based on polymer.	Silica nanoparticles used in plastic prevent gas and oxygen penetration and hence increases the shelf-life of products.
Nutritional drink	Oat chocolate Nutritional drink mix, toddler health.	300 nm iron nanoparticles (Sun Active Fe).	Increased bioavailability and reactivity of iron nanoparticles.
Nutritional supplement	Mycrohydrin powder	A complex of silica mineral hydride in the form of molecular cages that has a diameter of 1–5 nm.	Improved efficacy and bioavailability. Also shows strong antioxidant property due to the release of H-ion in presence of moisture.

surface area, faster rate of mass transfer, enhanced chemical and biological activity, higher reaction rate of enzymes, penetrability, quantum properties, etc., of NPs play a leading role in increasing the range of their applications. Different types of NPs based on bulk material used, have been developed including metallic, carbon based, polymer based, ceramic, semiconductor based, and lipid-based NPs. The nanomaterials made from these tiny particles are generally used as food additives that enhance the shelf-life by preserving the foods from contaminants. Nanomaterials and nanoscale food additives in forms of preservatives, antimicrobials sensors, flavoring agent, packaging substances, encapsulated food components, and so forth are used to influence nutrient composition and improve product shelf-life, texture, flavor, and so on [4]. An outline of some relevant nanomaterials widely used in different areas of the food industry are shown in Table 20.2.

TABLE 20.2 Significance of Nanomaterials in Food Industry [33]

Area	Nanomaterials	Composition	Implementation
Food processing	Nano-sieves	Specific nanoparticles	Destruction of pathogens, Removal of contaminants.
	Nano-sensors	Aptasensors	Identify the presence of microorganisms, control of food spoilage.
	Nanoemulsions	Gum arabica or modified starch, soy, caseinate.	Improves stability and storage of foods.
Food packaging	Nanoparticles	Ag, ZnO, Mg, SiO_2	Antioxidant, Antibacterial
	Nanocomposite	Fe-Cr/Al_2O_3 Ni/Al_2O_3	Increased shelf-life during storage, Protects food from microbial attack.
	Nanofilms	Bioactive compounds	Biodegradable in nature, excellent barrier properties.
Food supplements	Nanocochleates	Coiled nanoparticles	Increase the nutritive value of food.
	Nanomicelles	Aquanova, novasol	Liquid carrier, improved solubility.

Source: Adapted from Ref. [33]

The classification of nanomaterials is based on size, shape, arrangement, and attributes like high surface volume ratio. These features clearly manifest the eminent physiochemical characteristics like solubility, diffusivity, bioavailability, color, optics, strength, toxicity, magnetism, and thermodynamics [14]. On the basis of above features, nanomaterials are categorized into three forms –organic, inorganic, or combination of both.

The biodegradability of organic nanomaterial is much higher than inorganic nanomaterial, therefore the former is often used in the preparation of bio nanocomposites blending with a biodegradable polymer. The significance of inorganic nanomaterial is mainly in food packaging as an antimicrobial agent. NPs used for the synthesis of organic nanomaterial are polymeric NPs, lipid-based NPs, liposomes, solid lipid NPs and/or carbon-based NPs, while the synthesis of inorganic nanomaterials include the NPs having metallic nanostructure [37]. These particles are of silicon dioxide (SiO_2), titanium dioxide, silver, zinc, iron, or their oxides, and have solid form at room temperature thus possess spherical or non-spherical shape. Several nanostructures display specific benefits due to their structural organization itself. Nanotubes self-assembled from the hydrolyzed protein of milk like alpha-lactalbumin can act as nanocarriers of supplements and nutrients. Micelles of lipids and proteins like casein can be utilized as vehicles for protection, entrapment, and delivery of nutraceuticals.

Polysaccharide protein solutions separate into phases spontaneously with nanodroplets which remain dispersed in the continuous phase. Granules of starch expand upon heating and hydration and can be recrystallized and can find immense applications [32]. Dextran and other extrusion products can contribute in encapsulation of bio-actives. These nanostructures, along with increasing the quality and nutrient content of food can also enrich the taste, consistency, and texture of food. Ice creams and mayonnaise containing low fat nanostructures were found to be creamier compared to their alternatives containing full fat. Nanodroplets of around 100 nm are formed in homogenized milk. In dairy industry structures of fat globules, casein micelles and whey proteins at the nanoscale are primarily used to prepare butter emulsions, complex liquids like milk, foams for whipped creams and ice creams, solid cheese and yogurt gel networks [32, 34].

20.3 CONTRIBUTION OF NANOTECHNOLOGY (NT) IN FOOD INDUSTRY

The continuous rise in consumer demands for healthy, safe, and quality foods instigate the research to search an alternative technology that improve the food quality and safety as well as disturbing least the nutritional balance of the food. The market of food requires technologies that are vital to maintain leadership of the market in the industry of food processing to manufacture fresh, convenient, flavorful, and authentic food products, and the most effective solution to it is "nanotechnology." The significance of NT in the food industry was first described by United States Department of Agriculture (USDA) in an article published in the year 2003 [8].

NT has created a major industrial outbreak and provides an array of opportunities related to the structures, materials, or systems in different industries like pharmaceuticals, agriculture, food, etc. NT plays an important role in the food system for the development of new products such as nutraceuticals and functional foods. NT intwined with food science creates new possibilities and is now becoming a rising sector throughout the globe. Therefore, both developed and developing countries are taking interest in adapting this technology as it not just strengthened the food processing but also other aspects of food products [4].

NT provides effective food solutions for different areas like production, processing, packaging, quality, safety, and security (Figure 20.1). There are two main types of application of NT in the food industry, i.e., nanostructured ingredients and nano-sensing. Nanostructures can be used as food additives, nutrient carriers, antimicrobial agents in food processing as well as fillers in food packaging while nano-sensing is used to attain appropriate evaluation of quality and safety of foods. The functional materials development, nanoscale and microscale processing development, development of product, and design of methods and instrumentation in food production may get benefited from NT. Nanoparticle based materials can be less toxic, have high thermal and pressure stability, improve food processing and security and environmentally friendly. Numerous researchers, organizations, and industries are arising with new methods, techniques, and products, which have a direct NT application in science of food [6].

The most remarkable discoveries in NT that can significantly enhance food processing and food products are in the following areas:

- Health promotion through bioactive compound delivery;
- Ingredients with enhanced functionality, with novel flavor and texture;
- Novel packaging materials;
- Materials with microbial-repellent characteristics;
- Advanced tools of nanosensors for food safety, shelf-life, and quality monitoring.

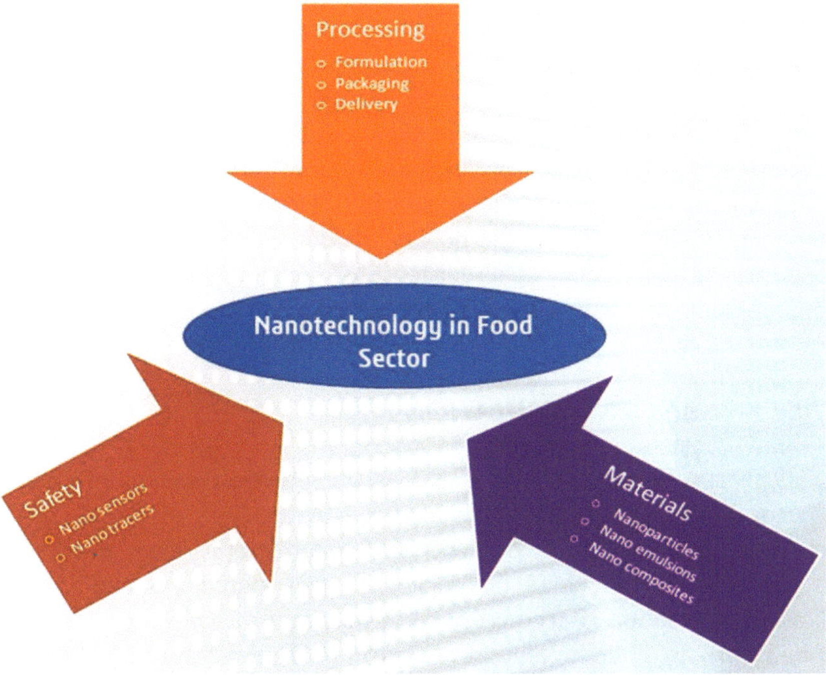

FIGURE 20.1 An overview of the significance of nanotechnology in the food sector.

20.4 APPLICATIONS OF NANOTECHNOLOGY (NT) IN FOOD PROCESSING

The term "Food Processing" is the fusion of several steps from raw material collection, processing, preservation, packaging to storage in order to make the food into consumable form. Apart from these steps, processing also involves removal of toxins, elimination of pathogenic microbes, enzyme deactivation and provides better marketing and distribution of food.

These practices are developed as such that they maintain the nutritional balance and quality of food as well as also provide protection from the contamination and spoilage. There are various conventional methods available for the processing of foods, but there is a huge opportunity for improvements in most of the operations by the intervention of NT-based applications [27]. The significance of NT in the processing of food can be realized by taking into consideration of its role in food products improvement in terms of appearance, texture, nutritional value, taste, and the shelf-life. The use of nanomaterials in food processing include delivery of nutrients, improves gelation and viscosity, addition of nutraceuticals, fortification of micronutrients and encapsulation of flavors.

In food processing, NT can either be used directly or indirectly. In direct usage, nanosized material are directly assimilated into the food by its clear declaration while indirect usage includes the application of nanosized material in packaging material and nanosensors [28]. Development of nanostructured food ingredients maintain the quality of food in terms of nutrition, taste, texture, and consistency. The existence of contaminants, mycotoxins, and microorganisms can be easily and accurately proved by nanosensors.

20.4.1 NANOENCAPSULATION

Nanoencapsulation can be defined as the practical use or relevance of encapsulation to the extent of nanometer in the form of films, layers, or micro-diffusion. These ingredients are of nanoscopic scale and make up a covering over the food in order to protect ingredients and/or flavors. Nanoencapsulation is accomplished in the form of nano-capsules in food. Lipid, carbohydrate, protein, cyclodextrin (CD), dendrimers, and non-ionic surfactant vesicles or niosome based nanoencapsulation systems have been widely investigated in food processing [1, 9, 10, 18, 19, 23].

Nano-capsules impart several advantages in food processing like easy handling, prevent oxidation, improves stability, retention of volatile compounds, protection from environmental factors, increase vitamin and fat solubility, carries flavors, improves bioavailability, constant delivery of active ingredients and improved efficacy [25]. The reduction in the size of ingredients to nanoscale can also alter some properties of food like texture by providing large surface area. The enhanced interfacial interaction can lead to a change of functional properties of the system. The incorporation of more than one water-in-oil-in-water emulsions of nanosized lipids results in enhancement of spreadability and stability of foods. A recent example of this technique is low fat mayonnaise that needs no additional stabilizers and thickeners but imparts a suitable texture, taste, and mouthfeel. The formulation of low-fat food products could have a good application of nanoemulsions [7].

Nanocapsules also preserve the food by trapping the undesirable component and odor in food during the delivery of that component to its target, hence supports food preservation. Nanocapsules have the ability to increase bioavailability by carrying its component in the form of food supplements via the gastrointestinal tract (GIT). They can be used to increase the nutritional value by delivering the health supplements and micronutrients in the food. They ensure the correct and exact delivery of components to the target even if the conditions are unsuitable [16].

An example of nano-based carrier is liposome, which controls the delivery of different components like nutraceuticals, enzymes, micronutrients, antimicrobial agents, and additives within the system. Another example is zein fibers coated with gallic acid to prevent degradation of lipids in the system prior to the target delivery. The efficiency of lipid-based encapsulation methods is much better than other encapsulation methods because they provide better solubility and specificity to the components associated with its encapsulation [13].

20.4.2 NANOEMULSIONS

Nanoemulsion is an emulsion containing dispersed globules having a diameter of 500 nm or less. Thus, they are prepared by incorporation of function ingredients in the form of droplets (100 nm–500 nm) in combination with high pressure valve homogenizers or microfluidizers [16]. The utilization of nanoemulsions is for the processing of food products for sweeteners, salad dressings, flavored oils, etc. They liberate distinct flavors after getting stimulated with different stimulants like heat, pH, ultrasonic waves, etc. They maintain the characteristic flavor of product by preserving them from undesirable reactions such as oxidation and enzymatic reactions. Mainly two energy approaches are used for the production of nanoemulsions – high energy approach and Low energy approach. High energy methods include high pressure and high-speed techniques like ultra sound method, high pressure homogenization while low energy methods consists of membrane emulsification, solvent displacement, spontaneous emulsification, phase, and emulsion inversion point [27].

Mainly three types of nanoemulsions are formed depending upon their compositions–oil in water nanoemulsion, water in oil nanoemulsion and bi-continuous nanoemulsion (Figure 20.2). In all these nanoemulsion types, the interface of the emulsion is stabilized by using an appropriate combination of co-surfactants and/or surfactants. The three main components for the processing of nanoemulsions are oil, surfactant/co-surfactant and aqueous phase. The selection of oil is very important as it greatly affects the other ingredients, especially in case of o/w emulsion. Generally, the oil having potential solubility has been selected for the preparation of nanoemulsions. Some suitable oil phases are Captex 355, Myritol 318, IPM, modified vegetable oils, etc. [5].

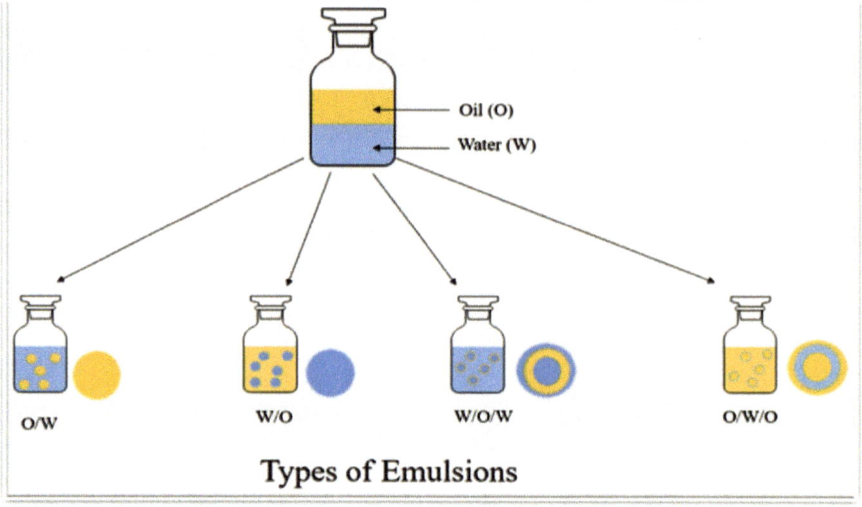

FIGURE 20.2 Different types of nanoemulsions used in the food industry.

Surfactant used for the formulation of nanoemulsions cannot be easily chosen. It can be ionic or non-ionic in nature, but ionic surfactant should be avoided due to its toxicity effects. Various surfactants are mostly used such as lecithin, poloxamers, and polysorbate 80. Co-surfactants lower down the oil-water interfacial tension to obtain nanoemulsion. Co-surfactants penetrate into the surfactant monolayer providing additional fluidity to interfacial film and thus distracting the liquid crystalline phases. Alcohols of intermediate chain length such as hexanol, pentanol, and octanol are some co surfactants preferably used to produce nanoemulsion [31].

Nanoemulsions are formed with lipophilic components which mix completely with the oil droplets and influenced with different properties like molecular weight, solubility, hydrophobicity, and melting point. Nanoemulsions are droplets smaller in size and high heat stability than any other conventional emulsion, therefore are preferred more than conventional emulsion. For example, citral, β-carotene, capsaicin, tributyrin, flaxseed oil, fat-soluble vitamins and coenzyme Q are formulated for production of nanoemulsions and have high stability to gravitational separation and droplet agglomeration. They also provide large surface area resulting in increased access and thus rate of digestive enzymes in GIT and easy absorption of food components. Lipase is an example of an enzyme that can readily digest the nanoemulsion in GIT. Nanoemulsions can also be used as an antimicrobial agent in food processing as they are very effective on gram-positive bacteria. Tributyl phosphate, soybean, and nonionic surfactants are the examples of nanoemulsions that reduces the scope of spoilage by analyzing the presence and growth of microbes [36].

20.4.3 INORGANIC NANOPARTICLES (NPS)

Inorganic NPs of metals and their oxides like iron oxide, silver, silicon, titanium dioxide, or zinc oxide have been extensively used in food processing. These nanoforms are present in amorphous or crystalline form at optimal temperatures and can be non-spherical or spherical with varying sizes and surface chemistries depending on conditions of fabrication and material used. These NPs show variation in their ability to dissolve in different solutions under different conditions and chemical reactiveness which contributes in their gastrointestinal fate and toxicity [30]. Zinc oxide NPs are used as supplements of zinc and as functional foods. Iron oxide NPs can be utilized as bioavailable iron and also as food colorants. Titanium oxide NPs contribute optical properties to foods like enhanced brightness and lightness and are used as functional ingredients. In general, the size of particles used as lightening agents ranges from 100 to 300 nm. NPs of silicon dioxide are used in certain food powders as anticaking agents and thus improves the properties of flow, for example in icing sugar, salts, powdered milk, spices, and dry mixture. They are amorphous solids of spherical shape and usually lying-in range between 100 and 1,000 nm in range but can also be smaller in size [24].

20.5 APPLICATIONS OF NANOTECHNOLOGY (NT) IN FOOD PACKAGING

The demand of fresh, healthy, and ready to eat food has been increased from last few decades due to the changing lifestyle and fast urbanization. The maintenance of wholesomeness of food is a bit complex in terms of its processing, packaging, storage, shipment, distribution, safety of food and environment and economic feasibility. Among all these factors, packaging stands at an important place as it performs several different functions in food. Among all these criteria, environmental safety is largely affected by the packaging material used to pack the food especially plastic polymers [3]. As a response to challenges associated with food safety and environmental issues, a need has been generated to develop the packaging material that can work as a possible outcome of these issues.

NT in packaging is categorized based on the purpose of the application. These forms of packaging technology are usually biodegradable having high sustainability and eco-friendly nature. NPs used for food packaging are generated either from animals and plant sources in the form of proteins, lipids, and carbohydrates, or metal particles. For example, zein from maize, collagen from animal, zinc oxide and carbon atoms are used in the synthesis of nanotubes and NPs for packaging material [21].

20.5.1 NANOSENSORS

Nanosensors have been made with the combination of different nanostructured materials and bio receptors designed in an integrated system. These bioanalytical devices have quick detection power, integrity, and low cost, thus gaining a lot of attention in the food industry. Nanosensors have the potential to be integrated with an array of analytes due to their high sensitivity and specificity. These devices have a high surface-to-volume ratio and excellent optical and electric properties due to conjugation with various types of nanomaterials such as carbon nanotubes, NPs (metallic, non-metallic, and metal oxide), semiconductor NPs, nanorods, nanowires, nanobiofilms, nanofibers, and quantum dots [22].

The prime application of nanosensors is to detect the presence of toxins, changes in color, and production of gases which leads to food spoilage. Nanosensors attached with packaging material also provide a better tracking system for the internal or external conditions of foods during storage and distribution. Time-temperature integrator and gas detector are the two most common nanosensors used for packaging in the food industry. The metals used for the production of gas sensors are platinum, palladium, and gold. Gold-based NPs also contribute sometimes to detect aflatoxin B1 toxin in milk [7].

Apart from food packaging, nanosensors can also be used for other purposes in the food industry like flavor detection, the presence of pesticide residues, and the presence of carcinogens in food. These nanosensors include array biosensors, nanoparticle-based sensors, electronic nose, electronic tongue, nano-test strips, etc. In some cases, the packaging is made of DNA and single walled carbon nanotubes that greatly improves the sensitivity of the sensors.

20.6 THREATS OF FOOD NANOTECHNOLOGY (NT)

NT in the food sector provides numerous applications that help to improve different properties of foods and provides several benefits from processing to packaging, preservation to quality. Though this technology has tremendous applications and potential, but public opinion is pretty contradictory. Food safety obtained from the direct or indirect use of different nanomaterials is a big topic for debate among scientists and researchers. Even some scientists have been issued warnings regarding the use of nanomaterials and about health diseases affiliated with their use. Nanomaterials can be potentially dangerous because they are minute in size and can easily penetrate and accumulate within the cell, causing health hazards [12]. Nanostructures can have high toxicity towards human and the environment due to their mobility and reactive nature. Nanotubes can also cause damage to the human body because they are hard to remove from the body due to their minute size, e.g., carbon nanotubes can even cause damage to lungs [20]. NPs can also cause toxicity to humans and leads to severe health diseases such as increased oxidative stress, cancer, mutation in DNA structure or even death.

NPs present in GIT intervene the normal functioning of GIT due to their large surface area. Larger the surface area, the higher the adsorption of surface-active compounds in GIT. This leads to the denaturation of many globular proteins and reduces the activity of many enzymes present in it. Therefore, nanoparticles at high levels can reduce the extent or rate of digestion of lipid, starch, or protein in the GIT [11]. Metal NPs, mainly obtained from silver, zinc, etc., can also possess toxicity by affecting the cellular components (such as proteins, nucleic acids, or lipids) essential for the maintenance of biochemical processes. Such actions of mechanisms are leadingly important for NPs of inorganic type which are absorbed by cells of the intestine, as most organic NPs are digested before getting absorbed [8].

Though various food and agricultural products of NT are distributed in the markets and environment, yet legal authorities throughout the globe are still struggling to make standards or rules in order to reduce the risks and threats associated with NT. Risk evaluation and consumer perception are significant topics in NT to be considered in analogous with the growth and application. Essential risk evaluation of nanotechnologies and their applications in food products necessitates progressive research in the fields of classifying NPs, dosage levels, impact, and effects, regulation, and consumer acquaintance.

20.7 SUMMARY

The intervention of NT has bought a revolution in the food sector. It has incorporated significant changes in different areas of the food industry like processing, preservation, packaging, quality, and safety of foods. These changes not only enhance the shelf-life of foods with better management of spoilage but also provided an improved tracking and tracing system in order to maintain quality and safety. Many large food industries pay in millions to enhance and progress nano-systems which can help food preservation in a better way. NT also overcomes the problem of food security in many countries by controlling the food wastage and spoilage, hence increase the availability of food among the common masses. Accuracy and integrity of manufacture dealing with nanoscale devices are the prime challenges of using NT. Upon mastering this peculiar technology, it can lead to valuable opportunities in an array of fields. Nanoscale materials include NPs, nanoemulsions, nanotubes, nanocomposites, nanosensors, etc., and perform several beneficial functions in several fields, from medicine to food production. These nanomaterials and devices focus on the accuracy in delivery of food processing units as well as enhance the properties of foods like texture, taste, consistency, and shelf-life. In the use of NT in the food sector, benefits are numerous, but several drawbacks also persist. NPs can cause toxicity and other health issues due to their small size and high bioavailability. NPs obtained from metals can detect the presence of bacteria accurately but also cause toxicity to the environment and can harm the beneficial microbial flora of our intestine. Proper care must be taken when new nano-systems are being designed so that they become both lowly toxic for food as well as environment friendly.

KEYWORDS

- **food industry**
- **food processing**
- **food science**
- **gastrointestinal tract**
- **nanoencapsulation**
- **nanoparticles**
- **nano-sensing**
- **nanotechnology**

REFERENCES

1. Aher, N., Banerjee, S., Bhansali, S., Yadav, R., Shidore, M., Mhaske, S., & Khandare, J., (2013). Poly (ethylene glycol) versus dendrimer prodrug conjugates: Influence of prodrug architecture in cellular uptake and transferrin mediated targeting. *Journal of Biomedical Nanotechnology, 9*(5), 776–789.
2. Aschberger, K., Gottardo, S., Amenta, V., Arena, M., Botelho, M. F., Bouwmeester, H., Brandhoff, P., et al., (2015). Nanomaterials in food: Current and future applications and regulatory aspects. *Journal of Physics: Conference Series, 617*, E-article 012032.
3. Atares, L., & Chiralt, A., (2016). Essential oils as additives in biodegradable films and coatings for active food packaging. *Trends in Food Science & Technology, 48*, 51–62.
4. Chellaram, C., Murugaboopathi, G., John, A. A., Sivakumar, R., Ganesan, S., Krithika, S., & Priya, G., (2014). Significance of nanotechnology in food industry. *APCBEE Procedia, 8*, 109–113.
5. Chime, S. A., Kenechukwu, F. C., & Attama, A. A., (2014). Chapter 3: Nano emulsions-advances in formulation, characterization and applications in drug delivery. In: *Application of Nanotechnology in Drug Delivery* (pp. 77–111). Intech Open.
6. Dasgupta, N., & Ranjan, S., (2018). Nanotechnology in food sector; Chapter 1. In: Dasgupta, N., & Ranjan, S., (eds.), *An Introduction to Food Grade Nanoemulsions* (Vol. 1, pp. 1–18). Singapore: Springer.
7. Dera, M. W., & Teseme, W. B., (2020). Review on the application of food nanotechnology in food processing. *American Journal of Engineering and Technology Management, 5*(2), 41–47.
8. EFSA (European Food Safety Authority), (2016). Safety assessment of the substance zinc oxide, nanoparticles, for use in food contact materials. *EFSA Journal, 14*, 4401–4408.
9. Fathi, M., Donsi, F., & McClements, D. J., (2018). Protein-based delivery systems for the nanoencapsulation of food ingredients. *Comprehensive Reviews in Food Science and Food Safety, 17*, 920–936.

10. Fathi, M., Mozafari, M. R., & Mohebbi, M., (2012). Nanoencapsulation of food ingredients using lipid-based delivery systems. *Trends in Food Science & Technology, 23*(1), 13–27.
11. Frohlich, E. E., & Frohlich, E., (2016). Cytotoxicity of nanoparticles contained in food on intestinal cells and the gut microbiota. *International Journal of Molecular Sciences, 17*, 1–22.
12. Golmohammadi, H., Morales-Narváez, E., Naghdi, T., & Merkoçi, A., (2017). Nanocellulose in sensing and biosensing. *Chemistry of Materials, 29*(13), 5426–5446.
13. Guhan-Nath, S., Sam-Aaron, I., Allwyn-Sundar, R. A., & Ranganathan, T. V., (2014). Recent Innovations in nanotechnology in food processing and its various applications: A review. *International Journal of Pharmaceutical Sciences Review and Research, 29*(2), 116–124.
14. Gupta, A., Eral, H. B., Hatton, T. A., & Doyle, P. S., (2016). Nanoemulsions: Formation, properties and applications. *Soft Matter, 12*, 2826–2841.
15. Hamad, A. F., Han, J. H., Kim, B. C., & Rather, I. A., (2018). The intertwine of nanotechnology with the food industry. *Saudi Journal of Biological Sciences, 25*, 27–30.
16. He, X., & Hwang, H. M., (2016). Nanotechnology in food science: Functionality, applicability, and safety assessment. *Journal of Food and Drug Analysis, 24*(4), 671–681.
17. Jafarali, K. M., Jayakumar, C., & Jashbhai, B., (2013). Nutrition in food and science, potential of nanotechnology in functional foods. *Journal of Food and Agriculture Emirates, 25*(1), 10–19.
18. Jafari, S. M., (2019). Biopolymer nanostructures for food encapsulation purposes; Chapter 20. In: Yousefi, M., Orojzadeh, P., & Jafari, S. M., (eds.), *Nanoencapsulation in the Food Industry* (Vol. 1; pp. 607–625) London – UK: Academic Press.
19. Jafari, S. M., (2019). Lipid-based nanostructures for food encapsulation purposes; Chapter 11. In: Matos, P. M., Pando, D., & Gutiérrez, G., (eds.), *Nanoencapsulation of Food Ingredients by Niosomes* (Vol. 2, pp. 447–481). London – UK: Academic Press.
20. Krishnan, U., Barabara, S., Arifullah, M., Abdul-Nabi, N. N., & Antonella, P., (2018). Relevance of nanotechnology in food processing industries. *International Journal of Agriculture Sciences, 10*(7), 5730–5733.
21. Kulkarni, S. K., (2015). Synthesis of nanomaterials, part II: Chemical methods. In: Kulkarni, S. K., (ed.), *Nanotechnology: Principles and Practices* (pp. 77–109). Cham: Springer.
22. Kuswandi, B., & Moradi, M., (2019). Improvement of food packaging based on functional nanomaterial; Chapter 16. In: Siddiquee, S., Melvin, G. J. H., & Rahman, M. M., (eds.), *Nanotechnology: Applications in Energy, Drug and Food* (pp. 309–344). Cham – Switzerland, Springer.
23. Mazzobre, M. F., Elizalde, B. E., Dos, S. C., Cevallos, P. A. P., & Buera, M. P., (2010). Nanoencapsulation of food ingredients in cyclodextrins: Effect of water interactions and ligand structure. In: Shahidi, F., Smith, J., & Charter, E., (eds.), *Functional Food Product Development* (pp. 24–38). New York: Wiley Online Library.
24. McClements, D. J., & Xiao, H., (2017). Is nano safe in foods? Establishing the factors impacting the gastrointestinal fate and toxicity of organic and inorganic food-grade nanoparticles. *NPJ. Science of Food, 1*(6), 1–13.
25. Meghani, N., Dave, S., & Kumar, A., (2019). Nanofood and its application: A review. *BAOJ Nanotechnology, 5*(1), 1–16.

26. Momin, J. K., & Joshi, B. H., (2015). Nanotechnology in foods. In: Rai, M., Ribeiro, C., Mattoso, L., & Duran, N., (eds.), *Nanotechnologies in Food and Agriculture* (pp. 3–24). Cham: Springer.
27. Pradhan, N., Singh, S., Ojha, N., Shrivastava, A., Barla, A., Rai, V., & Bose, S., (2015). Facets of nanotechnology as seen in food processing, packaging, and preservation industry. *BioMed Research International, 2015*, 1–17.
28. Sahoo, M., Vishwakarma, S., Panigrahi, C., & Kumar, J., (2021). Nanotechnology: Current applications and future scope in food. *Food Frontiers, 2*, 3–22.
29. Samal, D., (2017). Use of nanotechnology in food industry: A review. *International Journal of Environment, Agriculture and Biotechnology, 2*(4), 22070–22078.
30. Satyanarayana, T., & Reddy, S. S., (2018). Review on chemical and physical synthesis methods of nanomaterials. *International Journal for Research in Applied Science & Engineering Technology, 6*, 2885–2889.
31. Savardekar, P., & Bajaj, A., (2016). Nanoemulsions: A review. *International Journal of Research in Pharmacy and Chemistry, 6*(2), 312–322.
32. Sekhon, B. S., (2010). Food nanotechnology – overview. *Nanotechnology, Science and Applications, 3*, 1–15.
33. Shafiq, M., Anjum, S., Hano, C., Anjum, I., & Abbasi, B. H., (2020). An overview of the applications of nanomaterials and nanodevices in the food industry. *Foods, 9*(148), 1–27.
34. Singh, H., (2016). Nanotechnology applications in functional foods: Opportunities and challenges. *Preventive Nutrition and Food Science, 21*(1), 1–8.
35. Singh, T., Shukla, S., Kumar, P., Wahla, V., Bajpai, V. K., & Rather, I. A., (2017). Application of nanotechnology in food science: Perception and overview. *Frontal Microbiology, 8*, 1–7.
36. Wang, Y., Zhang, Q., Zhang, C. L., & Li, P., (2012). Characterization and cooperative antimicrobial properties of chitosan/ nano-ZnO composite nanofibrous membranes. *Food Chemistry, 132*(1), 419–427.
37. Yu, H., Park, J. Y., Kwon, C. W., Hong, S. C., Park, K. M., & Chang, P. S., (2018). Overview of nanotechnology in food science: Preparative methods, practical applications, and safety. *Journal of Chemistry, 2018*, 1–10.

INDEX

C

F

J

K

L

M

N

O

W

X

Y

Z